Topics in
Current Physics

38

Topics in Current Physics

Founded by Helmut K. V. Lotsch

Coherent Radiation Sources

Edited by A. W. Sáenz and H. Überall

With Contributions by
J. U. Andersen   V. V. Beloshitsky   B. L. Berman
E. Bonderup   R. A. Carrigan, Jr.   S. Datz
W. M. Gibson   L. Ya. Kolesnikov   G. D. Kovalenko
M. A. Kumakhov   E. Laegsgaard   A. L. Rubashkin
A. W. Sáenz   H. Überall   G. B. Yodh

With 118 Figures

Springer-Verlag Berlin Heidelberg GmbH

Professor Albert W. Sáenz

Naval Research Laboratory, Washington, DC 20375, USA and
Department of Physics, Catholic University of America,
Washington, DC 20064, USA

Professor Herbert Überall

Department of Physics, Catholic University of America,
Washington, DC 20064, USA

ISBN 978-3-642-51187-5   ISBN 978-3-642-51185-1 (eBook)
DOI 10.1007/978-3-642-51185-1

Library of Congress Cataloging in Publication Data. Main entry under title: Coherent radiation sources. (Topics in current physics ; 38). Includes bibliographies. 1. Radiation sources. 2. Coherence (Optics). 3. Bremsstrahlung. 4. Channeling (Physics). I. Sáenz, Albert William, 1923–. II. Überall, Herbert, 1931–. III. Andersen, J. U. IV. Series. QC476.S6C64 1985 537 85-9851

© Springer-Verlag Berlin Heidelberg 1985
Originally published by Springer-Verlag Berlin Heidelberg in 1985.

2153/3150-543210

The editors of this book wish to dedicate it to

Hans A. Bethe
(Pioneer of bremsstrahlung)

Jens Lindhard
(Pioneer of channeling)

Preface

Energetic electromagnetic radiation finds frequent uses in science (e.g., for experiments in nuclear and elementary - particle physics), in technology (for materials testing), and in medicine (for medical X-rays). The most common method of generating such radiation is via the process of "bremsstrahlung" (a German term coined by A. Sommerfeld, meaning "braking radiation") in which a beam of electrons is directed into matter (e.g., a metal target), losing energy during its collisions with the atoms and releasing this energy in the form of emitted radiation.

The character of such radiation may be drastically changed by the use of a target with periodic structure (most commonly, a crystal target). The coherent waves emitted from individual crystal atoms interfere with each other, monochromatizing and polarizing the radiation and often increasing its intensity manifold, thereby creating a powerful radiation source of high quality for purposes of scientific and technical applications. This is true both for the well - established "coherent bremsstrahlung" process in which the interfering radiation is emitted while the electrons cross a succession of crystal planes, as well as for the more recently discovered process of "channeling radiation" (generating radiation of even higher intensity, but lower energy) in which the radiation is emitted while the electrons propagate along a crystal plane, or a crystal axis, in an oscillatory fashion.

In this book we have assembled an authoritative collection of review articles, written by either the original expounders of the relevant phenomena, or by investigators who carried out original experiments or who were involved in the latest phases of coherent - radiation research and its applications. We hope that the book will prove useful to readers interested in this fascinating subject.

Washington, April 1985   *A.W. Sáenz • H. Überall*

Contents

List of Contributors

Andersen, Jens U.

Institute of Physics, University of Aarhus, DK-8000 Aarhus C, Denmark

Beloshitsky, Vladimir V.

Institute of Nuclear Physics, Moscow State University, Moscow 117234, USSR

Berman, Barry L.

Lawrence Livermore National Laboratory, University of California,
Livermore, CA 94550, USA   and
Department of Physics, George Washington University,
Washington, DC 20052, USA

Bonderup, Ejvind

Institute of Physics, University of Aarhus, DK-8000 Aarhus C, Denmark

Carrigan, Richard A., Jr.

Fermi National Accelerator Laboratory, Box 500,
Batavia, IL 60510, USA

Datz, Sheldon

Oak Ridge National Laboratory, Oak Ridge, TN 37830, USA

Gibson, Walter M.

Department of Physics and Institute for Particle-Solid Interactions, State
University of New York at Albany, 1400 Washington Avenue,
Albany, NY 12222, USA

Kolesnikov, Leonid Ya.

Institute of Physics and Technology, The Ukrainian Academy of Sciences,
Kharkov 310108, USSR

Kovalenko, Grigorij D.

Institute of Physics and Technology, The Ukrainian Academy of Sciences,
Kharkov 310108, USSR

Kumakhov, Muradin A.

Institute of Nuclear Physics, Moscow State University, Moscow 117234, USSR

Laegsgaard, Erik

Institute of Physics, University of Aarhus, DK-8000 Aarhus C, Denmark

Rubashkin, Alexander L.

Institute of Physics and Technology, The Ukrainian Academy of Sciences, Kharkov 310108, USSR

Sáenz, Albert W.

Naval Research Laboratory, Washington, DC 20375, USA   and
Department of Physics, Catholic University of America,
Washington, DC 20064, USA

Überall, Herbert

Department of Physics, Catholic University of America,
Washington, DC 20064, USA

Yodh, Gaurang B.

Department of Physics and Astronomy, University of Maryland,
College Park, MD 20742, USA

1. Introduction

A. W. Sáenz and H. Überall

With 1 Figure

Coherent radiation is given off by charged particles when they traverse a periodic array of electric or magnetic fields. The resulting interference phenomena
are often found to affect decisively the character of the emitted radiation, for
example, by increasing its intensity by orders of magnitude, monochromatizing its
spectrum, or polarizing the radiation. These effects have provided coherent radiation sources for many scientific and technical purposes.

The periodic structures involved can be either man-made (such as an alternating
periodic magnetic field), or of natural origin (i.e., crystals). The former structures have recently been extensively discussed in the technical domain as possible
sources of microwave, X-ray, and light radiation. In this connection we mention the
Smith-Purcell effect [1.1], in which an electron beam grazes an optical grating on
a metal surface and radiates due to the fluctuating induced charges on the grating;
the radiation of an electron beam passing along a corrugated-metal "slow-wave structure"; the "undulator" [1.2] and the "free-electron laser" (proposed by *Madey*
[1.3] and built by *Elias* et al. [1.4]), in which the electron beam passes through
an alternating periodic magnetic field in cavities and emits coherent microwave
radiation; and a related device known as the "wiggler", which has been used at the
Stanford Synchrotron Radiation Laboratory to increase and harden the X-ray intensity of SPEAR's synchrotron radiation.

This book is mainly concerned with coherent radiation emitted by a beam of electrons (or positrons) when traversing natural periodic targets, e.g., crystals. The
only exception is the last chapter, on transition radiation, in which charged particles pass through a stack of alternating laminated foils. Both crystal and transi-
tion-radiation sources have enjoyed numerous applications for scientific purposes,
such as serving as coherent, tunable, polarized monochromatic sources of gamma rays
for nuclear- and particle-physics experiments, and as a means of high-energy particle identification and energy measurement.

Coherent bremsstrahlung, the crystal-enhanced, monochromatized, and polarized
variant of ordinary bremsstrahlung (radiation from electrons when passing by atoms
of matter) was first mentioned by *Williams* [1.5]. The detailed theoretical discussion of this effect by *Überall* [1.6,7] has given rise to extensive experimental activity verifying and studying the coherent-bremsstrahlung phenomenon and utilizing

the resulting quasi-monochromatic, polarized radiation as a source for scientific
experiments. The basis of this effect lies in the fact that the recoil generated
in the radiation process is taken up not by a single atom (as in ordinary bremsstrahlung), but with a certain probability by the crystal as a whole. This is similar to the Mössbauer effect [1.8] in which nuclear radiation is emitted by a nucleus
which is part of a crystal. Chapter 2 of this book (the chapters being essentially
in chronological order as far as crystal-radiation phenomena are concerned) deals
with the theory of coherent bremsstrahlung, and Chap.3 (written by members of a
laboratory with a presently operative coherent-bremsstrahlung beam) presents a review of coherent-bremsstrahlung experiments.

Charged particles traversing a crystal have been found to exhibit a "channeling"
phenomenon as first discussed by *Lindhard* [1.9], which was initially verified for
heavy particles. Electrons and positrons have later also been found to be subject
to channeling; the electrons, e.g., can get trapped by, and oscillate back and
forth across, a crystal plane while moving alongside it ("planar channeling"), or
they may execute a helical motion around and along a crystal axis ("axial channeling"). Chapter 4 presents a discussion of the channeling phenomenon, and specifically of one of its more recently studied aspects, namely, charged-particle channeling through a mechanically bent crystal, which can cause a bending of the particle beam. This chapter may serve to familiarize the reader with the channeling
phenomenon, in preparation for the following three chapters on channeling radiation.

An electron trapped by and undulating along a crystal plane or axis will, due
to its small mass, have a high probability of radiating in this accelerated type
of motion. The ensuing "channeling radiation" was discussed and analyzed in a de-
finitive fashion by *Kumakhov* [1.10], and has been observed by *Berman* et al. at the
Lawrence Livermore Laboratory [1.11], as well as at many other laboratories. As in
the case of coherent bremsstrahlung, one obtains here a series of quasi-monochromatic, polarized peaks in the radiation spectrum, albeit at much lower energies.

Spectral peaks for coherent bremsstrahlung are depicted in Fig.1.1a and b, and
for channeling radiation in Fig.1.1c. The coherent-bremsstrahlung peaks [1.12] correspond to 15- MeV electrons incident along the <100> direction of a silicon crystal
(Fig. 1.1a), having a fundamental peak at ~3 MeV with overtones, or incident at 1°
from the <110> direction (Fig.1.1b) having a fundamental peak at ~0.1 MeV with overtones. These are examples of type-B and type-A coherent bremsstrahlung, respectively
(Chap.2); note the vastly higher intensity of type-A peaks. Figure 1.1c shows the
calculated spectrum [1.13] of the channeling radiation of 56- MeV electrons channeled in the (110) plane of silicon. The radiation peaks lie here between 42 and
128 keV, an energy range typical for channeling radiation. If Figs.1.1b and c were
drawn to the same scale, the highest of the channeling-radiation peaks would be at
least an order of magnitude above those of the type-A coherent-bremsstrahlung peaks
(but with latter being located at higher energy). In turn, the type-A peaks are an

2

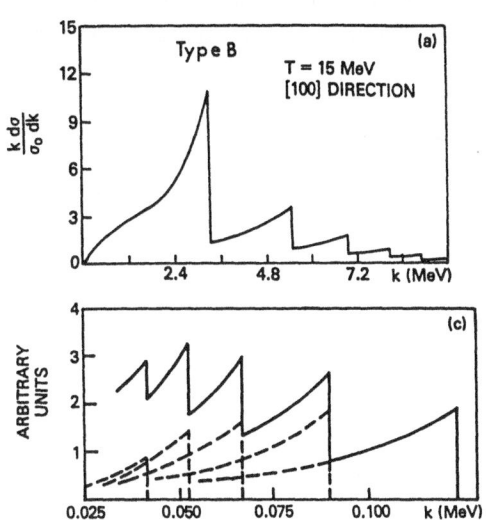
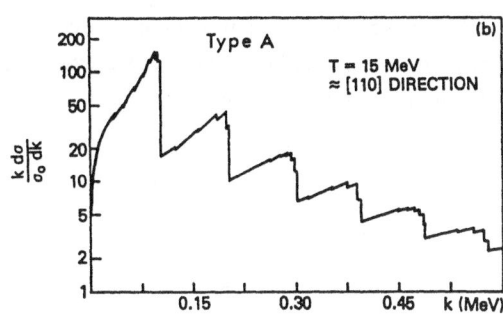

Fig.1.1a-c. Peaks in the spectrum of coherent bremsstrahlung, (a) of type B and (b) of type A, from 15-MeV electrons, and in the spectrum of channeling radiation (c) from 56- MeV electrons, all in a silicon crystal target

order of magnitude higher than the type-B coherent-bremsstrahlung peaks, but with the latter again being located at higher energies.

The similarity of coherent-bremsstrahlung and of channeling-radiation spectral peaks, evident in Fig.1.1, raises the question of a possible connection between the two radiation processes. After some initial controversy, this question was answered affirmatively by *Andersen* et al. [1.14]. They described electrons planarly (respectively, axially) channeled in crystals by one-dimensional (respectively, two-dimensional) Bloch-type wave functions, which correspond either to "bound" states of the transverse motion lying within narrow energy bands, or to "free" states with wide energy bands lying in the continuum. Transitions between free states of transverse motion lead to conventional coherent bremsstrahlung, which was previously described in the Born approximation [1.6,7]. Transitions between bound states produce Kumakhov's channeling radiation [1.10]. In this description, a third type of coherent radiation exists, namely, that due to free-bound transitions. These have now also been observed experimentally [1.15], and are discussed theoretically in Chap.2.

The topic of channeling radiation is reviewed in Chaps.5-7 of this book. Chapter 5 discusses the classical theory of channeling radiation, originally expounded by *Kumakhov* [1.10]. Chapter 6 treats channeling radiation quantum mechanically, and Chap.7 describes the early experimental history of channeling radiation, continuing from there up to the latest experimental results.

The last chapter of this book, Chap.8, discusses transition radiation, which unlike the radiations considered in the preceding chapters, is not produced in crystals, but in laminated-foil targets when they are traversed by high-energy charged particles. Transition radiation has also been utilized for nuclear- and particle-physics experiments, primarily as a means of energy measurements for high-energy particles, and it also represents a potential coherent radiation source.

While scientific applications of coherent radiation sources have been numerous, many technical applications also appear feasible and have been discussed at various places in the literature.

References

1.1 S.J. Smith, E.M. Purcell: Phys. Rev. **92**, 1069 (1953)
1.2 H. Motz: J. Appl. Phys. **22**, 527 (1951)
1.3 J.M. Madey: Appl. Phys. **42**, 1906 (1971)
1.4 L. Elias, W. Fairbank, J. Madey, H.A. Schwettmann, T. Smith: Phys. Rev. Lett. **33**, 717 (1976)
1.5 E.J. Williams: K. Dan. Vidensk. Selsk. Mat. Fys. Medd. 13. No.4 (1935)
1.6 H. Überall: Phys. Rev. **103**, 1055 (1956)
1.7 H. Überall: Z. Naturforsch. **17a**, 332 (1962)
1.8 R. Mössbauer: Z. Phys. **151**, 124 (1958); Naturwiss. **45**, 538 (1958)
1.9 J. Lindhard: K. Dan. Vidensk. Selsk. Mat. Fys. Medd. **34**, No.14 (1965)
1.10 M.A. Kumakhov: Phys. Lett. **57A**, 17 (1976) and subsequent papers
1.11 See, e.g., M.J. Alguard, R.L. Swent, R.H. Pantell, B.L. Berman, S.D. Bloom, S. Datz: Phys. Rev. Lett. **42**, 1148 (1979)
1.12 A.W. Sáenz, H. Überall: Phys. Rev. **B25**, 4418 (1982)
1.13 A.W. Sáenz, H. Überall, A. Nagl: Nucl. Phys. **A372**, 90 (1981)
1.14 J.U. Andersen, K.R. Eriksen, E. Laegsgaard: Phys. Scr. **24**, 588 (1981)
1.15 J.U. Andersen, E. Bonderup, E. Laegsgaard, B.B. Marsh, A.H. Sørensen: Nucl. Instr. Methods **194**, 209 (1982)

2. Theory of Coherent Bremsstrahlung

A. W. Sáenz and H. Überall

With 7 Figures

Two souls, alas, are dwelling in my breast!

Goethe, Faust I, 1112.

In this chapter we review theoretically the phenomenon of coherent bremsstrahlung (CB) of electrons traversing a crystal. This process has been studied for more than thirty years, but it now appears in a new light after the realization by *Andersen* [2.1] that CB and the more recently discovered process of channeling radiation (CR) constitute two different but related aspects of one and the same physical phenomenon. These two types of radation may be described in a unified fashion by the use of Bloch functions for the electrons in the crystal. CB kinematics is discussed from this viewpoint, a theoretical calculation of low-energy CB intensities is given, and a model calculation for free-bound transitions is carried out as an example of the link between CB and CR phenomena.

2.1 Overview

The term "(ordinary) bremsstrahlung" designates the radiation emitted by a charged particle (usually an electron or positron) during its passage near an atom or molecule, or more specifically, the nuclei thereof. When electrons (or positrons) traverse a crystal with their direction of incidence close to a direction of major crystal symmetry, the elementary bremsstrahlung transition amplitudes may add up with definite **phases,** leading to constructive interference; the corresponding radiation is referred to as "coherent bremsstrahlung" (CB). It exhibits a spectrum with a series of peaks, quite different from that of ordinary bremsstrahlung generated in an amorphous or polycrystalline target.

Coherent bremsstrahlung is a subject with a long history, the earlier parts of which are covered in three major reviews by *Diambrini-Palazzi* [2.2], *Timm* [2.3], and *Ter-Mikaelian* [2.4], and in other relevant summaries [2.5,6]. The earliest mention of the subject goes back to *Williams* [2.7,8], and further early work is due to *Ferretti* [2.9], *Landau* and *Pomeranchuk* [2.10], *Ter-Mikaelian* [2.11], *Dyson* and *Überall* [2.12], and *Purcell* [2.13]. *Frisch* [2.14] contributed physical arguments elucidating the interference phenomenon.

While much of the previous work was based on the Weizsäcker-Williams approximation, the first quantitative calculation of CB spectra [2.15] and polarization

[2.16,17], based on the first Born approximation, was given by *Uberall* in a form in which it could be compared with experiment. In [2.15-17], whose results were intended to apply at high electron energies ($\gtrsim$ several 100 MeV), various approximations appropriate to such energies were also made.

Since, for reasons of unitarity, the Born approximation is expected to fail at very high energies [2.9], second-Born-approximation effects in the crystal potentials, or semiclassical approximations, have been considered by *Schiff* [2.18], *Ferretti* and *Gamberini* [2.19], and especially by *Akhiezer* et al. [2.20,21]. In general, second-Born effects should become considerable for high atomic numbers, for photon energies near the high-frequency limit, and for low incident electron energies. They introduce a difference between the coherent bremsstrahlung emitted by electrons and that emitted by positrons (which in the first Born approximation are the same), increasing the radiation intensity for positrons and decreasing that for electrons. It was shown, however [2.21], that CB results calculated on the basis of the first Born approximation have a validity that exceeds that of the nominal applicability of this approximation. For example, the first-Born-approximation cross-section expressions remain valid even if the usual condition of applicability $RZe^2/d\Theta \ll 1$ is not satisfied, where d is the lattice plane spacing, Θ the angle between the incident electron momentum and a (major) crystal axis, and R the atomic screening length, and fast particles ($E \gg m$) are considered. Indeed, the numerous experimental studies of high-energy coherent bremsstrahlung that have been carried out [2.22,23] since 1960 have all given results in essential agreement with the first-Born-approximation theory [2.2-4]. According to *Akhiezer* et al. [2.21] it is necessary to satisfy the condition $\Theta \gg (Ze^2/Ed)^{\frac{1}{2}} \equiv \Theta_c$ in order for this theory to be applicable for a calculation of CB, where E ($\gg m$) is the incident electron energy. The conclusions of these authors were intended to apply at such high electron energies that their quasi-classical approach was valid. It does not necessarily follow that the condition $\Theta \gg \Theta_c$ needs to be satisfied in order for the first-Born-approximation approach to CB to be valid at much lower electron energies. For example, *Komar* et al. [2.24] have successfully interpreted by a first-Born-approximation analysis their CB experiments with electrons of 7-10 MeV incident at $\Theta = 0$ along the <100> axis of Si.

It is interesting to note that the opposite condition, namely $\Theta \lesssim \Theta_c$, plays an important role in the phenomenon of channeling radiation (CR) [2.25,26], which has a close connection with CB (to be discussed below), the condition being applicable at electron energies high enough so that a classical description is meaningful. The term "channeling" describes the phenomenon in which a beam of charged particles with an initial direction making a small angle Θ with a major crystal axis or set of crystal planes will propagate along such an axis (axial channeling), or along a crystal plane or between two such planes (planar channeling), guided by the resultant screened Coulomb force exerted on it by the atoms of the axes or planes, and thus achieve large penetration depths. According to *Lindhard* [2.27], many aspects

6

of channeling can be understood by replacing this resultant force by an average force due to continuously charged strings and planes, and he showed that when classical mechanics is applicable the particles of the beam are trapped into a channeling state if $\Theta \lesssim \Theta_c$, where Θ_c is termed the "Lindhard critical angle".

Quantum mechanically, a particle in such a channeling state is described by a wave function which, in a certain sense, is transversely "bound", and has a nonzero probability of emitting spontaneous radiation, i.e., CR, in allowed transitions to channeling states of lower transverse energy (see Sect.2.2 for further details; CR is discussed *in extenso* in Chap.6).

In order to calculate more accurately (beyond the first Born approximation) the emission of radiation by charged particles, e.g., electrons, traversing a crystal, one evaluates the first-order matrix element for radiative transitions [2.28],

$$J_\lambda = \int_{\text{crystal}} \psi_{\mathbf{p}'}^{\dagger}(\mathbf{r})(\boldsymbol{\alpha} \cdot \boldsymbol{\epsilon}_\lambda^*) \, e^{-i\mathbf{k}\cdot\mathbf{r}} \psi_{\mathbf{p}}(\mathbf{r}) d^3 r \quad , \tag{2.1}$$

where $\mathbf{p}$, $\mathbf{p}'$, and $\mathbf{k}$ denote, respectively, the momenta of the initial and final electron, and of the photon, whose polarization vector is denoted by $\hat{\epsilon}_\lambda$; $\boldsymbol{\alpha}$ is a three-vector whose components are the usual 4×4 Dirac matrices, the electron wave functions $\psi_{\mathbf{p}}(\mathbf{r}), \psi_{\mathbf{p}'}(\mathbf{r})$ are taken as four-components spinors which are of Bloch form, and the integration in (2.1) is taken over the volume of the crystal. Thus [2.18,29],

$$\psi_{\mathbf{p}}(\mathbf{r}) = e^{i\mathbf{p}\cdot\mathbf{r}} u_{\mathbf{p}}(\mathbf{r}) \quad , \quad u_{\mathbf{p}}(\mathbf{r} + \mathbf{L}) = u_{\mathbf{p}}(\mathbf{r}) \quad , \tag{2.2}$$

where $\mathbf{L}$ is a vector of the direct lattice, and similarly for $\psi_{\mathbf{p}'}(\mathbf{r})$.

In this more exact approach, the connection between CB and CR becomes obvious [2.1]. If the Bloch functions in (2.1) are states of high enough energy, it is expected that these functions can be usefully expanded into a plane wave plus correction terms representing the distortion of the plane wave by the periodic crystal potential. The first Born approximation to J_λ results from keeping only the correction terms proportional to the electronic charge e [2.30]. On the other hand, this perturbation expansion is useless when calculating the emission of CR, since this radiation involves transitions between transversely bound states which can be approximated by Bloch states in the tight-binding approximation, but not by distorted plane waves. We thus see that CB and CR are two different aspects of the same phenomenon [2.1], depending on whether the Bloch functions of the electrons are close to free (f) plane-wave states or correspond to channeling states (b).

The radiative transitions of charged particles in a crystal may thus be classified as free-free (ff) transitions, which represent the conventional (coherent) bremsstrahlung (CB), bound-bound transitions (bb), representing channeling radiation (CR) with its various line spectra, and a new type of phenomenon, namely, free-bound (fb) transitions. The three types of radiation may be experimentally distinguished (as will be discussed in Sect.2.5) by kinematical effects, namely, by the dependence of their transition energies on the angle of incidence on the crystal. In summary, it is seen that Bloch functions represent electrons in both bound channeling states and in (quasi-)free continuum states in the lattice potential, and permit an evaluation of the radiation matrix element (2.1) that describes both CB (ff) and CR (bb), as well as fb radiation, in a unified fashion. Incidentally, Bloch-wave channeling [2.31] and channeling radiation [2.32,33] have also been discussed by researchers in the field of electron microscopy.

The principal applications that have been made of CB so far have mainly been for scientific purposes, and have made use of the quasi-monochromatic and highly polarized nature of CB peaks [2.34-37]. For example, CB has been used in elementary-particle physics (e.g., the processes $\gamma p \rightarrow n\pi^+$ [2.38-42], $\gamma p \rightarrow p\pi^0$ [2.40,43], or ρ^0 photoproduction from hydrogen and carbon [2.44]) and in nuclear physics (e.g., proton photoproduction in ^{7}Li [2.45]), employing the CB sources that have been set up at a number of high-energy electron accelerator laboratories such as Frascati, DESY, Cornell, Cambridge, Tokyo, and Kharkov.

While originally [2.11,15] CB was viewed as a high-energy phenomenon which would be significant for electron energies of ~200 MeV or more, a number of low-energy experiments that were subsequently carried out at Leningrad [2.46], NRL [2.47], and Livermore [2.48] have demonstrated the persistence of the coherent-bremsstrahlung effect for energies as low as 35 keV [2.46], although the height of the coherence peaks at such low energies is much reduced. CB as a side-product of experiments on planar [2.30] or axial [2.49] channeling radiation has also been observed in Si crystals for electrons in the few-MeV region by the Aarhus group, and in Si and gold crystals by a group at Illinois [2.50].

The present review is organized as follows. In Sect.2.2, we present a discussion of the kinematics of CB and CR for relativistic longitudinal and nonrelativistic transverse motion of the electrons (with respect to a crystal channel), employing a unified viewpoint for the two phenomena. In Sect.2.3, we discuss a theory [2.51] of CB which is based on the first nonvanishing Born approximation, but which is exact in other respects. In particular, it does not use the small-angle and high-energy approximations, and is believed to be valid down to 1-MeV electron energies for low-Z crystals ($Z \lesssim 14$). In Sect.2.4, some numerical predictions of this theory in the 15-30-MeV electron-energy range are shown and are compared to a CB experiment at these energies. Section 2.5 contains a more detailed overview of ff, fb, and bb kinematical relationships, and a model calculation of fb transition intensities as an example of the link between CB and CR phenomena. Finally, in Sect.2.6, our results and the present status of CB research are summarized.

2.2 General Remarks on CB and CR; Kinematics

In this section, we consider the process

$$e + crystal \rightarrow e' + \gamma + crystal \quad , \tag{2.3}$$

where the electrons, assumed to be relativistic, are described by the relativistic Bloch functions (2.2). The cross section for emission of the photon in (2.3) is given in perturbation theory (to lowest order in powers of $e \equiv 1/137^{\frac{1}{2}}$) by [2.28]

$$d\sigma = \frac{2\pi}{v/V} |H_{if}|^2 \delta(E - E' - k) \frac{Vd^3k}{(2\pi)^3} \frac{Vd^3p'}{(2\pi)^3} \quad . \tag{2.4a}$$

Here E, E', and k are the energies of the initial electron, the final electron, and the photon, respectively, v is the relative speed of the incident electron, V the volume of the crystal, and

$$|H_{if}|^2 = \frac{2\pi e^2}{kV} |J_\lambda|^2 \quad , \tag{2.4b}$$

with J_λ the radiative matrix element of (2.1). The integral over the crystal in (2.1) can be written as a sum of integrals over individual unit cells, and the Bloch-periodicity property in (2.2) allows us to write J_λ as a linear combination of integrals over a single unit cell V_0:

$$J_\lambda = \sum_{\mathbf{L}} \exp[i(\mathbf{p} - \mathbf{p}' - \mathbf{k}) \cdot \mathbf{L}] \int_{V_0} \exp[i(\mathbf{p} - \mathbf{p}' - \mathbf{k}) \cdot \mathbf{r}] u_{\mathbf{p}'}^\dagger \alpha_\lambda u_{\mathbf{p}} d^3r \quad , \tag{2.4c}$$

where the summation runs over all the lattice vectors **L** of the crystal. The transition probability therefore contains the factor

$$\left| \sum_{\mathbf{L}} \exp[i(\mathbf{p} - \mathbf{p}' - \mathbf{k}) \cdot \mathbf{L}] \right|^2 \cong \frac{2\pi}{V_0} N \sum_{\mathbf{G}} \delta(\mathbf{p} - \mathbf{p}' - \mathbf{k} - \mathbf{G}) \quad , \tag{2.5}$$

where the sum runs over all reciprocal-lattice vectors $\mathbf{G} = 2\pi\mathbf{g}$ and N, the number of unit cells in the crystal, is, as usual, assumed to be large. Equations (2.4a and 5) give rise to the energy and momentum conservation laws (Fig.2.1)

$$E = E' + k \quad , \tag{2.6a}$$

$$\mathbf{p} = \mathbf{p}' + \mathbf{k} + \mathbf{G} \quad . \tag{2.6b}$$

The z axis will be taken parallel to a major crystal axis, and the direction of the incident electron momentum **p** will be assumed to make a small angle Θ with the positive z direction. The projections of **p** on the xz and yz planes are denoted by θ_x, θ_y, respectively. The crystal planes that are almost parallel to **p**, i.e., the xz planes or the (shaded) yz planes, each consist, in reality, of a two-dimensional array of lattice points, but as argued by *Lindhard* [2.27], they may in some cases be replaced by continuous planes, especially for fast electrons with $|\mathbf{p}_\perp| \ll p$. At

9

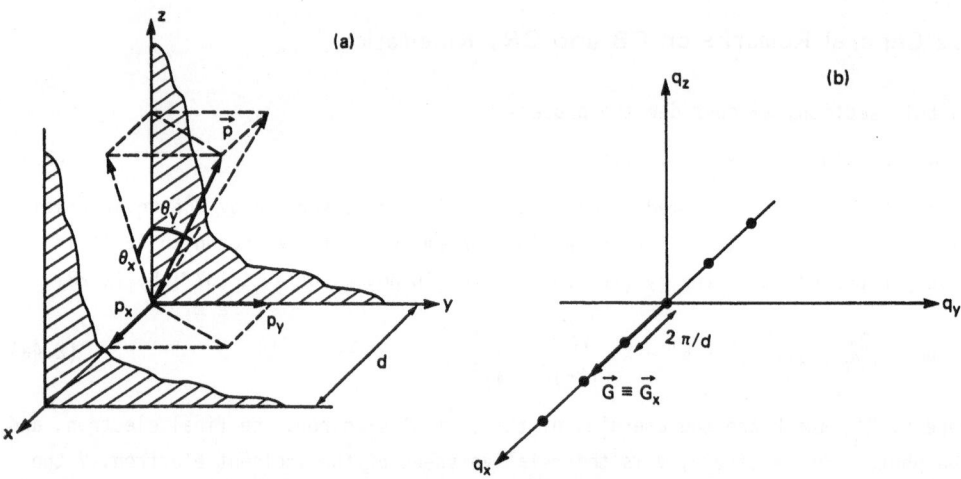

Fig.2.1. (a) Geometry of incident electron beam, crystal axes and planes. (b) One-dimensional reciprocal lattice

ultra-relativistic electron or positron energies, i.e., in the quasi-classical regime, as well as at lower electron energies in the quantum-mechanical regime (e.g., below 10 MeV) there occurs: (1) planar channeling along the yz planes if $\theta_x \lesssim \theta_0$, $\theta_y \gtrsim \theta_0$; (2) planar channeling along the xz planes if $\theta_y \lesssim \theta_0$, $\theta_x \gtrsim \theta_0$; (3) axial channeling if $\theta_x \lesssim \theta_0$, $\theta_y \lesssim \theta_0$. Here $\theta_0 \sim \Theta_c$ in the quasi-classical regime, but generally θ_0 has to be estimated from numerically computed band spectra in the quantum-mechanical regime. CB can occur in all three cases in the latter regime, sometimes spectrally superposed on CR [2.52].

Figure 2.1 corresponds to case (1), when the yz planes involved are the (110) planes of Si. In this example, we replace these atomic planes by continuous charge distributions, shown as cross-hatched planes in Fig.2.1a. The corresponding one-dimensional reciprocal lattice consists of the reciprocal lattice points spaced along the q_x axis, as shown in Fig.2.1b. In this case, the electron "sees" the one-dimensional periodic potential in the x direction arising from the continuous charge distribution in these planes, leading to a one-dimensional band structure for the transverse electron energy levels, as discussed below.

Figure 2.2 corresponds to case (3): axial channeling along the z axis. In this case, one should apply Lindhard's continuum approximation only to the chains of atoms parallel to the z direction, replacing them by continuous strings (Fig.2.2a). The zx and yz crystal planes may then be viewed as consisting of parallel strings [2.52], each plane thus resembling a harp. This picture may be termed the "harp model" of the crystal. For example, if the <001> direction in a Si crystal is taken as the z axis, then the strings are as shown in Fig.2.2a. In this case, the relevant direct lattice is a 2D square lattice, whose points are the intersection points of the strings with the xy plane. The corresponding 2D reciprocal lattice is shown

10

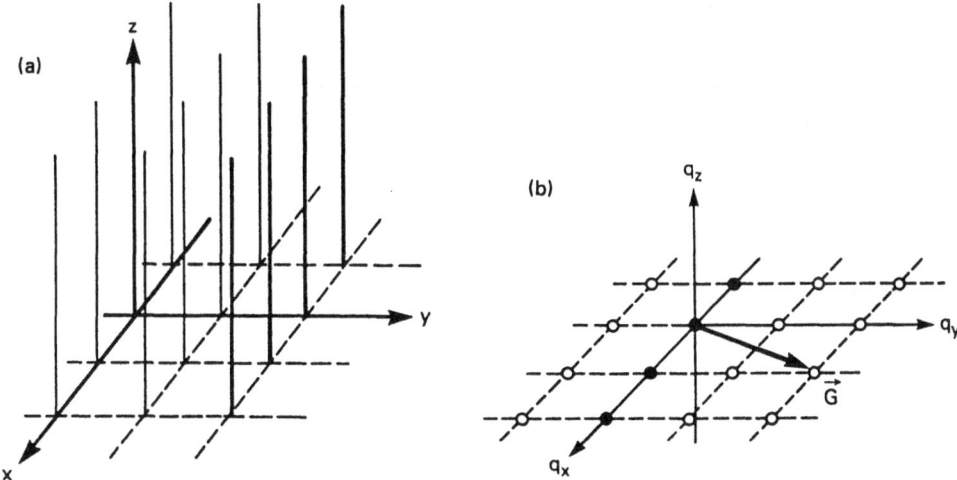

Fig.2.2. (a) Harp model of a crystal. (b) Two-dimensional reciprocal lattice

in Fig.2.2b. In this example, the x and y axes in Fig.2.2 form an angle of 45°
with the <100> and <010> crystallographic axes, respectively, and the lattice spacing of the direct lattice is equal to $d/\sqrt{8}$, d being the atomic spacing along the
strings. The electron "sees" a two-dimensional periodic potential in the xy plane,
with the periods of the direct lattice (Fig.2.2a). This leads to a transverse electron energy spectrum of the two-dimensional band type, as discussed below.

We now give a simple heuristic derivation of the harp model starting from the
Dirac equation governing the motion of an electron in the three-dimensional potential $V(\mathbf{r})$ of the lattice, which for simplicity we take to have cubic structure
[2.53]. We write this equation in the form

$$(i\boldsymbol{\alpha} \cdot \boldsymbol{\nabla} - E - \beta m)\psi(\mathbf{r}) = V(\mathbf{r})\psi(\mathbf{r}) \quad , \tag{2.7}$$

where we express $\psi(\mathbf{r})$ as

$$\psi(\mathbf{r}) = \exp(ip_z z)v_p(\mathbf{r}) \quad , \tag{2.8a}$$

$$v_p(x,y,z + a) = v_p(x,y,z) \quad , \tag{2.8b}$$

and where (2.8b) follows from the Bloch-function property of $\psi(\mathbf{r})$, (2.2). Since we
will be concerned with the situation when the electron motion is relativistic,
with momentum $\mathbf{p}$ almost parallel to the z axis, we can expect that $v_p(\mathbf{r})$ will only
be weakly z-dependent.

We introduce a two-dimensional spinor $w_p(\mathbf{r}_\perp)$, with $\mathbf{r}_\perp = (x,y)$, by

$$w_p(\mathbf{r}_\perp) = \frac{1}{a} \int_0^a v_p(\mathbf{r}) \, dz \quad , \tag{2.9a}$$

and the "harp potential"

11

$$V_h(\mathbf{r}_\perp) = (1/a) \int_0^a V(\mathbf{r})\, dz \quad . \tag{2.9b}$$

We now multiply both sides of (2.7) by $\exp(-ip_z z)$ and average both sides over a period a in the z direction. Making the additional approximation

$$\frac{1}{a} \int_0^a V(\mathbf{r}) v_p(\mathbf{r})\, dz \cong V_h(\mathbf{r}_\perp) w_p(\mathbf{r}_\perp) \quad , \tag{2.9c}$$

where we have used (2.9a,b) and the remark after (2.8b) on the weak z dependence of $v_p(\mathbf{r})$, we obtain:

$$(i\boldsymbol{\alpha} \cdot \nabla_\perp - \alpha_z p_z + E - \beta m) w_p(\mathbf{r}_\perp) = V_h(\mathbf{r}_\perp) w_p(\mathbf{r}_\perp) \quad , \tag{2.10}$$

which is the Dirac equation for a two-dimensional lattice of the harp type. In the derivation of (2.10), we have also used the fact that the average of $\partial v_p(\mathbf{r})/\partial z$ over the interval $0 \leq z \leq a$ is zero by (2.8b) (assuming that v_p is sufficiently smooth).

Separating the spinor $w_p(\mathbf{r}_\perp)$ into large and small components,

$$w_p(\mathbf{r}_\perp) = \begin{pmatrix} w_p^>(\mathbf{r}_\perp) \\ w_p^<(\mathbf{r}_\perp) \end{pmatrix} \quad , \tag{2.11a}$$

leads to the Pauli-type equation

$$(i\boldsymbol{\sigma} \cdot \nabla_\perp - \sigma_z p_z)(E - V_h + m)^{-1}(i\boldsymbol{\sigma} \cdot \nabla_\perp - \sigma_z p_z) w_p^>(\mathbf{r}_\perp) = (E - V_h - m) w_p^>(\mathbf{r}_\perp) \quad . \tag{2.11b}$$

We now keep in mind that in the experiments we want to describe, the electron energy is $E > 1$ MeV while the potential $V_h(\mathbf{r}_\perp)$ is only $\lesssim 100$ eV deep [2.49]. With the ensuing simplifications, there results

$$\left[-\frac{1}{2m\gamma} \nabla_\perp^2 + V_h(\mathbf{r}_\perp) \right] w_p^>(\mathbf{r}_\perp) = \varepsilon w_p^>(\mathbf{r}_\perp) \quad , \tag{2.12a}$$

where

$$\gamma = E/m \quad , \tag{2.12b}$$

$$\varepsilon = \frac{(E^2 - m^2) - p_z^2}{2E} \quad . \tag{2.12c}$$

Equation (2.12a) is the two-dimensional Schrödinger equation for the non-relativistic transverse motion of the electron (with relativistic mass γm) in the harp potential of the lattice. Hence, this equation has a solution $w_p^>(\mathbf{r}_\perp)$ which is a two-dimensional Schrödinger-Bloch function:

$$w_p^>(\mathbf{r}_\perp) = \exp(i\mathbf{p} \cdot \mathbf{r}_\perp) u_p(\mathbf{r}_\perp) \tag{2.13a}$$

$$u_p(\mathbf{r}_\perp + \mathbf{L}_\perp) = u_p(\mathbf{r}_\perp) \quad , \tag{2.13b}$$

with $L_\perp$ any vector of the relevant two-dimensional lattice (intersection of the harp strings with the xy plane).

The above procedure can be repeated for the case of continuous lattice *planes*, Fig.2.1a, in which case one averages the crystal potential over these planes and obtains a one-dimensional Schrödinger equation analogous to (2.12a), its solutions being Bloch functions of the one-dimensional transverse motion.

Figure 2.3 shows the calculated [2.1,30] energy eigenvalues ε^R of one-dimensional Schrödinger-Bloch functions corresponding to the periodically spaced lattice planes of Fig.2.1a, for 4-MeV electrons propagating in directions almost parallel to the (100), (110), and (111) planes of a Si crystal as indicated, plotted vs. θ_x. These eigenvalues pertain to the rest system of longitudinal electron motion, $p_z = 0$. They are depicted in terms of a reduced Brillouin-zone scheme, and show a number of "bound" states (corresponding to the electrons being channeled by a lattice plane), as well as continuum levels (corresponding to the electrons crossing lattice planes). As the energy increases, it is seen that the band gaps get narrower and the levels approach the parabolas $p_x^2/2m$ of free transverse motion in the rest system of longitudinal electron motion, $p_z = 0$. As mentioned in Sect.2.1, CR and CB emission arise, respectively, from bb and ff transitions, which can be quantitatively computed from

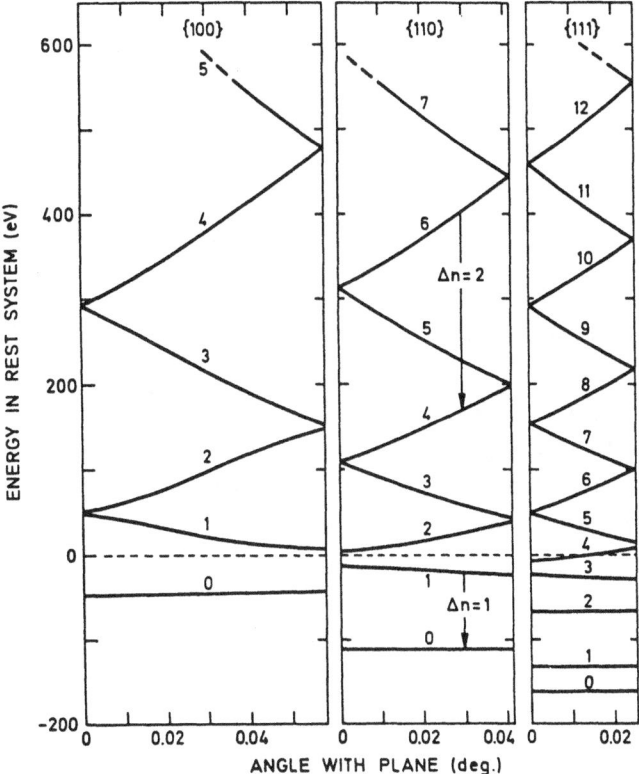

Fig.2.3. Levels of transverse motion (in the rest system) between continuous lattice planes (Fig.2.1a) for 4-MeV electrons in Si crystals, for the lattice planes (100), (110), and (111) as indicated [2.30]

13

these band schemes and a knowledge of the relevant Bloch functions. In addition, there exists a new type of CB, corresponding to fb transitions.

We will now discuss the kinematics for CB and CR, respectively, for free electrons in a three-dimensional lattice and electrons described in the context of the harp model.

For the free-electron case, which is relevant when analyzing CB in the first Born approximation, one finds from (2.6) that the energy of a CB photon is given by

$$k = \frac{\mathbf{p} \cdot \mathbf{G} - \frac{1}{2} \mathbf{G}^2}{E + \mathbf{G} \cdot \hat{k} - \mathbf{p} \cdot \hat{k}} \quad . \tag{2.14a}$$

For a given photon emission angle relative to $\mathbf{p}$ and $\mathbf{G}$, (2.14a) predicts a series of discrete photon energies, specified by the value of $\mathbf{G}$. There are general restrictions on which vectors $\mathbf{G}$ contribute significantly to the CB spectrum, dictated by the values of momentum transfer $\mathbf{q} \equiv \mathbf{p} - \mathbf{p}' - \mathbf{k}$ for which the relevant matrix element is large, e.g., $\mathbf{G}$ must lie in a pancake-shaped region [2.3,15] given by

$$\delta \leqslant q_\ell \lesssim 2\delta \quad ,$$
$$0 \leqslant q_t \lesssim 2k/E \quad . \tag{2.14b}$$

Here $q_\ell = \hat{p} \cdot \mathbf{q}$, $q_t = (q^2 - q_\ell^2)^{\frac{1}{2}}$, and δ is the minimum momentum transfer $\delta = p - p' - k$, where, for $E \gg m$,

$$\delta \cong \frac{k/E}{2(E - k)} \quad . \tag{2.14c}$$

For electrons of energies from one to tens of MeV, one has $p/\gamma^2 \gg G$, which allows 2.50 the simplication

$$k \cong \frac{\mathbf{p} \cdot \mathbf{G}}{E - \mathbf{p} \cdot \hat{k}} \tag{2.14d}$$

of (2.14a) appropriate to CB.

Alternatively, one may consider the case of Fig.2.1a, where the initial direction of the electron is almost parallel to both the yz and xz planes ($\theta_x, \theta_y \ll 1$), and hence also to the crystal axis z. One may additionally assume the photon to be emitted in a direction close to forward ($k_z = k \cos\theta_k$, $\theta_k \ll 1$), which is usually the case for $E \gg m$, and to lie in the lowest part of the spectrum ($k \ll p$), which is also the case for the CB and the prominent CR peaks [2.51]. Writing

$$\mathbf{p} = p_z \hat{z} + \mathbf{p}_\perp \quad , \tag{2.15a}$$

and effecting similar decompositions for $\mathbf{p}$, $\mathbf{k}$, and $\mathbf{G}$, we can express (2.6b) in the form

$$p_z = p_z' + k_z \quad , \tag{2.15b}$$

14

$$\mathbf{p}_\perp = \mathbf{p}'_\perp + \mathbf{G} + \mathbf{k}_\perp \quad , \tag{2.15c}$$

remembering that for a harp potential, considered in the following, one has $G_z = 0$.

One may approximately separate the electron energies into longitudinal and transverse parts,

$$E \cong E_z + E_\perp \quad , \tag{2.16a}$$

where

$$E_z = (p_z^2 + m^2)^{\frac{1}{2}} \quad , \tag{2.16b}$$

and where for θ_x, $\theta_y \ll 1$ the transverse energy $E_\perp(\mathbf{p}_\perp)$ is nonrelativistic, and for free electrons is given by

$$E_\perp(\mathbf{p}_\perp) = p_\perp^2/2\gamma m \quad , \tag{2.16c}$$

while for electrons bound in a two-dimensinal lattice of harp strings, $E_\perp(\mathbf{p}_\perp)$ generally has to be computed numerically. Under the further assumption that $k_\perp/k \ll \gamma^{-1}$ (satisfied if $\theta_k \ll 250$ mrad for 2-MeV, $\ll 100$ mrad for 5-MeV, $\ll 25$ mrad for 20-MeV or $\ll 10$ mrad for 50-MeV electrons), one obtains from (2.6a):

$$E_\perp(\mathbf{p}_\perp) \cong E_\perp(\mathbf{p}_\perp - \mathbf{G}) + (k/2\gamma^2) \quad , \tag{2.16d}$$

which determines the photon energy of the transition in the laboratory system:

$$k = 2\gamma^2[E_\perp(\mathbf{p}_\perp) - E_\perp(\mathbf{p}_\perp - \mathbf{G})] \quad . \tag{2.16e}$$

This equation is appropriate for CR and CB for the harp model, under the assumptions mentioned after (2.14d).

Equations (2.14d and 16e) can be reconciled with each other if the appropriate specializations are made. Specializing $m\gamma$ times the square bracket in (2.16e) to the case of free electrons, and using $G_z = 0$ as assumed in the derivation of (2.14d) one obtains $\mathbf{p}_\perp \cdot \mathbf{G} - G^2/2$, i.e., the numerator of (2.14a). The remaining factor $2\gamma/m$ of (2.16e) can be shown to arise from $(E - \mathbf{p} \cdot \hat{\mathbf{k}})^{-1}$ of (2.14d), using the simplifying assumptions made after this equation. Note, however, that even in its general form, the denominator of (2.14c) represents the well-known factor $(1 - \beta\cos\theta_{pk})^{-1}$, where θ_{pk} is the angle between $\hat{\mathbf{p}}$ and $\hat{\mathbf{k}}$, which gives rise to the forward emission of high-energy bremsstrahlung, or to the "Doppler shift" of CR photons from the $\lesssim 50$-eV region of the transversely bound states [2.49] into the $\lesssim 100$-keV region of photon energies observed in the laboratory system [2.54].

2.3 Born-Approximation Theory of Coherent Bremsstrahlung

In the following, we shall adopt a Born-approximation approach, but retain the exact three dimensional nature of the lattice potential. The consequences of including reciprocal-lattice vectors with $G_z \neq 0$ will be pointed out. In addition, the high-energy and small-angle approximation of high-energy CB studies [2.2,15] will not be made here [2.51]. Our formulas are expected to be accurate only for the lighter elements and for electron kinetic energies $T \gtrsim 1$ MeV. This is a reasonable expectation since, e.g., *Andersen* et al. [2.30] showed that the relevant Born-approximation result was accurate for 4-MeV electrons, and, on the other hand, the measurements of *Rester* and *Dance* [2.55] at $T = 1$ MeV are in reasonably good agreement with the Bethe-Heitler formula of ordinary bremsstrahlung for $Z = 13$.

Our discussion will be confined to perfect crystals with high enough Debye temperature, e.g., diamond or silicon. In such cases, a substantial fraction of bremsstrahlung events occur without transfer of energy to the crystal (zero-phonon processes), just as in the Mössbauer effect. In events of this type, the crystal as a whole takes up the recoil. By "coherent bremsstrahlung" we will mean free-free one-photon emission by a charged particle in a crystal in which the phonon occupation numbers are unchanged. Bremsstrahlung emission processes in crystals in which these occupation numbers change can be expected to be largely incoherent; such non-zero phonon processes are background effects of no interest in this review.

For crystals with a cubic space lattice, such as diamond or silicon, the reciprocal-lattice vectors are of the form

$$\mathbf{G} = (2\pi/a) \sum_{i=1}^{3} n_i \hat{e}_i \ , \tag{2.17a}$$

where $\hat{e}_i$ is a unit vector along the ith cubic axis, a is the length of the edge of the primitive unit cell, and the n_i are integers.

For a given longitudinal recoil momentum

$$G_\ell = \mathbf{G} \cdot \hat{p} \ , \tag{2.17b}$$

the photon energy has to be smaller than or equal to an energy k_G depending on G_ℓ, above which the CB spectrum cuts off more or less abruptly. This effect leads to the well-known peaks in the CB spectrum, and is due to the existence of the minimum momentum transfer δ (2.14b), which must be $\leq G_\ell$. From (2.14a), one has

$$k \leq k_G \cong \frac{pG_\ell}{E - p + G_\ell} \ , \tag{2.17c}$$

assuming that $p \gg G_\ell$.

Two types of CB can be distinguished:

Type A: $\delta \ll 2\pi/a$. In this case, which applies to photon energies $k \ll E$ (typically $x \ll 0.1$ where $x = k/T$), and which is prominent in the high-energy regime [2.15] there are significant contributions to coherent bremsstrahlung only from the reciprocal-lattice plane approximately normal to the direction of incidence that contains the origin ($G_z = 0$). Then, since $G_\ell = G \sin\Theta \cong G\Theta$, where Θ is the angle of the incident electron direction with a lattice axis, one has

$$\Theta \geqslant \delta/G \quad , \tag{2.17d}$$

and this condition leads to the well-known coherent-bremsstrahlung minimum for $\Theta = 0$. The reader is referred to [Ref.2.15, Fig.1] for a geometrical overview of this situation.

Type B: $\delta \gtrsim 2\pi/a$. In this case, the reciprocal-lattice planes approximately normal to the direction of incidence and spaced successively from the $G_z = 0$ plane (but excluding the latter plane) will furnish the prominent coherent-bremsstrahlung peaks, and the angle of incidence $\Theta = 0$ is permitted for coherent-bremsstrahlung production. This situation will mainly be encountered in the low-energy case, and the peaks will be typically situated at $x \gtrsim 0.1$ [2.51].

Applying the standard rules of quantum electrodynamics and using Van Hove's time-dependent formalism [2.56,57], one finds that in the lowest nonvanishing Born approximation the coherent-bremsstrahlung cross section, differential with respect to **p'** and **k** and summed over photon polarizations, is given by

$$\frac{d^2\sigma_{coh}}{d^3\mathbf{p'}d^3\mathbf{k}} = \frac{8\pi\sigma_0 N}{V_0}\,\delta(E - E' - k)\,\frac{m^2}{pE'k}\sum_{\mathbf{G}}\delta(\mathbf{p} - \mathbf{p'} - \mathbf{k} - \mathbf{G})$$

$$\times \;|\mathscr{S}(\mathbf{G})|^2\,\frac{[1 - F(\mathbf{G})]^2}{G^4}\left(\frac{p'^2\sin^2\theta'(4E'^2 - G^2)}{(E'k - \mathbf{p'}\cdot\mathbf{k})^2}\right.$$

$$+ \frac{p^2\sin^2\theta(4E^2 - G^2)}{(Ek - \mathbf{p}\cdot\mathbf{k})^2} - \frac{2}{(E'k - \mathbf{p'}\cdot\mathbf{k})(Ek - \mathbf{p}\cdot\mathbf{k})} \tag{2.18}$$

$$\times \left[p'p\,\sin\theta'\sin\theta\cos\phi(4E'E - G^2 - 2k^2)\right.$$

$$\left.\left.+ k^2(p'^2\sin^2\theta' + p^2\sin^2\theta)\right]\right)$$

for unpolarized incident electrons. Here the sum over **G** runs over all reciprocal vectors; further, $\sigma_0 = Z^2\alpha(e^2/mc^2)^2$, with $\alpha = 1/137$, and Z is the nuclear charge of each atom in the crystal, and we have assumed for simplicity that only one species of atom is present; N is the number of unit cells in the crystal, V_0 the volume of a unit cell, θ the angle between **p** and **k**, θ' the angle between **p'** and **k**, and ϕ the angle between the planes (**p**,**k**) and (**p'**,**k**). The quantity $[1 - F(\mathbf{G})]/G^2$ is proportional to the Fourier transform of the screened nuclear Coulomb field, $[1 - F(\mathbf{G})]$ taking into account the effect of screening by the atomic electrons. The quantity

$$\mathscr{S}(\mathbf{G}) = \sum_{j=1}^{\nu}\,[\exp W_j(\mathbf{G})]\exp(-2\pi i\boldsymbol{\rho}_j\cdot\mathbf{G}) \tag{2.19}$$

describes the interference effect of the atoms in a unit cell and also takes account of the thermal motion of the lattice by means of Debye-Waller-type factors

$\exp[-W_j(\mathbf{G})]$ pertaining to each of the atoms $j = 1,\ldots,\nu$ in a unit cell, $\boldsymbol{\rho}_j$ being the position vector of the jth of these atoms.

For crystals of the diamond-type structure ($\nu = 8$), and assuming as usual that all the $W_j(\mathbf{G})$'s are equal, with $W(\mathbf{G})$ their common value, the factor $|\mathscr{S}(\mathbf{G})|^2$ in (2.18) is equal to $\exp[-2W(\mathbf{G})]$ times a number equal to 64 if all the $n_i (i = 1,2,3)$ are even and their sum is a multiple of 4, equal to 32 if all the n_i are odd, and equal to zero otherwise. The CB calculations for crystals of the diamond type, namely Si crystals [2.58], reported in the next section were carried out under this assumption. The reciprocal lattice for crystals of this structure is shown, e.g., in [2.2].

The cross sections for CB production, differential with respect to the photon direction and to the photon energy, are both of experimental interest and can be obtained from (2.18) without making any approximations. This can be done in a very simple way, as pointed out in [2.51] for the case of the cross section differential with respect to photon energy, but no mention of this fact occurs in the earlier literature.

We will only consider the latter cross section here:

$$\frac{d\sigma_{coh}}{dk} = \iint d^3p' \, d\hat{k} \, \frac{d^2\sigma_{coh}}{d^3p'd\hat{k}} \quad . \tag{2.20}$$

After performing the trivial integration over $\mathbf{p}'$, the integration over $\hat{k}$ can be carried out by a simple trick whose main point is that $\mathbf{p} + \mathbf{G}$ (*not* $\mathbf{p}$) is taken as polar axis for each $\mathbf{G}$. In more detail, let $\mathbf{k}$ make an angle θ_1 with $\mathbf{p} + \mathbf{G}$ and let its azimuthal angle, with respect to an arbitrary direction perpendicular to $\mathbf{p} + \mathbf{G}$, be ϕ_1. We first integrate over θ_1 using the familiar formula for changes of variables in delta functions. We are left with an integral over ϕ_1 which is elementary and are thus led to the following expression [2.59]:

$$\frac{d\sigma_{coh}}{dk} = \frac{4\pi^2\sigma_0 N}{V_0} \frac{m^2}{pk^2} \sum_{\mathbf{G}}{}' \frac{|\mathscr{S}(\mathbf{G})|^2[1 - F(\mathbf{G})]^2}{G^4|\mathbf{p} - \mathbf{G}|} I(\mathbf{G}) \quad , \tag{2.21a}$$

where

$$I(\mathbf{G}) = \sum_{r=1}^{4} I_r(\mathbf{G}) \quad , \tag{2.21b}$$

with

$$I_1(\mathbf{G}) = \frac{[p_\perp(\mathbf{G})]^2(4E^2 - G^2)}{\varepsilon_G^2} \quad ,$$

$$I_2(\mathbf{G}) = (4E'^2 - G^2)\left(-1 + \frac{2E}{\rho_G} - \frac{m^2(E - \eta_G)}{\rho_G^3}\right) \quad ,$$

$$I_3(\mathbf{G}) = 2\frac{4EE' - G^2 + 2k^2}{\varepsilon_G}\left[(E - \varepsilon_G)\left(\frac{E'}{\rho_G} - 1\right) - \frac{\sigma_G}{\rho_G}\right] \quad ,$$

18

$$I_4(\mathbf{G}) = \frac{2k^2}{\varepsilon_G} \left(\eta_G + E' + \frac{[p_\perp(\mathbf{G})]^2 - m^2}{\rho_G} \right) \quad . \tag{2.21c}$$

Here,

$$p_\perp(\mathbf{G}) = [\,|\mathbf{p} - \mathbf{G}|^2 - (E' - \varepsilon_G)^2]^{\frac{1}{2}} \quad ,$$

$$\varepsilon_G = \rho_G/2k \quad ,$$

$$\eta_G = \frac{\sigma_G(E) - \varepsilon_G}{|\mathbf{p} - \mathbf{G}|} \quad ,$$

$$\rho_G = \left[(E' - \eta_G)^2 - \frac{G^2 p^2 - (\mathbf{G} \cdot \mathbf{p})^2}{|\mathbf{p} - \mathbf{G}|^2} \left(1 - \frac{(E' - \varepsilon_G)^2}{|\mathbf{p} - \mathbf{G}|^2} \right) \right]^{\frac{1}{2}} \quad ,$$

$$\sigma_G = \mathbf{p} \cdot (\mathbf{p} - \mathbf{G}) \quad . \tag{2.21d}$$

Notice, in particular, that $p_\perp(\mathbf{G})$ is the magnitude of the part of $\mathbf{p}'$, which is perpendicular to $\hat{\mathbf{k}}$ for given $\mathbf{p}$, $\mathbf{G}$. The prime in the sum in (2.21a) means that one sums over all those Gs such that

$$|E - \sigma_G| < |\mathbf{p} - \mathbf{G}| \quad , \tag{2.21e}$$

which is equivalent to requiring that $|\hat{\mathbf{k}} \cdot (\mathbf{p} - \mathbf{G})| \leqslant |\mathbf{p} - \mathbf{G}|$ for the $\mathbf{p}$, $\mathbf{G}$ considered, implying an inequality expressed approximately by (2.17c).

Although each term of the series in (2.21a) is a lengthy combination of elementary functions, the series can be easily evaluated numerically on a fast electronic computer. Selected results of such evaluations will be discussed in Sect.2.4.

Mozley and *DeWire* [2.60] have pointed out that an axial collimation of the emitted radiation has the effect of additionally monochromatizing the CB peaks. The photons of energies sufficiently close to those at the CB peaks are emitted in directions almost parallel to $\mathbf{p}$, (2.14c), and hence are affected very little by the collimation, while photon intensities at energies far below the peaks are cut down drastically because they emerge at large angles. This effect was verified experimentally in the GeV region by *Tsuru* et al. [2.61]. In contrast with the difficulties of verification at these high energies, it should be relatively straightforward to establish the effect experimentally at electron energies of $\lesssim 10$ MeV, where a rather moderate axial collimation is predicted to have spectacular effects (Sect. 2.4).

Let θ_c denote the collimation angle for photons around the incident electron direction, and let

$$\psi_c \equiv \theta_c/(m/E) \quad . \tag{2.22a}$$

Then, for given values of G_ℓ and θ_c, only photons with energies k satisfying the inequality

$$k_G - \Delta k_G \leqslant k \leqslant k_G \tag{2.22b}$$

will pass through the collimator, where

$$\Delta k_G \cong \frac{k_G(1 - k_G/E)\psi_c^2}{1 + (1 - k_G/E)\psi_c^2} \quad . \tag{2.22c}$$

Equation (2.22c) is an approximation valid for $m/E \ll 1$. The numerical results on axially collimated coherent bremsstrahlung reported in Sect.2.4 are based on exact kinematical formulas and on a formula for the spectrum of the resultant radiation which is an exact consequence of (2.18). The latter formula is analogous to (2.21a) but much more complicated, and hence will be omitted. Again, the photon spectrum predicted by this formula was readily evaluated on a fast electronic computer.

2.4 Numerical Results

In this section, numerical results will be discussed that were obtained from the theory of Sect.2.3. They all refer to thin Si single-crystal targets at room temperature. We show only the coherent part of the bremsstrahlung cross section, and plot the quantity

$$(x/\sigma_0) \frac{d\sigma_{coh}}{dx} \tag{2.23}$$

vs. $x = k/T$, where T is the kinetic energy of the incident electron. The values of $F(\mathbf{G})$ used in these numerical calculations were obtained from the analytical expression for the atomic scattering factor of Si given by *Bonham* and *Strand* [2.62]. The values of $W(\mathbf{G})$ used in our numerical work correspond to the room-temperature value $B \equiv (8\pi^2/3)<u^2> = 0.45\ \text{Å}^2$ derived from X-ray measurements [2.63], where $<u^2>$ is the mean-square vibrational displacement of the atoms from their equilibrium positions.

Coulomb multiple-scattering effects have not been included, but are known to have no overwhelming effect. We have estimated that for electrons with $T = 10$ MeV, crystals with $d = 5$-$10\ \mu$m are adequate to keep the half-angle of divergence of the electron beam caused by multiple scattering below 1-$2°$, which will be seen in the following to be sufficient to preserve the appearance of the CB peaks.

Figure 2.4 illustrates CB of type B, where the most prominent peaks originate from the successive reciprocal-lattice planes that lie approximately normal to the direction of incidence, except for the plane containing the origin. Figure 2.4a shows the CB spectrum for electrons of $T = 15$ MeV, incident exactly parallel to the <100> direction, without photon collimation (dashed curve), and with the emitted bremsstrahlung collimated by $\theta_c = 0.5°$ (solid curve). The two lowest and most prominent peaks are seen to occur at $x = 0.22$ and 0.36, i.e., at $k = 3.3$ and 5.4 MeV, respectively. Each of the peaks in Fig.2.4a drops off abruptly, since above the cut-

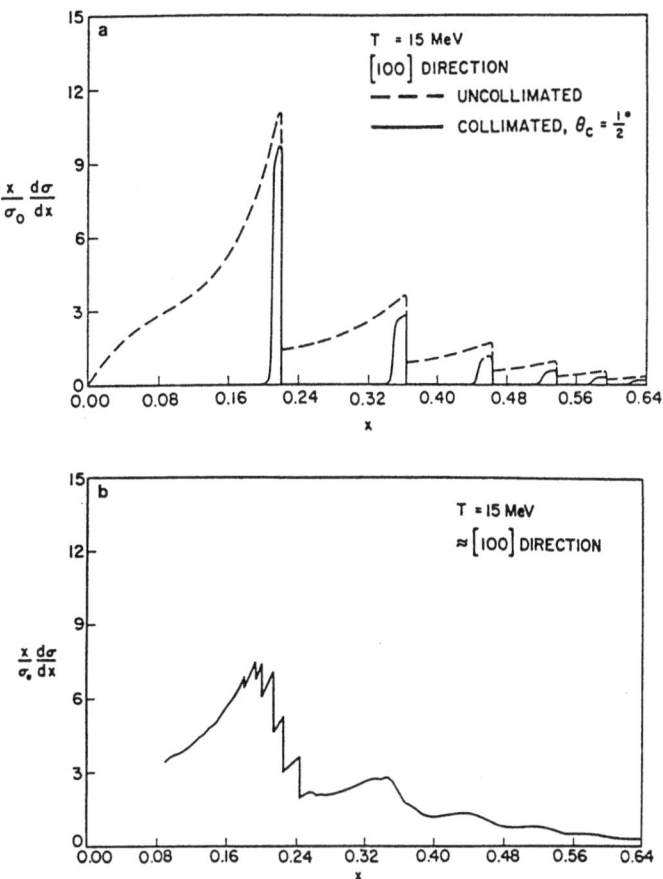

Fig.2.4. (a) Coherent part of bremsstrahlung spectrum of T = 15-MeV electrons incident along the <100> direction of a Si crystal at room temperature: without photon
collimation (---), and with photon collimation of $\theta_c = 1/2°$ (——). (b) As (a) (un-
collimated), but with the electrons incident in the (001) plane at an angle of
$\Theta = 2°$ with the <100> direction

off energy, the contribution of an entire reciprocal lattice plane disappears in the
present case.

If now the beam is misset somewhat from the <100> direction, namely, if it is
taken to be incident in the (001) plane and makes an angle of $\Theta = 2°$ with the <100>
direction, then as shown in Fig.2.4b the peaks "melt", i.e., spread out, showing a
fine structure. This structure comes about since now the cutoff frequencies, corresponding to the reciprocal-lattice points in each contributing reciprocal plane,
are slightly different. The numerical calculations depicted in this figure, as well
as similar calculations not shown here, demonstrate that at the low electron energies considered in this section a half-angle of beam divergence of less than 2°,
e.g., as results from multiple scattering, will not substantially affect the overall
peak structure of the coherent bremsstrahlung spectrum at these energies.

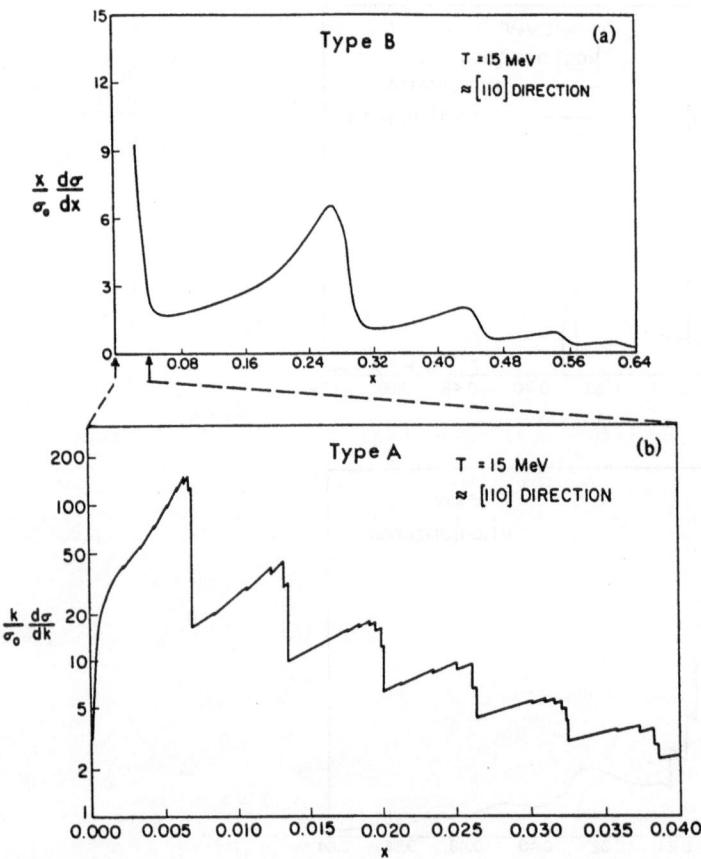

Fig.2.5a,b. Coherent Bremsstrahlung spectrum of T = 15 MeV electrons incident in the (001) plane of a Si crystal at room temperature, making an angle of $\Theta = 1^\circ$ with the <110> direction: (a) high-energy-type portion (B-type CB) and (b) low-energy portion of the spectrum (A-type CB)

The following discussion will include coherent peaks of type A, each of the most prominent ones being generated by contributions of points in a reciprocal plane almost orthogonal to the direction of incidence, containing the origin, and whose normal makes a small but nonzero angle Θ with the direction of incidence. These peaks are located in the low-energy portion of the spectrum. We first consider electrons with T = 15 MeV incident on a Si crystal in the (001) plane and making an angle of 1° with the <110> direction. Figure 2.5a, which depicts the high-energy portion of the spectrum, shows a series of "melted" type-B peaks (again consisting of small spikes that have been averaged in the figure) and a large rise of the spectral intensity at low energies. The latter consists of a series of much more closely spaced type-A peaks, the first of which occurs at about 100 keV, as shown in Fig.2.5b. From our previous discussion, we know that all the latter peaks disappear for $\Theta \to 0$, causing the well-known dip in the coherent bremsstrahlung intensity at

22

the origin when plotted versus Θ. The observed minimum of this dip is, of course, furnished by the incoherent bremsstrahlung contribution which is independent of Θ.

Two further remarks are in order regarding Fig.2.5. As mentioned, the theoretical results of this section were obtained using a full 3D lattice, and hence a full 3D reciprocal lattice. If only a 1D reciprocal lattice (such as leads to the Brillouin zone of Fig.2.3), or the 2D reciprocal lattice of the harp model (Fig.2.2) had been used, where $G_z = 0$, the type-B peaks of Fig.2.5a (as well as of Fig.2.4) would have been lost. The harp model thus only describes the more prominent lower-energy CB peaks of A type, but it and the continuous-plane model of Fig.2.1 additionally describe the CR peaks (if bound-state effects are included), which are found at even lower energies ($k \lesssim 10$ keV, or $x \lesssim 0.0007$ for 15 MeV electrons [2.54]). Such CB peaks would appear in the very lowest part of the spectrum of Fig.2.5b, but they were of course not predicted by our Born-approximation calculation.

Figure 2.6 corresponds closely to the geometry of the experiment of *Walker* et al. [2.48]. We assume that a beam of electrons (or positrons) with T = 28 MeV is incident in the (110) plane of a Si crystal, misset from the <111> direction by $\Theta = 1^{\circ}$. Figure 2.6a exhibits the calculated type-A peaks of the coherent spectrum, in which the lowest and most prominent peak is located at k = 560 keV, close to the experimental value of ~0.5 MeV [2.48]. Note that this coherent structure lies on top of a much larger, (presumably) incoherent-bremsstrahlung contribution in these experiments, as seen from Fig.2.6b, where the measured spectra of forward-emitted CB [2.48] are shown for the situation of Fig.2.6a but with Θ = 50.5 min of arc (top), and for a random beam direction far from the <111> axis and from major planes (bottom).

2.5 Kinematics of CB and CR. Theory of fb Transition Intensities

In this section we will first study the kinematics of ff and fb transitions, both of which can be considered as CB, and trivially, the kinematics of bb transitions, for the concrete example of electrons of kinetic energy T = 3.5 MeV incident on a Si crystal in a direction close to the <100> axis. For this case, experiments by *Andersen* et al. [2.49] on fb and bb transitions have succeeded in determining a set of five bound states that are present in this situation for axial channeling. Table 2.1 lists the energies of the bound states as obtained from the experimental CR transition energies (in the laboratory system) shown in [Ref.2.49, Fig.10]. Plots of photon transition energies can now be obtained using this information and (2.16e). The characteristic variations of these energies under changes of the incidence angles θ_x and θ_y (Fig.2.1) clearly indentify a given transition peak as corresponding to an ff, fb, or bb transition. Indeed, bb transitions between two narrow energy bands of the 2D harp structure yield photon energies which are essentially angle-

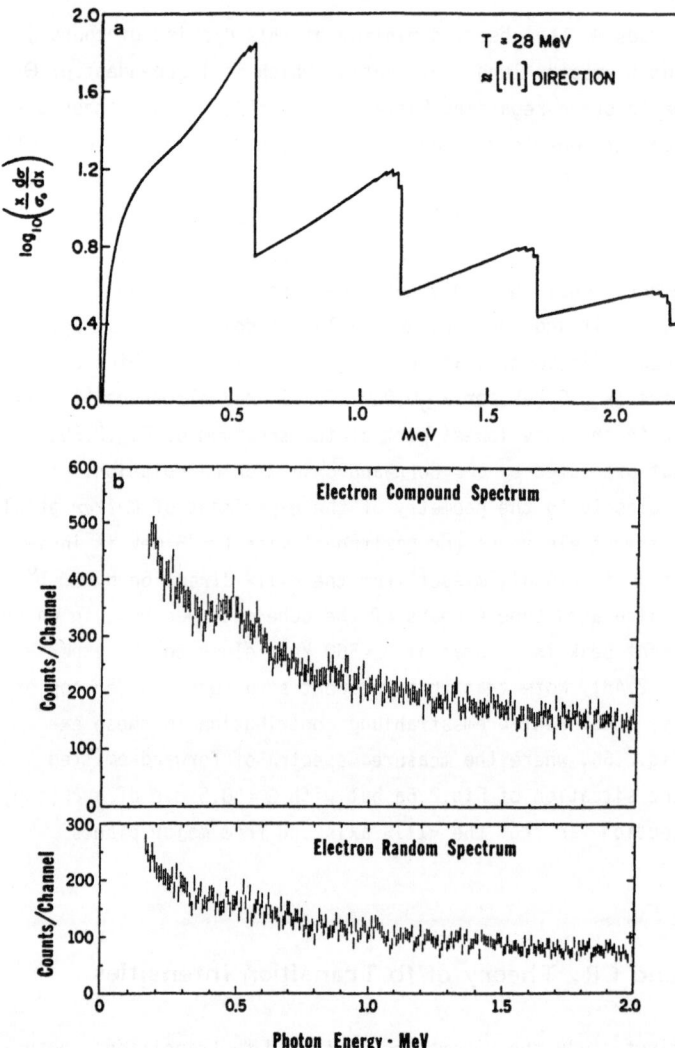

<u>Fig.2.6.</u> (**a**) Low-energy portion of the coherent bremsstrahlung spectrum of T = 28-MeV electrons incident in the (110) plane of a Si crystal at room temperature, making an angle of Θ = 1° with the 111 direction. (**b**) Measured spectra of forward-emitted CB [2.48], for case (**a**) with Θ = 50.5 min of arc (*top*), and for a random beam direction far from the <111> axis and from major planes (*bottom*)

independent. On the other hand, fb transitions from initial states of high enough transverse energy show a parabolic dependence on $\Theta \cong (\theta_x^2 + \theta_y^2)^{\frac{1}{2}}$, since such initial energies are well approximated (in the laboratory system) by [2.49]

$$E_\perp(\mathbf{p_\perp}) \cong \frac{p^2\Theta^2}{2\gamma m} + U_0 \quad , \tag{2.24}$$

where U_0 (= -13.89 eV in the present case [2.49]) represents an average lattice potential. Finally, ff transitions (i.e., CB proper) yield photon energies depending approximately linearly on θ_x, θ_y, which follows if one expresses them as the differ-

24

Table 2.1. Experimental bound-state energies of transverse motion in the laboratory system for 3.5 MeV electrons incident close to the <100> axis of a Si crystal

Bound state	$\varepsilon_{n\ell}$ [eV]
1s	-43.0
2p	-16.2
2s	-10.6
3d	- 6.5
3p	- 4.1

ence of two parabolas. Numerous experimental examples of this linear dependence of CB peaks on transverse electron energy have been studied in [2.50].

For T = 3.5 MeV electrons, the transition energies of emitted CR and CB photons in the laboratory frame are depicted vs. θ_y in Fig.2.7, for fixed values of θ_x.

a) The horizontal lines in this figure represent bb transitions (as shown in Table 2.1), corresponding to $\theta_x = 0$, cf. [Ref.2.49, Fig.10].

b) For fb transitions, (2.16e) can be shown to reduce to

$$k = 2\gamma^2\left(\frac{p_\perp^2}{2\gamma m} + U_0 + |\varepsilon|\right) \quad , \qquad\qquad (2.25a)$$

as expected, if the fb transition takes place between a quasi-free state with transverse momentum $p_\perp$, with an energy $E_\perp(p_\perp)$ as in (2.24), and a bound state (narrow band) of the harp model, with an energy $\varepsilon < 0$ in the laboratory frame. The parabolic curves in the upper (lower) portion of Fig.2.7 depict the energies of the fb transitions to the five bound levels listed in Table 2.1 for $\theta_x = 0.35°$ ($\theta_x = 0°$), as predicted by (2.25a) with $n_x = 8$.

c) Photon energies of various ff transitions, predicted by (2.16e and 24) are also shown in Fig.2.7. The straight lines on the upper right-hand side of this figure depict these ff photon energies vs. θ_y for $\theta_x = 0.35°$, $n_x = 8$, $n_y = 1, \ldots, 5$, where n_x and n_y are the integers determining the components

$$G_x = \frac{2\pi}{a} n_x \quad , \quad G_y = \frac{2\pi}{a} n_y \qquad\qquad (2.25b)$$

of $\mathbf{G}$ in the x and y directions, with $a = d/\sqrt{8}$, d being the atomic spacing along the harp strings as before.

We conclude this section with a summary of cross section and polarization formulas for fb transitions. The corresponding radiation can be viewed as intermediate between coherent bremsstrahlung and channeling radiation.

The formulas were derived on the basis of the following simple model, which is accurate enough for our purposes. The initial electron state ψ_i is taken to be a plane wave, and the final electron state ψ_f as transversely bound in the field of

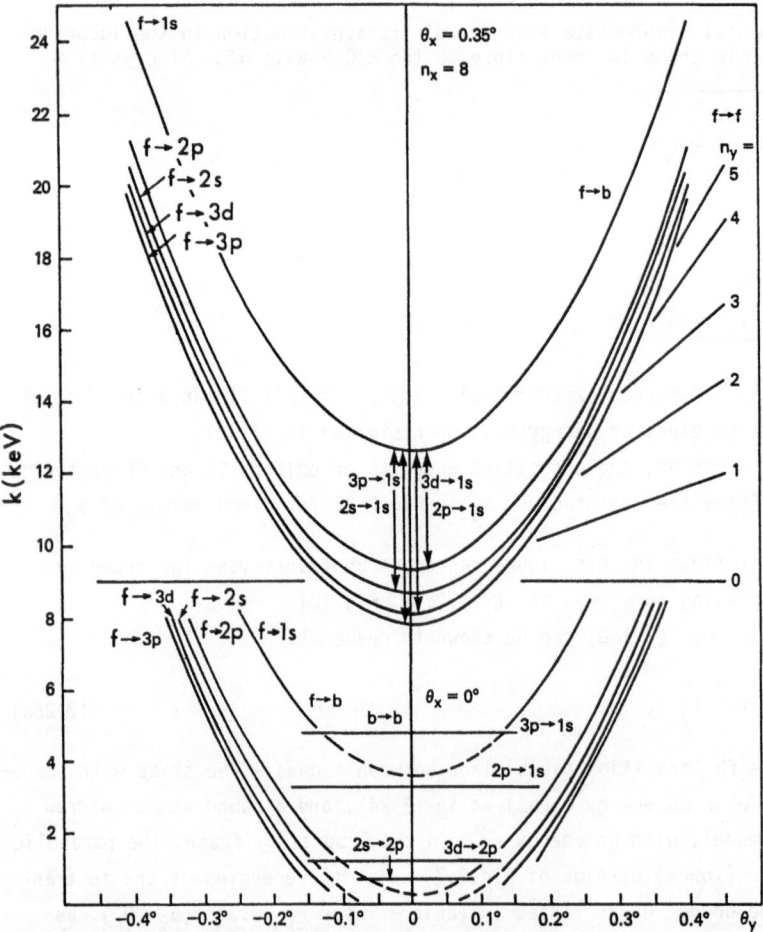

Fig.2.7. Laboratory transition energies of CB and CR photons from $T = 3.5$-MeV electrons incident on Si close to the <100> direction, as a function of θ_y, for fixed values $n_x = 8$ and $\theta_x = 0.35°$ (*upper part*), or $n_x = 0$ and $\theta_x = 0°$ (*lower part*). Vertical arrows between the $f \rightarrow b$ curves of the *upper part* indicate a spacing equal to the corresponding photon energies of $b \rightarrow b$ transitions in the *lower part*

a single harp string. (The same results are obtained with more effort by taking ψ_f as the product of a plane wave and a 2D Bloch wave function given by the lowest-order tight-binding approximation.) Choosing the z axis in the direction of the harp strings, we write

$$\psi_i = C_i \left(\frac{u}{\frac{\boldsymbol{\sigma} \cdot \mathbf{p} u}{E + m}} \right) \exp(i\mathbf{p} \cdot \mathbf{r}) \quad , \tag{2.26a}$$

$$\psi_f = C_f \left(\frac{u}{\frac{\boldsymbol{\sigma} \cdot \mathbf{p}_0 u}{E' + m}} \right) \exp(i p'_z z)\varphi_{n\ell}(\mathbf{r}_\perp) \quad . \tag{2.26b}$$

26

Here, C_i, C_f are normalization constants; u is the two-component spinor $\binom{1}{0}[\binom{0}{1}]$ when the initial electron spin points in the +z [-z] direction in the rest frame; σ has its usual meaning, $\mathbf{p}$ is the initial electron momentum vector, $\mathbf{p}_0 = (-i\partial/x,$ $-i\partial/\partial_y, p_z')$, with p_z' the z component of the final electron momentum, and E, E' are the initial and final electron energies, respectively. The transverse bound-state wave function $\varphi_{n\ell}(\mathbf{r}_\perp)$ has the form

$$\varphi_{n\ell}(\mathbf{r}_\perp) = (2\pi)^{-\frac{1}{2}} e^{\pm i\ell\phi} R_{n\ell}(r_\perp) \quad , \tag{2.26c}$$

where $\mathbf{r}_\perp = (x,y,0)$; $(r_\perp,\phi)$ are polar coordinates in the x, y plane, and $R_{n\ell}(r_\perp)$ satisfies a radial Schrödinger equation with the electron rest mass replaced by E and with the potential $V_{st}(r_\perp)$ corresponding to a single harp string. Naturally, $\int |\varphi_{n\ell}(\mathbf{r}_\perp)|^2 d\mathbf{r}_\perp = 1$.

We denote by $d^2\sigma_\lambda/dk\, d\Omega_k$ the cross section *per harp string of length L*, differential with respect to the photon energy k and photon solid angle $d\Omega_k$, for emission of a photon of polarization $\hat{\epsilon}_\lambda$ in the fb transition $\psi_i \to \psi_f$. To lowest order in e^2, this cross section is found to be

$$\frac{1}{L}\frac{d^2\sigma_\lambda}{dk d\Omega_k} = \frac{e^2}{4}\frac{k}{pE'}\frac{|I_{n\ell}(\kappa)|^2}{1-(q_z/E_z')\hat{k}_z}\left\{\frac{E'+m}{E+m}p^2(\hat{\epsilon}_\lambda\cdot\epsilon_\lambda^*)\right. \tag{2.27a}$$

$$\left. +\frac{E+m}{E'+m}q^2(\hat{\epsilon}_\lambda\cdot\hat{\epsilon}_\lambda^*)+2[(\hat{\epsilon}_\lambda\cdot\mathbf{p})(\hat{\epsilon}_\lambda^*\cdot\mathbf{q})+(\hat{\epsilon}_\lambda^*\cdot\mathbf{p})(\hat{\epsilon}_\lambda\cdot\mathbf{q})-(\hat{\epsilon}_\lambda\cdot\hat{\epsilon}_\lambda^*)(\mathbf{p}\cdot\mathbf{q})]\right\}\delta(k-k_{n\ell}) \quad ,$$

where

$$E_z' = (m^2 + q_z^2)^{\frac{1}{2}} \tag{2.27b}$$

and $\mathbf{q} = \mathbf{p} - \mathbf{k}$; $k_{n\ell}$ is the photon energy in the laboratory frame for the fb transition:

$$k_{n\ell} \cong \frac{p_\perp^2/2E + |\epsilon_{n\ell}|}{1 - \beta_z\hat{k}_z} \quad , \tag{2.28a}$$

[which is essentially the same as (2.25a), except for the empirical U_0], where $\beta_z = p_z/E$, and

$$I_{n\ell}(\kappa) = \int_0^\infty J_\ell(\kappa\rho)R_{n\ell}(\rho)\rho d\rho \quad . \tag{2.28b}$$

Here, J_ℓ is a Bessel function of the first kind of order ℓ, and $\kappa = |\mathbf{p}_\perp - \mathbf{k}_\perp|$, with $\mathbf{p}_\perp$, $\mathbf{k}_\perp$ the components of $\mathbf{p}$, $\mathbf{k}$ perpendicular to the z direction.

From (2.27), the corresponding cross section summed over photon polarizations is

$$\frac{1}{L}\frac{d^2\sigma}{dk d\Omega_k} = \frac{1}{L}\sum_{\lambda=1}^{2}\frac{d^2\sigma_\lambda}{dk d\Omega_k} = \frac{e^2}{2}\frac{k}{pE'}\frac{|I_{n\ell}(\kappa)|^2}{1-(q_z/E_z')\hat{k}_z}\left[\frac{E'+m}{E+m}p^2\right.$$

$$\left. +\frac{E+m}{E'+m}q^2 - 2(\hat{k}\cdot\mathbf{p})(\hat{k}\cdot\mathbf{q})\right]\delta(k-k_{n\ell}) \quad , \tag{2.29}$$

27

and the linear photon polarization is

$$P \equiv \frac{(d^2\sigma_1/dkd\Omega_k) - (d^2\sigma_2/dkd\Omega_k)}{d^2\sigma/dkd\Omega_k}$$

$$= \frac{2[(\hat{\varepsilon}_1 \cdot \mathbf{p})(\hat{\varepsilon}_1 \cdot \mathbf{q}) - (\hat{\varepsilon}_2 \cdot \mathbf{p})(\hat{\varepsilon}_2 \cdot \mathbf{q})]}{\frac{E' + m}{E + m} \mathbf{p}^2 + \frac{E + m}{E' + m} \mathbf{q}^2 - 2(\hat{k} \cdot \mathbf{p})(\hat{k} \cdot \mathbf{q})} \quad . \tag{2.30}$$

Some important aspects of fb transitions can be deduced from these results. Notice first that the cross sections (2.27 and 29) are proportional to $|I_{n\ell}(\kappa)|^2$, i.e., to the population of the state $\varphi_{n\ell}$ which would be created by an incident plane wave in the crystal, with transverse momentum vectpr $\mathbf{p}_\perp - \mathbf{k}_\perp$ ($\cong \mathbf{p}$ in the Aarhus experiments [2.49]). Notice also that the cross sections are small at $\hat{k} = \hat{p}$ (forward emission). In our approximation, the fb transition rates are the same whether the incident electrons are polarized or not. As expected, zero circular polarization for the fb photons is predicted here.

Detailed theoretical discussions of the energy and angular distributions of fb photons from a Si crystal will be given in a subsequent publication.

2.6 Summary

Ten years ago, the subject of coherent bremsstrahlung appeared to be well understood, well studied, and well utilized, with few of its aspects left open to further exploration. Most details of the theory had been worked out [2.2,3] (at least in the first Born approximation), numerous high-energy experiments had been carried out, confirming quantitatively the detailed theory (including, e.g., the predicted polarization and the Mozley-DeWire collimation effect [2.60,61]), low-energy experiments [2.24,48] and theory [2.64] showed the existence of CB effects even in the region of electron energies of a few MeV and below, and research applications of the polarized, quasi-monochromatic CB peaks had been made [2.38,45]. The subject thus appeared to be as good as closed, when the discovery of channeling radiation [2.26,65], which was partly motivated by the study of modifications in the low-energy positions of CB spectra of electrons or positrons associated with their channeling [2.48], suddenly brought renewed interest to it. The interest increased when it became clear [2.1,30] that CB and CR are closely related phenomena and in fact can be described in a unified fashion by the use of Bloch wave functions for the electrons or positrons traversing the crystal [2.18,30,33].

Although in recent years CR, rather than CB, has occupied most of the attention of experimentalists in the field, interesting new CB results have emerged as by-products of these efforts. The physical situation may be viewed by an extrapolation of Fig.2.5. The type-B CB peaks for 15 MeV electrons are located in the 4-10 MeV region of Fig.2.5a. If the low-energy portion of the figure is blown up, as in Fig. 2.5b, one perceives the type-A CB peaks located in the 100-600-keV region. Channeling-radiation peaks would now be located in the lowest-energy portion of this figure, i.e., in the $\leq$100 keV region, to the left of the lowest-energy peak in Fig. 2.5b (Chap.1, Fig.1.1c). They have shapes quite similar to those of the CB peaks in Fig.2.5a and b, with a sharp cutoff at the high-energy side (e.g., [Ref.2.54, Fig.5]). Hence, we predict that CR experiments in the 100 keV region, when extended to higher photon energies, will readily encounter CB peaks of the fb and ff types. The intensities of the peaks will, however, successively decrease, being highest for CR, lower for CB of type A, and lower still for CB of type B.

CB has been studied in conjunction with CR measurements by the Aarhus [2.49] and Illinois [2.50] groups. This unified experimental approach likewise symbolizes the unity of CR and CB phenomena, and one may consider it a source of great satisfaction that at the present time, the channeling effects of electrons or positrons in crystals, whose existence had once been viewed [2.48,66] as an annoyingly unknown factor potentially affecting the otherwise known phenomenon of CB, have been understood and recognized as contributor and complement to CB radiation of these particles when they transverse a crystal.

Acknowledgments. We are indebted to Drs. B. Buxton, S. Datz, D.J. Nagel, A. Nagl, and G. Temmer for enlightening discussions. We also wish to thank Professor Yu.S. Korobochko for helpful contributions. One of us (H. Überall) acknowledges partial support from the National Science Foundation at Catholic University.

References

2.1 J.U. Andersen: Nucl. Instrum. Methods **170**, 1 (1980)
2.2 G. Diambrini Palazzi: High-Energy Bremsstrahlung and Electron Pair Production in Thin Crystals. Rev. Mod. Phys. **40**, 611 (1968)
2.3 U. Timm: Coherent Bremsstrahlung of Electrons in Crystals. Fortschr. Phys. **17**, 765 (1969)
2.4 M.L. Ter-Mikaelian: *High-Energy Electromagnetic Processes in Condensed Media* (Wiley-Interscience, New York 1972) pp.34-113 and 391-410
2.5 U. Timm: Grundlagen des Überall-Effekts. Deutsches Elektronen-Synchrotron Report DESY 64/9 (Exp.), Hamburg, June 1964 (unpublished)
2.6 A.P. Komar, Yu.S. Korobochko: Bremsstrahlung of Relativistic Electrons in Crystals. Fiz. Tverd. Tela **13**, 245 (1971) [English transl.: Sov. Phys-Solid State **13**, 192 (1971)]
2.7 E.J. Williams: Phys. Rev. **45**, 729 (1934)
2.8 E.J. Williams: K. Dan. Vidensk. Selsk. Mat. Fys. Medd. **13**, No. 4 (1935)
2.9 B. Ferretti: Nuovo Cimento **7**, 118 (1950)
2.10 L. Landau, I. Pomeranchuk: Dokl. Akad. Nauk. SSSR **92**, 535, 735 (1953)
2.11 M.L. Ter-Mikaelian: Zh. Eksp. Teor. Fiz. **25**, 296 (1953)
2.12 F.J. Dyson, H. Überall: Phys. Rev. **99**, 604 (1955)
2.13 E.M. Purcell: unpublished (1955)
2.14 O.R. Frisch: Acta Phys. Austriaca **12**, 331 (1959)
2.15 H. Überall: Phys. Rev. **103**, 1055 (1956)

2.16 H. Überall: Phys. Rev. **107**, 223 (1957)
2.17 H. Überall: Z. Naturforsch. **17**a, 332 (1962)
2.18 L.I. Schiff: Phys. Rev. **117**, 1394 (1960)
2.19 B. Ferretti, G. Gamberini: Lett. Nuovo Cimento 3, 113 (1970)
2.20 A.I. Akhiezer, P.I. Fomin, N.F. Shul'ga: Pis'ma Zh. Eksp. Teor. Fiz. **13**, 713 (1971) [English transl.: JETP Lett. **13**, 506 (1971)]
2.21 A.I. Akhiezer, V.F. Boldyshev, N.F. Shul'ga: Fiz. Elem. Chastits At. Yadra **10**, 51 (1979) [English transl.: Sov. J.Part. Nucl. **10**, 19 (1979)]
2.22 G. Bologna, G. Diambrini Palazzi, G.P. Murtas: Phys. Rev. Lett. **4**, 134 (1960)
2.23 G. Barbiellini, G. Bologna, G. Diambrini Palazzi, G.P. Murtas: Phys. Rev. Lett. **8**, 454 (1962)
2.24 A.P. Komar, Yu.S. Korobochko, V.I. Mineev, A.F. Petrochenko: Zh. Tekh. Fiz. **41**, 807 (9171) [English transl.: Sov. Phys.-Tech.Phys. **16**, 631 (1971)]
2.25 A.A. Vorobiev, V.V. Kaplin, S.A. Vorobiev: Nucl. Instrum. Methods **127**, 265 (1977); *note*, however, that this work represents a controversial approach to the subject of channeling radiation
2.26 M.A. Khumakov: Phys. Lett. **57**A, 17 (1976)
2.27 J. Lindhard: K. Dan. Vidensk. Selsk. Mat. Fys. Medd. **34**, No. 14 (1965)
2.28 See, e.g., W. Heitler: *The Quantum Theory of Radiation*, 3rd. ed. (Clarendon, Oxford 1954)
2.29 Although a need for the use of Bloch functions was indicated by Schiff [2.18], he incorrectly proceeded to employ them in place of plane-wave states in a first-order Born matrix element, rather than in the fundamental radiation matrix element of (2.1)
2.30 J.U. Andersen, K.R. Eriksen, E. Laegsgaard: Phys. Scr. **24**, 588 (1981)
2.31 A. Howie: Philos. Mag. **14**, 223 (1966)
2.32 C.J. Humphreys: Electr. Microsc. **4**, 68 (1980)
2.33 D.M. Bird, F.B. Buxton: Proc. Roy. Soc. (London) A**379**, 459 (1982)
2.34 L. Criegee, G. Lutz, H.D. Schulz, U. Timm, W. Zimmermann: Phys. Rev. Lett. **16**, 1031 (1966)
2.35 V.G. Gorbenko, Yu.V. Zhebrovskii, L.Yu. Kolesnikov, I.I. Miroshnichenko, L.M. Romas'ko, A.L. Rubashkin, P.V. Sorokin: Yad. Fiz. **11**, 1044 (1970) [English transl.: Sov. J. Nucl. Phys. **11**, 580 (1970)]
2.36 M. Kobayashi, S. Hiromatsu, K. Kondo, S.I. Kurokawa, T. Nishikawa, H. Yoshida: J. Phys. Soc. Japan **36**, 1 (1974)
2.37 V.G. Gorbenko, Yu.V. Zhebrovskii, N.A. Kovalenko, L.Ya. Kolesnikov, I.I. Miroshnichenko, A.L. Rubashkin, S.V. Shalatskii: Yad. Fiz. **24**, 961 (1976) [English transl.: Sov. J. Nucl. Phys. **24**, 503 (1976)]
2.38 C. Geweniger, P. Heide, U. Kötz, R.A. Lewis, P. Schmüser, H.J. Skronn, H. Wahl, K. Wegener: Phys. Lett. **29**B, 41 (1969)
2.39 Z. Bar-Yam, J. de Pagter, J. Dowd, W. Kern, D. Luckey, L.S. Osborne: Phys. Rev. Lett. **25**, 1053 (1970)
2.40 V.G. Gorbenko, Yu.V. Zhebrovskii, I.M. Karnaukhov, L.Ya. Kolesnikov. A.L. Rubashkin, P.V. Sorokin, Yu.N. Telegin: Yad. Fiz. **30**, 136 (1979) [English transl.: Sov. J. Nucl. Phys. **30**, 70 (1979)]
2.41 V.A. Get'man, V.G. Gorbenko, A.Ya. Derkach, Yu.V. Zhebrovskii, I.M. Karnaukhov, L.Ya. Kolesnikov, V.S. Kuz'menko, A.A. Lukhanin, Yu.N. Ranyuk, A.L. Rubashkin, V.M. Sanin, P.V. Sorokin, E.A. Sporov, Yu.N. Telegin, S.V. Shalatskii, V.F. Grushin: Yad. Fiz. **31**, 930 (1980) [English transl.: Sov. J. Nucl. Phys. **31**, 480 (1980)]
2.42 V.A. Get'man, V.G. Gorbenko, V.F. Grushin, A.Ya. Derkach, Yu.V. Zhebrovskii, I.M. Karnaukhov, L.Ya. Kolesnikov, A.A. Luchanin, A.L. Rubashkin, V.M. Sanin, P.V. Sorokin, E.A. Sporov, Yu.N. Telegin, S.V. Shalatskii: Yad. Fiz. **32**, 1008 (1980) [English transl.: Sov. J. Nucl. Phys. **32**, 521 (1980)]
2.43 V.B. Ganenko, V.G. Gorbenko. L.M. Derkach, Yu.V. Zhebrovskii, L.Ya. Kolesnikov, I.I. Miroshnichenko, V.I. Nikiforov, A.L. Rubashkin, V.M. Sanin, P.V. Sorokin, S.V. Shalatskii: Yad. Fiz. **23**, 310 (1976) [English transl.: Sov. J. Nucl. Phys. **23**, 162 (1976)]
2.44 G. Diambrini-Palazzi, G. McClellan, N. Mistry, P. Mostek, H. Ogren, J. Swartz, R. Talman: Phys. Rev. Lett. **25**, 478 (1970)

2.45 Yu.P. Antuf'ev, Yu.V. Zhebrovskii, L.Ya. Kolesnikov, V.S. Kuz'menko, I.I. Miroshnichenko, A.L. Rubashkin, P.V. Sorokin: Yad. Fiz. **9**, 680 (1969) [English transl.: Sov. J. Nucl. Phys. **9**, 394 (1969)]
2.46 Yu. S. Korobotchko, V.F. Kosmach, V.I. Mineev: Zh. Eksp. Teor. Fiz. **48**, 1248 (1965) [English transl.: Sov. Phys.-JETP **21**, 834 (1965)]
2.47 T.F. Godlove, M.E. Toms: U.S. Naval Research Laboratory, Nuclear Physics Division Annual Report 1969 (unpublished), pp. 96-97
2.48 R.L. Walker, B.L. Berman, S.D. Bloom: Phys. Rev. A**11**, 736 (1975)
2.49 J.U. Andersen, E. Bonderup, E. Laegsgaard, B.B. Marsh, A.H. Sørensen: Nucl. Instrum. Meth. **194**, 209 (1982)
2.50 J.E. Watson, J. Koehler: Phys. Rev. A**24**, 861 (1981); Phys. Rev. B**25**, 3079 (1982)
2.51 A.W. Sáenz, H. Überall: Phys. Rev. B**25**, 4418 (1982)
2.52 V.V. Beloshitsky, M.A. Kumakhov: Radiat. Eff. **56**, 25 (1981)
2.53 The legitimacy of replacing a three-dimensional lattice potential by a suitably averaged potential to describe the quantum-mechanical motion of electrons in crystals is treated by M.V. Berry: J. Phys. C**4**, 697 (1971) in a more rigorous manner than our discussion in the present section
2.54 A.W. Sáenz, H. Überall, A. Nagl: Nucl. Phys. A**372**, 90 (1981)
2.55 D.H. Rester, W.E. Dance: Phys. Rev. **161**, 85 (1967)
2.56 L. Van Hove: Phys. Rev. **95**, 249 (1954)
2.57 In particular, Eq. (2.7) in A.W. Sáenz: Phys. Rev. **119**, 1542 (1960), was used [a factor i is missing in the second exponential in that equation]
2.58 See, e.g., C. Kittel: *Introduction to Solid State Physics*, 5th ed. (Wiley, New York 1976) pp.25, 69
2.59 This simple method for evaluating $d\sigma_{coh}/dk$ exactly from Eq. (2.18) was pointed out by one of us (A.W.S.) in unpublished NRL work, in which Eqs. (2.21a-d) and the exact formulas for CB alluded to in the last paragraph of this section were first derived. The simplicity of this method is to be contrasted with the complicated procedure used in the literature (see, e.g., [2.15-17] to calculate $d\sigma_{coh}/dk$ from the Bethe-Heitler formula using the high-energy and small-angle approximation
2.60 R.F. Mozley, J. DeWire: Nuovo Cimento **27**, 1281 (1963)
2.61 T. Tsuru, S. Kurokawa, T. Nishikawa, S. Suzuki, T. Katayama, M. Kobayashi, K. Kondo: Phys. Rev. Lett. **27**, 609 (1971)
2.62 R.A. Bonham, T.G. Strand: J. Chem. Phys. **39**, 2200 (1963)
2.63 *International Tables for X-Ray Crystallography*, ed. by K. Lonsdale (Kynoch Press, Birmingham 1968) Vol.III, p.237
2.64 Yu.S. Korobochko: Zh. Tekh. Fiz. **36**, 1394 (1966) [English transl.: Sov. Phys.-Tech. Phys. **11**, 1041 (1967)]
2.65 R.W. Terhune, R.H. Pantell: Appl. Phys. Lett. **30**, 265 (1977)
2.66 R.L. Walker, B.L. Berman, R.C. Der, R.M. Kavanagh, J.M. Khan: Phys. Rev. Lett. **25**, 5 (1970)

3. Coherent Bremsstrahlung – Experiment

G. D. Kovalenko, L. Ya. Kolesnikov, and A. L. Rubashkin
With 21 Figures

A review of experimental research performed at different laboratories on coherent
bremsstrahlung of electrons in crystals is presented. Methods for producing and monitoring quasi-monochromatic, linearly polarized photon beams at cyclic and linear
electron accelerators are considered. Spectral and orientation dependences of bremsstrahlung intensity and polarization in crystals are presented and discussed.

3.1 History

In studies of the interactions of high-energy particles with matter, the effect of
the structure of the medium is generally neglected and the interaction cross section is assumed to be a sum of cross sections for the separate atoms. However, there
are some cases when the medium structure must be taken into account, the phenomenon
of coherent bremsstrahlung (CB) being an example.

The periodicity of a medium gives rise to interference maxima in the CB intensity and polarization spectra. The effect of medium periodicity on high-energy electron bremsstrahlung was pointed out about 50 years ago by *Williams* [3.1]. Using the
method of virtual quanta, *Ferretti* [3.2] derived an expression for the coherent
bremsstrahlung cross section in an ideal crystal (neglecting thermal oscillations
and screening effects). He also pointed out the existence of certain initial electron
beam directions relative to the crystal axes in which the bremsstrahlung sharply increases. The modification of the Bethe-Heitler cross section for the case of ultra-
relativistic electron bremsstrahlung in condensed media was noted by *Landau* and
Pomeranchuk [3.3].

Ter-Mikaelian [3.4,5] studied the effects of the thermal oscillations of crystal-
lattice atoms and the screening of the nuclear Coulomb potential on the radiation
intensity in the crystal. He derived the exact formula for intensity-maximum positions in the coherent spectrum and obtained a criterion for the onset of interference effects. Independently of *Ter-Mikaelian*, *Dyson* and *Überall* [3.6] gave a criterion for the occurrence of interference effects and explained the influence of the
crystal lattice on the bremsstrahlung process. Later on, by using the Born approximation, *Überall* [3.7,8] calculated total coherent-bremsstrahlung cross sections

which could be conveniently compared with experimental data. He showed that for certain crystal orientations the coherent bremsstrahlung led to a linear photon polarization. Circular photon polarization was shown to occur in the case of polarized electron bremsstrahlung [3.9]. Similar conclusions were drawn by *Sekhposian* and *Ter-Mikaelian* [3.10] who used the method of virtual quanta.

The experiments by *Panofsky* and *Saxena* [3.11] did not confirm the effect of the crystal structure on the bremsstrahlung intensity. The enhancement of the soft-photon bremsstrahlung intensity in the crystal was first observed by *Frisch* and *Olson* [3.12] and some time later by *Saxena* [3.13]. An analysis of these experimental results has been given by *Schiff* [3.14].

The further progress in theoretical and experimental studies of CB is associated with experiments performed at the Frascati 1 GeV electron synchrotron by *Diambrini* and co-workers [3.15-22] during the years 1960-1962. Unexpectedly, in contrast to a continuous enhancement of γ-radiation intensity, as predicted by Überall, this group observed a spectrum with peaks whose origin was explained in terms of the crystal-lattice structure. Besides performing measurements, these authors calculated CB-intensity and polarization spectra.

In 1965, bremsstrahlung measurements were carried out in Tokyo for a silicon crystal at an electron energy E_0 of 0.7 GeV [3.23,24]. In 1966, similar experiments were performed for a diamond crystal and $E_0 = 6$ GeV at DESY (Deutsches Elektronen-Synchrotron) [3.25-27]. The DESY experiments were done for a crystal orientation that provided a higher photon polarization. The method of electron-positron pair production in a crystal, suggested by *Barbiellini* et al. [3.22], enabled the researchers at DESY to determine experimentally the degree of photon polarization at a coherent maximum.

In 1963, *Mozley* and *DeWire* [3.28] pointed out that the coherent quasi-monochromatic spectrum can be modified to a line spectrum by using tight collimation and thin single-crystal targets, thereby simultaneously increasing the beam polarization and reducing incoherent background. Theoretical estimates of this phenomenon were also made by *Bologna* [3.29] and *Lutz* [3.30]. Experimental evidence of this effect was obtained at the facilities at Hamburg [3.31], Erevan [3.32], and Tokyo [3.33].

As new electron accelerators were put into operation and the experimental and theoretical investigations on the physics of high-energy electromagnetic interactions proceeded at a rapidly increasing rate, beams of quasi-monochromatic, linearly polarized photons were obtained at several laboratories around the world. In 1968, a CB beam was obtained at the Kharkov 2-GeV electron linac [3.34-37]. Bremsstrahlung beams were produced at the 0.8-GeV electron synchrotron of the Tomsk Nuclear Physics Institute (1969) [3.38], the 6-GeV synchrotron of the Erevan Institute of Physics (1970) [3.32] as well as at the 6-GeV CEA synchrotron at Cornell University (1968) [3.39] with a subsequent gain in energy up to 10.8 GeV (1970) [3.40], at the 16-GeV Stanford linear accelerator (1971) [3.41], at the 5-GeV NINA synchrotron in Daresbury (1973) [3.42], and at the 2.5-GeV synchrotron of the University of Bonn [3.43].

The method based on target CB, and its use at modern proton accelerators at the Serpukhov Institute of High-Energy Physics [3.44], CERN, and FNAL [3.45,46] seems to be promising for the future. The CB from low-energy electrons was observed and analyzed in the work reported in [3.47,48]. Reviews of theoretical and experimental research on CB are given in [3.49-51].

3.2 General Formulas for Coherent Bremsstrahlung

3.2.1 Kinematics

According to the laws of conservation of energy and momentum, we have

$$\mathbf{p}_0 = \mathbf{p} + \mathbf{k} + \mathbf{q} \quad , \tag{3.1}$$

$$E_0 = E + k \quad , \tag{3.2}$$

where $\mathbf{p}_0$, E_0, $\mathbf{p}$, E, $\mathbf{k}$, k are the momentum and energy of the initial electron, final electron and photon, respectively; $\mathbf{q}$ is the recoil momentum of the nucleus. Owing to the large nuclear mass, the recoil energy in (3.2) may be neglected. Here, and henceforth, we use the system of units in which $m = c = \hbar = 1$.

Using momentum and angular relations, which are shown schematically in Fig.3.1, the longitudinal q_ℓ and transverse q_t components of the recoil momentum may be written approximately as

$$q_\ell = p_0 - p \cos\theta_e - k \cos\theta_k \quad , \tag{3.3}$$

$$q_t^2 = k^2\theta_k^2 + p^2\theta_e^2 + 2pk\theta_e\theta_k \cos\psi \quad , \tag{3.4}$$

(for small θ_e, θ_k), where θ_e, θ_k are the electron and photon emission angles relative to the direction of the initial electron, and ψ is the azimuthal angle between the planes $(\mathbf{p}_0, \mathbf{k})$ and $(\mathbf{p}_0, \mathbf{p})$.

Taking into account that the bremsstrahlung is directed strongly into the forward direction at high energies, we may derive from (3.1-4) the range of longitudinal and transverse components of the momentum transfer making the main contribution to CB:

$$\delta \lesssim q_\ell \lesssim 2\delta \quad , \qquad 0 \lesssim q_t \lesssim 2x \quad , \tag{3.5}$$

where the minimum recoil momentum is $\delta = q_\ell^{min} = x/2E_0(1-x)$, with $x = k/E_0$ the relative photon energy.

Thus, the kinematically permitted region of momentum transfer is a disc [called a "pancake" by *Überall*, see (2.14b,c)] with a thickness of about δ and a radius of ~ 1,

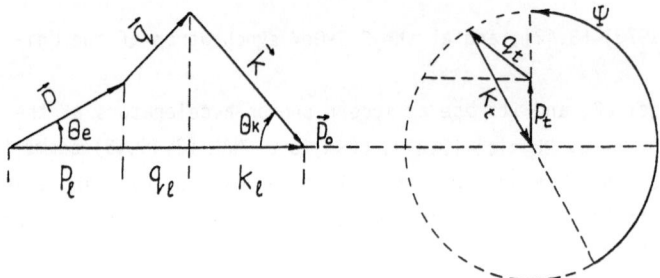

Fig.3.1. Momentum and angular relations in electron bremsstrahlung. Here, p_ℓ, q_ℓ, k_ℓ and p_t, q_t, k_t are, respectively, the longitudinal and transverse components of the momenta

which is axially symmetric with respect to the initial electron momentum $\mathbf{p_0}$ and which lies a distance δ from the origin of the interaction.

The minimum recoil momentum determines the effective longitudinal dimension of the interaction region, i.e., the coherence length

$$\ell_{coh} \sim 1/\delta \quad . \tag{3.6}$$

As the initial electron energy increases or the frequency of the emitted photon decreases, the size of the interaction region increases, and may exceed the lattice constant a. In this case, the number of atoms $N \sim \ell_{coh}/a$ located along the coherence length will participate coherently in the process and the bremsstrahlung intensity in a single crystal will increase N-fold in comparison with that in an amorphous medium [3.49-51].

3.2.2 Intensity and Polarization

To calculate the bremsstrahlung cross section $d\sigma_{cr}$ of electrons in a crystal, one should replace the nuclear electrostatic field potential $V(\mathbf{r})$ by the crystal-lattice potential

$$V_{cr} = \sum_{\mathbf{L}} V(\mathbf{r} + \mathbf{L}) \quad ,$$

where $\mathbf{L}$ is the lattice vector. This replacement leads to the multiplication of the well-known Bethe-Heitler cross section $d\sigma_{BH}$ by an additional factor $D(\mathbf{q}) = = |\sum_{\mathbf{L}} \exp(i\mathbf{q} \cdot \mathbf{L})|^2$

$$d\sigma_{cr} = D(\mathbf{q}) d\sigma_{BH} \quad . \tag{3.7}$$

For a simple cubic lattice with lattice spacing a, the diffraction factor in (3.7) has the form

$$D(\mathbf{q}) = \frac{(2\pi)^3}{\Delta} N \sum_{\mathbf{g}} \delta(\mathbf{q} - \mathbf{g}) \quad , \tag{3.8}$$

where N is the number of atoms in the crystal, $\Delta = a^3$ the volume of the unit cell of the direct lattice, and $\mathbf{g}$ denotes reciprocal-lattice vectors. The δ function causes the diffraction factor $D(\mathbf{q})$ to be nonzero only when $\mathbf{q} = \mathbf{g}$, i.e., when the momentum transfer coincides with the reciprocal-lattice vector.

For crystals with complex structure the factor $|S(\mathbf{g})|^2$ appears before the δ function in (3.8). This is the structure factor, which reflects the influence of atoms inside the cell on the magnitude of interference effects.

Taking into account the thermal lattice oscillations changes the expression for the bremsstrahlung cross section to the form

$$d\sigma_{cr} = d\sigma_{BH}\left\{\frac{(2\pi)^3}{\Delta} N \sum_{\mathbf{g}} |S(\mathbf{g})|^2 \exp(-\bar{A}g^2)\delta(\mathbf{q} - \mathbf{g}) + N[1 - \exp(1 - \bar{A}g^2)]\right\}$$

$$\equiv d\sigma^c + d\sigma^i \quad . \tag{3.9}$$

Here $\bar{A}$ is the r.m.s. displacement amplitude of an atom from its equilibrium position and $d\sigma^c$ and $d\sigma^i$ are the coherent and incoherent parts of the cross section, respectively.

After integrating the cross section over photon emission angles, the expression for the CB intensity may be written as

$$I(x,\theta,\alpha) = \frac{x}{N\sigma_0} \frac{d\sigma_{cr}}{dx} = [1 + (1 - x)^2](\Psi_1^c + \Psi_1^i) - \frac{2}{3}(1 - x)(\Psi_2^c + \Psi_2^i) \quad , \tag{3.10}$$

where $\Psi_{1,2}^c$ and $\Psi_{1,2}^i$ are the functions associated, respectively, with the coherent and incoherent parts of the bremsstrahlung spectrum; $\sigma_0 = 5.78 \times 10^{-28}Z^2 cm^2$, Z being the nuclear charge of the crystal atoms, θ is the polar angle between the momentum $\mathbf{p}_0$ and the chosen axis $\mathbf{b}_1$ and α the azimuthal angle between the planes $(\mathbf{p}_0, \mathbf{b}_1)$ and $(\mathbf{b}_1, \mathbf{b}_2)$ (Fig.3.2).

The functions $\Psi_{1,2}^i$ are weakly dependent on the parameter δ. For a diamond crystal, for example, we have $\Psi_1^i = 18.2$ and $\Psi_2^i = 17.4$.

The coherent part of the bremsstrahlung spectrum is given by

$$\Psi_1^c(\delta,\theta,\alpha) = \frac{(2\pi)^2}{va^3} 4\delta \sum_{\mathbf{g}} |S(\mathbf{g})|^2 \exp(-\bar{A}g^2) \frac{[1 - F(g^2)]^2}{g^4}$$

$$\times \frac{g_2^2 + g_3^2}{[g_1 + \theta(g_2 \cos\alpha + g_3 \sin\alpha)]^2} \quad , \tag{3.11}$$

$$\Psi_2^c(\delta,\theta,\alpha) = \frac{(2\pi)^2}{va^3} 24\delta^2 \sum_{\mathbf{g}} |S(\mathbf{g})|^2 \exp(-\bar{A}g^2) \frac{[1 - F(g^2)]^2}{g^4}$$

$$\frac{(g_2^2 + g_3^2)[g_1 + \theta(g_2 \cos\alpha + g_3 \sin\alpha) - \delta]}{[g_1 + \theta(g_2 \cos\alpha + g_3 \sin\alpha)]^4} \quad , \tag{3.12}$$

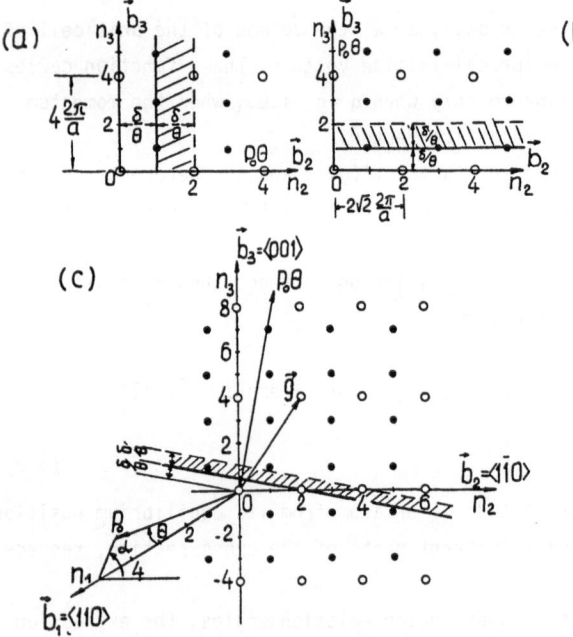

(a)   (b)   (c)

Fig.3.2a-c. The intersection of the diamond crystal (110) plane by a region of kinematically permitted momentum transfers. (a) $\alpha = 0$; (b) $\alpha = \pi/2$; (c) $0 < \alpha < \pi/2$. Structure-factor values $|S|^2 = 64$ are for reciprocal-lattice points shown by *open circles*, and $|S|^2 = 32$ for *full circles*

where ν is the number of atoms in the crystal cell, $g_{1,2,3}$ are the reciprocal-lattice-vector projections onto the crystallographic axes, and $[1 - F(g^2)]/g^4$ is a term taking into account the screening of the nuclear field by the atomic electrons.

The summation in (3.11,12) is performed over all reciprocal-lattice vectors **g** existing in the kinematically permitted region of recoil momenta (3.5). The coherent summation condition may be written in terms of the crystal orientation angles θ and α as:

$$\delta \leqslant [g_1 + \theta(g_2 \cos\alpha + g_3 \sin\alpha)] \lesssim 2\delta \quad . \tag{3.13}$$

By proper crystal orientation relative to the initial electron momentum p_0, one may fulfill the conditions for a number of reciprocal-lattice points to be present in the pancake (Fig.3.2a,b). In this case, the CB intensity is maximum and the relative energy x corresponding to the interference-maximum position in the bremsstrahlung spectrum is obtained from (3.13) ($g_1 = 0$):

$$x = 2\theta E_0(g_2 \cos\alpha + g_3 \sin\alpha)/[1 + 2\theta E_0(g_2 \cos\alpha + g_3 \sin\alpha)] \quad . \tag{3.14}$$

The interference condition **q** = **g** leads to a loss of symmetry of the bremsstrahlung process in a single-crystal radiator, and the photons emerge with a degree of polarization [3.8]

$$P(\delta,\theta,\alpha,\varphi) = 2(1 - x)\Psi_3^C(\delta,\theta,\alpha,\varphi)/I(x,\theta,\alpha) \quad ,$$

where

$$\Psi_3^c(\delta,\theta,\alpha,\varphi) = \frac{(2\pi)^2}{\nu a^3}\, 4\delta^3 \sum_{\mathbf{g}} |S(g^2)|^2 \exp(-\bar{A}g^2)\, \frac{[1 - F(g^2)]^2}{g^4}$$

$$\frac{(g_2^2 - g_3^2)\cos 2\varphi + 2g_2 g_3\,\sin 2\varphi}{[\theta(g_2\,\cos\alpha + g_3\,\sin\alpha)]^4} \quad . \tag{3.15}$$

Here φ is the angle between the plane ($\mathbf{p}_0$, $\mathbf{b}_2$) and the chosen polarization plane across the crystal axis $\mathbf{b}_1$.

Note that the maximum symmetry loss, and hence the highest CB polarization, are observed in the case when only a single reciprocal-lattice point contributes to the cross section.

This situation is illustrated by Fig.3.2c, which shows the intersection of the plane ($\mathbf{b}_2$, $\mathbf{b}_3$) of the diamond-crystal reciprocal lattice by the pancake. The sharp boundary of the pancake passes through the single point (020).

3.3 Experimental Equipment for Coherent Bremsstrahlung Studies

3.3.1 Experimental Layouts and Photon-Beam Formation Technique

In order to produce coherent bremsstrahlung and to study its properties, including the parameters of polarized photon beams, it is necessary to take into account certain properties of cyclic and linear electron accelerators.

Owing to a slow electron beam spill onto an internal crystal target (extraction time about 2-3 ms, duty cycle $\sim 10^{-1}$), electron synchrotrons allow one to use a pair γ-spectrometer with a coincidence technique.

Linear accelerators provide electron beams of high intensity (1 - 100 μA), small dimensions (1 - 3 mm), and low angular divergence ($\sim 10^{-4}$ rad), thereby enabling one to obtain photon beams of high intensity, polarization, and monochromaticity; however, they have a low duty cycle ($\sim 10^{-4}$) which complicates the use of the coincidence technique.

The first detailed CB experimental studies were performed at the 1-GeV electron synchrotron in Frascati [3-15.16] by *Diambrini* and his co-workers, who were the first to obtain (in 1962) a quasi-monochromatic linearly polarized photon beam with a diamond crystal. Figure 3.3 shows the experimental setup used by this group. The internal electron beam interacted with a $10 \times 5 \times 2$ mm^3 single-crystal diamond plate placed in a goniometer which permitted the rotation of the crystal target around vertical and horizontal axes. After cleaning (B), the collimated (C_1, C_2, C_3) CB beam entered the vacuum chamber of a pair spectrometer (S), after which its intensity was measured with a Wilson quantometer (Q). Electron-positron pairs produced in a converter (C) were detected by scintillation counters (A_1, A_2, A_3).

Fig.3.3

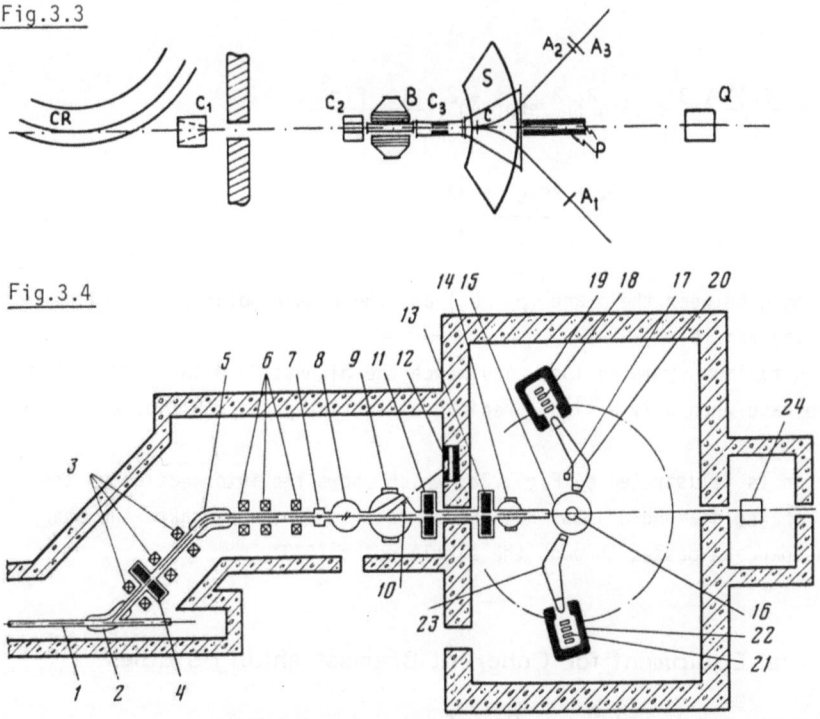

Fig.3.4

Fig.3.3. Experimental setup based by the Fascati group to produce coherent-bremsstrahlung beam. CR: crystal; C_1, C_2, C_3, P: collimators; S: pair spectrometer; A_1, A_2, A_3: scintillation counters; B: cleaning magnet; C: converter; Q: quantometer

Fig.3.4. Experimental arrangement used at the Kharkov linac. _1_: electron beam line; _2,5_: bending magnets for transport; _4_: collimator-monochromator; _3,6_: quadrupole lenses; _7,10_: secondary emission monitors; _8_: goniometer with diamond crystals; _9,14_: cleaning magnets; _11,13_: photon collimators; _12_: beam disposal; _15,16_: scattering chamber with liquid hydrogen and deuterium targets; _17_: Faraday cup; _18,21_: scintillation counter telescopes; _19,22_: counter shielding; _20,23_: magnetic spectrometers; _24_: quantometer

Bremsstrahlung and polarization experiments were performed with a diamond crystal at the Kharkov 2-GeV electron linac [3.34-37]. The experimental arrangement is shown schematically in Fig.3.4. After being transported through the formation system that consisted of two bending magnets (2,5) and quadrupole lenses (3,6), the electron beam of the Kharkov linac hit a single-crystal diamond target located in a goniometer (8). A collimator-monochromator (4) was used to set the width of the electron energy spectrum. The electron beam intensity was measured by secondary emission monitors (7,10) which were calibrated against the Faraday cup (17). The CB of electrons in the single-crystal radiator resulted in the production of a quasi-monochromatic, linearly polarized photon beam. Electrons passing through the radiator were analyzed by a secondary electron spectrometer (9) and were then deflected to the waste disposal (12). The photon beam formed by slit collimators (11,13) was

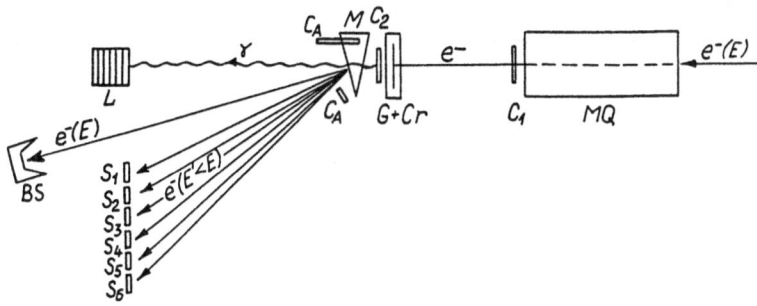

Fig.3.5. Experimental layout used to produce a tagged, linearly polarized photon beam at the Serpukhov proton accelerator. MQ: electron channel; C_1, C_2, C_A, S_1-S_6: scintillation counters; G + Cr: goniometer with crystal; M: bending magnet; L: photon shower detector; BS: primary electron beam absorber

cleaned by a magnet (14) and then entered the scattering chamber (15) with liquid hydrogen or deuterium targets (16). Nuclear reaction products were momentum-analyzed by magnetic spectrometers (20,23) and were detected by scintillation counter telescopes (18,21) placed within shielding (19,22). The total photon flux was measured by a quantometer (24).

Coherent bremsstrahlung from single crystals combined with photon tagging is used with electron beams from superhigh-energy proton accelerators.

In 1979, a beam of tagged linearly polarized photons was first obtained by the CB method [3.44] at the Serpukhov IHEP proton accelerator using a 31-GeV electron beam with an intensity of 4×10^4 particles per pulse, 1.7 s long, incident on a silicon crystal. A system for tagging (Fig.3.5) included a bending magnet M and scintillation counters S_1 - S_6 triggered by a master impulse from the C_1, C_2 counters monitoring the primary electron beam. The photons were detected by a lead shower counter L. Tagging channels covered a range of E_0-E = 8.2 - 27.1 GeV with a photon intensity I ~0.10 to 0.01 photons per electron and linear polarization $P_\gamma \sim (50 - 20\%)$, respectively. A goniometer G + Cr allowed the rotation of a single crystal, 70 mm in diameter, around horizontal and vertical axes within the range ~0.1 rad with an angular measurement accuracy of 2.5×10^{-5} rad.

3.3.2 Crystal Targets

It is common practice to use crystal plates of diamond, silicon, or germanium as CB radiators. Diamond-crystal targets are used more widely since they have a perfect lattice, a high Debye temperature and close packing of atoms, a comparatively low atomic number, and a relatively high radiation resistance.

Generally, a 15 - 20 carat natural high-purity diamond crystal is taken as the initial material, out of which diamond plates of different thicknesses ranging from 0.1 to 2 mm are cut. However, the high cost and difficulties in obtaining plates thinner than 0.1 mm and of larger dimensions (>10 × 10 mm^2) make diamond crystals

41

Table 3.1. Main characteristics of crystals used for CB production

Characteristics	Diamond	Silicon	Germanium
Atomic number	6	14	32
Atomic weight	12.01	28.09	72.59
Density [g/cm^3]	3.51	2.4	5.35
Lattice type	diamond	diamond	diamond
Lattice constant [Å]	3.568	5.4307	5.6576
Debye temperature [K]	1850	650	366
r.m.s. amplitude of thermal oscillations: for T = 293 K	127	270	304
for T = 77 K	110	145	117
Radiation length [g/cm^2]	43.3	22.2	12.08
Radiation length [cm]	12.34	9.25	2.26

somewhat disadvantageous. In the case of silicon and germanium crystals, it is possible to cut plates with a thickness of about 0.01 mm and with larger dimensions. The main parameters of crystals used for CB production are listed in Table 3.1.

3.3.3 Goniometers

To study the properties of CB and to obtain polarized photon beams, one needs goniometers which provide highly accurate crystal target orientation relative to the primary electron direction. Goniometer designs are specific for each accelerator. In view of the experimental techniques for CB and photon beam polarization measurements, we may formulate general requirements for goniometer design. It is desirable that the range of rotation angles should be ~0.1 rad and the accuracy of orientation angle setting should be within ~10^{-5} rad.

Figure 3.6a shows a diagram of a goniometer that is typical for most accelerators [3.52,53]. The frames A and B ensure the rotation of the internal frame C with a crystal around mutually perpendicular vertical (f_V) and horizontal (f_H) axes of the goniometer by angles Φ_V and Φ_H in the range ±8° with a setting accuracy of $\Delta\Phi = 5 \times 10^{-5}$ rad. The frame C has a third rotation axis f_A normal to f_H. By rotating the crystal around the axis f_A within the range 0 - 360°, the direction of the photon polarization vector can be changed. When the crystal axis b_1 coincides with the rotation axis f_A, the relation between the rotation angles Φ_H, Φ_V, and Φ_A and the crystal orientation angles θ, α, and φ (Fig.3.6b) may be written

$$\Phi_V = \Phi_V^0 + \theta\cos(\alpha + \varphi) \quad ,$$
$$\Phi_H = \Phi_H^0 + \theta\sin(\alpha + \varphi) \quad , \tag{3.16}$$

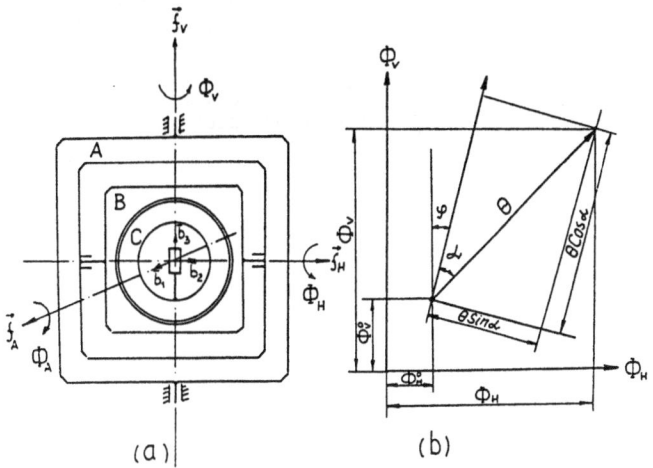

Fig.3.6. (a) Schematic
view of a goniometer.
(b) Relation between crystal orientation angles and
goniometer rotation angles

$$\Phi_A = \Phi_A^0 + \varphi \quad .$$ (3.16)

Here Φ_V^0, Φ_H^0, and Φ_A^0 are the angles at which θ, α, and φ, respectively, are equal to zero.

All goniometers are, as a rule, provided with a remote control, and the goniometer system at NINA operated automatically on-line with a computer [3.42]. The
SLAC goniometers and those used in pair γ-spectrometers for photon polarization
measurements in Hamburg and Erevan [3.41,42], have the azimuthal axis of rotation
along the direction of the electron momentum $\mathbf{p}_0$. If diamond or silicon crystals cut
along the (100) plane are used, it is not necessary to rotate the crystal around the
azimuthal axis Φ_A. The absence of the azimuthal axis of rotation is typical of the
goniometer setup for internal beams of electron synchrotrons [3.27,32,43].

3.4 Experimental Results and Discussion

3.4.1 Spectral Characteristics of Bremsstrahlung in Crystals

We mentioned above that, unlike the continuous Bethe-Heitler distribution, the spectrum of relativistic electron radiation in crystals is characterized by a peak-type
structure with a comparatively high degree of photon polarization. The intensities
and positions of the maxima in the spectrum depend on the initial particle energy,
crystal type, and the angles of crystal orientation relative to the electron beam
direction.

CB spectra were first measured with a pair spectrometer at electron synchrotrons
in Frascati [3.17-19] and Hamburg [3.25]. Here, particular attention was given to
measurements of electron radiation in diamond crystals at different initial particle
energies for different target orientations.

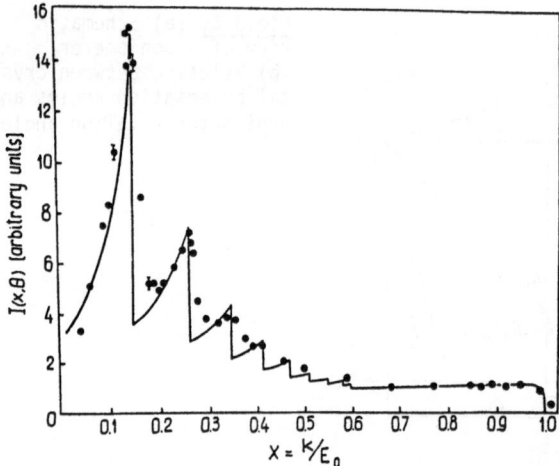

Fig.3.7. Bremsstrahlung intensity spectrum in a diamond crystal measured by the Frascati group ($E_0 = 1$ GeV, $\theta = 4.6 \pm 0.1$ mrad, $\alpha = 0^\circ$). The *solid curve* is the calculated spectrum

Figure 3.7 shows the bremsstrahlung intensity spectrum at $E_0 = 1$ GeV as measured by *Barbiellini* et al. [3.18]. A diamond plate, 2 mm thick, cut parallel to the axes $\mathbf{b}_1 = \langle 110 \rangle$, $\mathbf{b}_2 = \langle 1\bar{1}0 \rangle$, and $\mathbf{b}_3 = \langle 001 \rangle$, was used as a target. The electron beam was incident at $\theta = 4.6 \pm 0.1$ mrad to the $\mathbf{b}_1$ axis parallel to the $(\mathbf{b}_1, \mathbf{b}_2)$ plane ($\alpha = 0^\circ$). The spectrum shows several maxima predicted by the theory (solid line). Their positions versus photon energy may be obtained from (3.13), which, for the case considered, is

$$n_2 \sqrt{2} \cdot 2\pi/a = \delta/\theta \quad , \tag{3.17}$$

where n_2 is the row number along the axis $\mathbf{b}_2$ and the lattice constant $a = 922$ is in units of the electron Compton wavelength.

Figure 3.2a shows the intersection of the diamond reciprocal-lattice plane ($\mathbf{b}_2$, $\mathbf{b}_3$) with the pancake (the shaded region). The sharp boundary of the pancake contains the row of points with $n_2 = 1$, which is responsible for the first maximum in the radiation spectrum. As the photon energy increases, the minimum momentum transfer also increases and the boundary shifts to the right of the interaction origin. When the first row of points leaves the pancake, an abrupt drop in the coherent-bremsstrahlung intensity spectrum is observed. With a further increase in energy, the row of points with $n_2 = 2$ intersects the sharp boundary and a new intensity enhancement takes place having, however, a smaller amplitude than the first maximum, since the CB amplitude decreases with an increase in momentum transfer.

The radiation spectrum was calculated in [3.18] from (3.10-12) using the screening function of the form

$$[1 - F(g^2)]^2 \cdot g^{-4} = (\beta^2 + g^2)^{-2} \quad , \quad \beta = 111 \, z^{-1/3} \quad , \tag{3.18}$$

corresponding to an exponential potential.

44

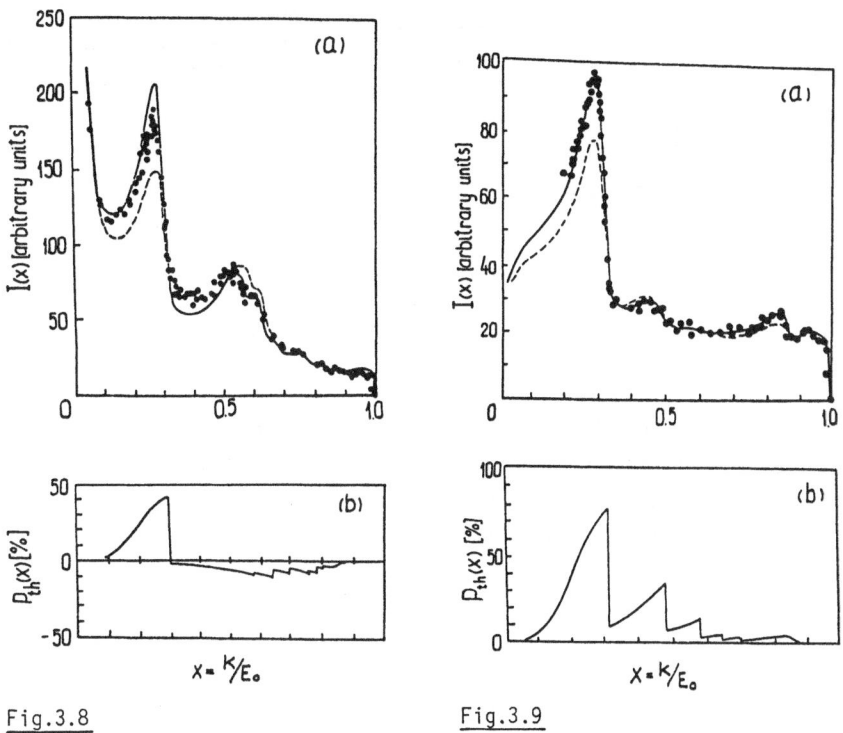

Fig.3.8   Fig.3.9

Fig.3.8a,b. CB intensity (**a**) and polarization (**b**) in a diamond crystal for $E_0 = 4.8$ GeV, $\theta = 3.44$ mrad, $\alpha = 0^\circ$ (DESY data). Theoretical spectra were calculated with a Hartree screening potential (——) and an exponential potential (----)

Fig.3.9a,b. The same as in Fig.3.8a,b for diamond-crystal orientation angles $\theta = 50$ mrad, $\alpha = 1.5^\circ$

The choice of a screening potential was discussed in detail by the DESY group [3.25], who studied radiation spectra from a diamond crystal at an electron energy of $E_0 = 4.8$ GeV. The diamond crystal, with dimensions $1 \times 4 \times 7$ mm^3, was cut parallel to the axes $\mathbf{b}_1 = <110>$, $\mathbf{b}_2 = <001>$, $\mathbf{b}_3 = <1\bar{1}0>$ (in the notation of the original paper). Coherent bremsstrahlung was studied for two crystal orientations relative to the incident particle momentum: (a) with a row of reciprocal-lattice points contributing to the CB cross section ($\alpha = 0^\circ$, $\pi/2$); and (b) with only one point [(020) in the DESY work] lying within the pancake, as shown in in Fig.3.2c ($\alpha \neq 0^\circ$, $\pi/2$).

The measured intensity spectra are shown in Figs.3.8 and 9a, respectively. The crystal orientation angles were $\theta = 3.44$ mrad, $\alpha = 0^\circ$, and $\theta = 50$ mrad, $\alpha = 1.5^\circ$, respectively. The curves in the figures show the intensity and polarization spectra calculated taking into account the initial divergence and multiple electron scattering in the crystal, and the photon collimation and energy resolution of the detecting equipment. The screening function was calculated using exponential (dashed curve) and Hartree (solid curve) potentials. The calculations with the Hartree-type potential provide a better description of the measured intensity spectra.

45

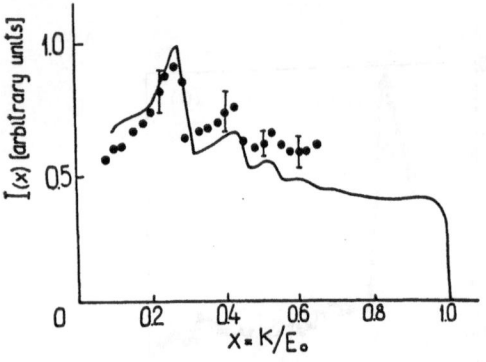

Fig.3.10. CB intensity in a silicon crystal, 0.24 mm thick. $E_0 = 1$ GeV, $\theta = 1.6 \times 10^{-2}$ rad

Spectral characteristics of electron radiation in silicon crystals at room temperature were studied by a group of Japanese physicists [3.23] at a synchrotron with a maximum beam energy of 720 MeV. Under similar experimental conditions, a silicon crystal, which has a lower Debye temperature and a larger unit cell volume, gives a lower photon intensity in the main interference maximum of the spectrum than the diamond crystal.

Figure 3.10 shows the bremsstrahlung spectrum of 1-GeV electrons from the Kharkov linac in a silicon crystal, 0.24 mm thick [3.54]. The measurements were performed using the secondary-electron method [3.55]. The momentum of the initial electron was fixed in the (001) plane at an angle of $\theta = 1.6 \times 10^{-2}$ rad to the crystal axis $b_1 = <110>$. The calculations (solid line) took into account electron beam divergence and multiple scattering in the crystal, as well as the energy and angular acceptances of the secondary-electron magnetic spectrometer. The spectrum was calculated with the Hartree-type screening potential. Bremsstrahlung spectra from germanium and niobium crystals at room temperature were also measured in this work.

The effect of crystal cooling on CB spectral characteristics has been studied by *Grishaev* et al. [3.56]. For diamond and silicon crystals having a sufficiently high Debye temperature, cooling should not lead to an appreciable increase of the CB cross section. Therefore, a germanium crystal having a low Debye temperature ($\theta_D = 366$ K) was used as a radiator. A 1-GeV electron beam with a divergence within 2×10^{-4} rad was incident at $\theta = 10^{-2}$ rad to the germanium crystal axis $b_1 = <111>$ in the (110) plane. The 0.165-mm-thick crystal was fixed in the goniometer so as to ensure a reliable thermal contact with a cryostat [3.57]. The measured bremsstrahlung spectra for two temperature values of 293 K (open circles) and 77 K (full circles) are depicted in Fig.3.11. In the region of the first interference maximum one can observe an intensity enhancement for the crystal cooled down to liquid nitrogen temperature by a factor of 1.2 in comparison with the bremsstrahlung at room temperature, which is in fair agreement with theoretical estimates (the solid curve is the calculation at 77 K).

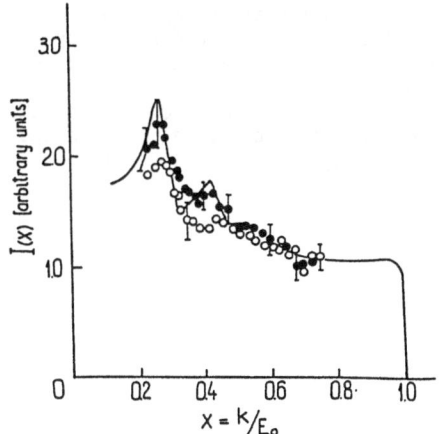

Fig.3.11. CB intensity in a germanium crystal, 0.165 mm thick: (●) T = 77 K, (○) T = 293 K. E_0 = 1 GeV, θ = 10^{-2} rad. The solid curve represents the calculation at T = 77 K

3.4.2 Orientation Dependence of Total Energy Flux and Photon Intensity

The orientation dependence of the total energy flux and the photon intensity at a given energy is important in the consideration of the coherent-bremsstrahlung process. This is above all due to the increase in total energy flux when the entrance angle at the crystal (relative to its axis or plane) of the electron is small. Figure 3.12 illustrates the total energy flux, measured with a Wilson quantometer, versus the angle θ for a diamond crystal, 2 mm thick, at α = π/2 and E_0 = 1.14 GeV [3.37]. The solid line represents the calculated dependence and the dash-dotted line shows the level of the incoherent part. The typical behavior of the total energy flux around θ = 0° can be used to determine the angles of initial orientation ϕ_V^0 and ϕ_H^0 (3.16).

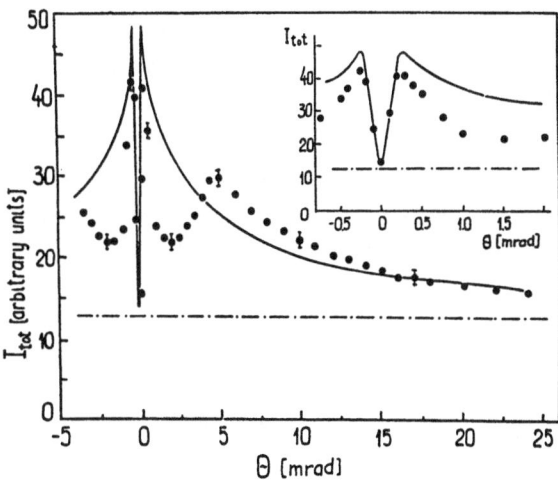

Fig.3.12. Total photon energy flux versus angle of diamond crystal orientation θ for E_0 = 1.14 GeV, α = π/2

47

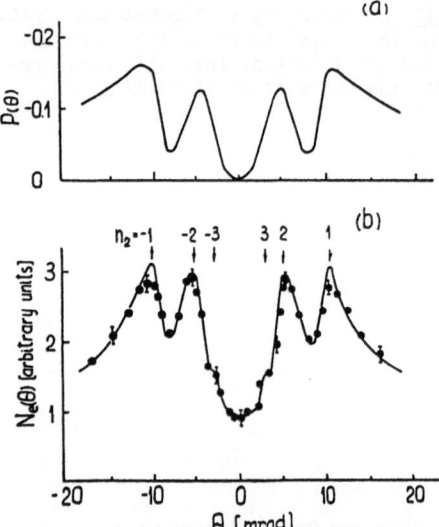

Fig.3.13a,b. Orientation dependence of CB polarization (**a**) and secondary-electron yields (**b**) from a 2 mm-thick diamond crystal, for $E_0 = 1.15$ GeV, $E = 0.8$ GeV, $\alpha = 0^\circ$, $\theta_c = 10^{-3}$rad, and $\Delta E/E = 4 \times 10^{-2}$. Here, the electron beam divergence was 10^{-4}rad and the r.m.s. angle of multiple scattering was 1.8×10^{-3}rad

The orientational intensity dependence at an assigned photon energy, similarly to bremsstrahlung energy spectra, shows several interference maxima. This fact allows one to use CB for the production of quasi-monochromatic, linearly polarized photon beams in crystals [3.24,58,59].

Below, we discuss the orientational CB intensity dependence at a fixed photon energy measured at the Kharkov 2-GeV electron linac [3.37,59,60]. Diamond crystals, 2, 0.3, and 0.08 mm thick, cut parallel to the crystal axes $\mathbf{b}_1 = <110>$, $\mathbf{b}_2 = <1\bar{1}0>$, and $\mathbf{b}_3 = <001>$, were used as targets. The dotted curve in Fig.3.13b shows the number of 800-MeV secondary electrons as a function of the orientation angle for a 2-mm-thick diamond crystal. The electron beam with $E_0 = 1.15$ GeV was directed along the (001) plane ($\alpha = 0^\circ$). The secondary-electron yield may be associated with the intensity of the photon-spectrum line with energy $k = E_0 - E = 350$ MeV. Interference maxima in the orientation dependence shown are due to successive contributions to the CB cross section from a row of reciprocal-lattice points of a diamond crystal with $n_2 = \pm 1, \pm 2, \pm 3$ along the axis $\mathbf{b}_2 = [1\bar{1}0]$.

The solid curves in Fig.3.13a,b show the CB intensity and polarization calculated using (3.10-13) and (3.15), assuming a Hartree-type screening potential. The electron beam divergence of 10^{-4}rad, the r.m.s. angle of multiple scattering in the crystal of 1.8×10^{-3}rad, the energy resolution of the secondary-electron magnetic spectrometer $\Delta E/E = 4 \times 10^{-2}$, and the collimation angle $\theta_c = 10^{-3}$rad were taken into account in the calculations.

The detection of reaction products from single pion photoproduction on nucleons and the lightest nuclei, e.g., $\gamma p \rightarrow p\pi^0$, $\gamma p \rightarrow n\pi^+$, $\gamma n \rightarrow p\pi^-$, $\gamma d \rightarrow \pi^0 d$, seems to be a rather promising method for studying the orientation dependence of the bremsstrahlung intensity in crystals.

For two-body reactions, the momentum and emission angle of one particle (e.g., $\pi^\pm$, p, or d), fixed by the magnetic spectrometer, determine the photon energy k, whereas the momentum and angle acceptances give the energy resolution $\Delta k/k$.

The yield of particles with fixed kinematic parameters is proportional to the photon-spectrum line intensity and to the reaction cross section, which may be written for linearly polarized photons as

$$d\sigma_{pol} = d\sigma_0(1 - P_\gamma \sum \cos 2\varphi) \quad . \tag{3.19}$$

Here, $d\sigma_0$ is the cross section for unpolarized photons, P_γ the degree of linear polarization, φ the angle between the reaction plane and the photon polarization vector, and $\sum = (d\sigma_\perp - d\sigma_\parallel)/(d\sigma_\perp + d\sigma_\parallel)$ the cross section asymmetry for normal and parallel directions of the polarization vector.

Thus, by measuring the yields for mutually perpendicular photon polarization vectors, one can eliminate the dependence of the cross section on P_γ, and the sum of yields can be unambiguously related to the photon-spectrum line intensity.

Figures 3.14a,c shows the sums of proton yields from the $\gamma p \rightarrow p\pi^0$ reaction [3.61] for perpendicular and parallel directions of the photon polarization vector, respectively, versus the orientation angles of a 2-mm-thick diamond crystal. The measurements were performed at a photon energy of 350 MeV and an initial electron beam

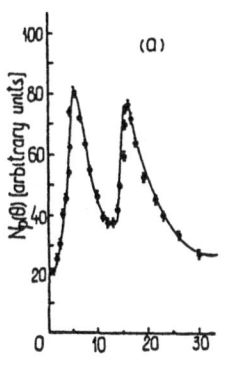
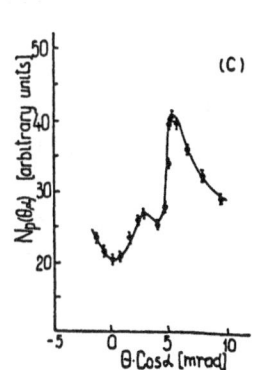
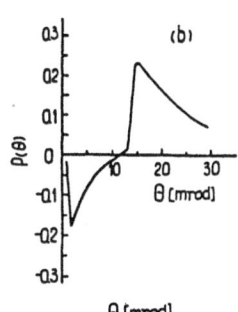
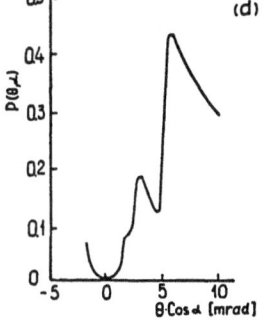

Fig.3.14a-d. Coherent-bremsstrahlung polarization P and proton yields $N_p = N_p^\perp + N_p^\parallel$ from the $\gamma p \rightarrow \pi^0 p$ reaction as functions of 2-mm thick diamond-crystal orientation angles. $E_0 = 1.15$ GeV, $k = 0.35$ GeV. (**a,b**): $\alpha = \pi/2$; (**c,d**): $\theta \sin\alpha = 75$ mrad

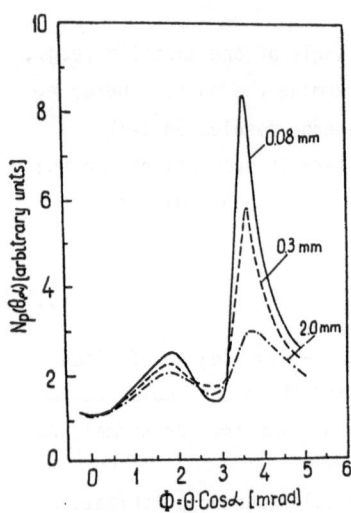

Fig.3.15. Proton yields $N_p^\perp = N_p^\parallel + N_p$ from the $\gamma p \to \pi^o p$ reaction versus orientation angles of diamond crystals of different thicknesses for $E_0 = 1.3$ GeV, $k = 0.33$ GeV, $\theta \sin\alpha = 75$ mrad, $\theta_c = 2 \times 10^{-4}$ rad

energy of 1.15 GeV. In the first case, the electron momentum was directed along the crystal plane $(1\bar{1}0)$ - $\alpha = \pi/2$ and the main contribution to the CB cross section was due to a row of reciprocal-lattice points indexed as n_3 along the axis $\mathbf{b}_3 = \langle 001 \rangle$.

Figure 3.14c illustrates the orientation dependence when the dominant intensity peak was due to a single reciprocal-lattice point (020). The intensity and polarization of the photon-spectrum line shown in Fig.3.14a-d by solid curves were calculated for $\theta_c = 5 \times 10^{-4}$ rad, $\Delta k/k = 0.05$ (the angles of divergence and multiple scattering of electrons were the same as in Fig.3.13).

The influence of the diamond-crystal thickness on the coherence effect (the ratio of the CB intensity in the interference maximum to the incoherent background), for the case of the single reciprocal-lattice point (020) contributing, is illustrated in Fig.3.15 [3.60]. At $E_0 = 1.3$ GeV, the protons from the $\gamma p \to p\pi^o$ reaction were detected for kinematic conditions corresponding to the photon energy $k = 330$ MeV. The curves are drawn through the experimental points, which were normalized to the level of the incoherent part of the bremsstrahlung for each value of the crystal thickness. The photon collimation angle θ_c was 2×10^{-4} rad. As the diamond-crystal thickness decreases, the coherence effect, and hence the photon polarization, increase. This is due to the decreasing r.m.s. angle of multiple scattering and, as will be shown below, to the reduction of the ratio of incoherent to coherent bremsstrahlung under tight photon beam collimation. The polarization for a diamond crystal, 0.08 mm thick, was calculated to be $P_\gamma = 0.92$.

3.4.3 Measurements of CB Polarization

The comparison of experimental CB intensity spectra with theoretical spectra that take into account the divergence and multiple scattering of the incident electrons as well as the photon beam collimation angles provides the set of parameters for

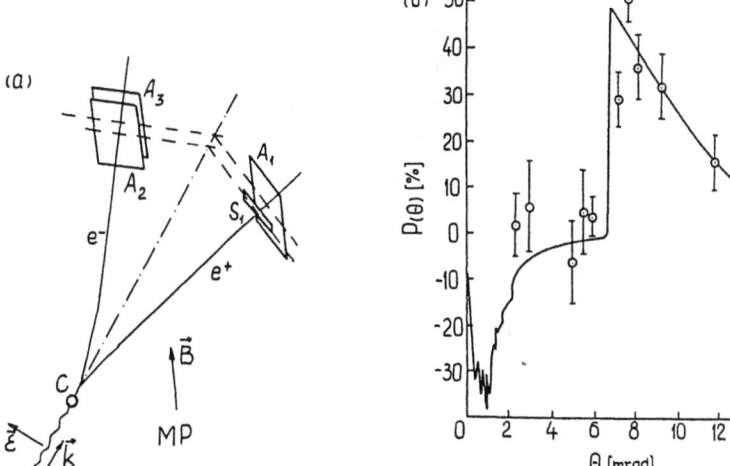

Fig.3.16. (a) Experimental setup for measuring the CB polarization at the Frascati synchrotron. Here, ϵ is the direction of the photon polarization vector, C the aluminum converter, B the magnetic field direction of the pair spectrometer, MP the median plane of the pair spectrometer, and S_1, A_1, A_2, A_3 are scintillation counters detecting e^+e^- pairs. (b) Polarization of 150-MeV photons as a function of the diamond orientation angle at $E_0 = 1$ GeV. Experimental points are normalized to the theoretical curve (——) at $\theta = 11.6$ mrad

calculating the photon polarization. However, direct measurements of the photon polarization in crystals are also of importance since they provide a useful tool to test CB theory.

The polarization of CB produced in a diamond crystal was first measured in Frascati [3.20] at $E_0 = 1$ GeV. The method of measuring the polarization is based on the fact that linearly polarized photons produce electron-positron pairs preferentially in the polarization plane [3.62]. Taking into account that the asymmetry coefficient

$$R = (d\sigma_\parallel - d\sigma_\perp)/(d\sigma_\parallel + d\sigma_\perp) \qquad (3.20)$$

($d\sigma_\parallel$ and $d\sigma_\perp$ being, respectively, the cross sections for pair production in the planes parallel and normal to the polarization plane) reaches its maximum value for equal energy distribution between the particles, *Barbiellini* et al. [3.20] measured symmetrical electron-positron pair production yields for 150-MeV photons in an aluminum radiator, 10^{-4} radiation lengths thick. The experimental layout is shown in Fig. 3.16a. Pair components were detected by scintillation counters S_1, A_1 and A_2, A_3. The vertical dimension of the S_1 counter defining the polarization plane was 0.5 cm, the remaining counters A_1, A_2, and A_3 being 10 cm high. The ratio of coincidence counts

$$\rho = S_1 A_2 A_3 / A_1 A_2 A_3 \qquad (3.21)$$

is related to the photon polarization P_γ and the asymmetry factor R by

$$\rho = \rho_0 (1 + R P_\gamma) \quad , \qquad (3.22)$$

51

where ρ_0 is the value of the ratio (3.21) for unpolarized photons. In the geometry of the experiment, R = 10.1%. Figure 3.16b shows the measured linear photon polarization as a function of the angle θ between the electron momentum $\mathbf{p}_1$ and the diamond-crystal axis $\mathbf{b}_1$ = <110>; $\mathbf{p}_1$ is parallel to the (110) plane which was oriented vertically in the experiment. The solid curve in the figure represents the theoretical predictions, which are in fair agreement with the experimental values.

Note that owing to the small production angles of the members of the pair, this procedure is applicable only up to energies of 500 MeV. This energy region may be extended to 1 GeV if wire spark chambers [3.63] are used instead of scintillation counters.

Barbiellini et al. [3.22] suggested a method of measuring linear photon polarization at energies above 1 GeV based on coherent electron-positron pair production in thin crystals and the correlation between the photon polarization vector and the reciprocal-lattice vector. For certain orientation angles of the crystal analyzer, one can observe the asymmetry $R_c = (d\sigma_\parallel - d\sigma_\perp)/(d\sigma_\parallel + d\sigma_\perp)$ in pair production relative to the $(\mathbf{k}, \mathbf{b}_1)$ plane which can be expressed in terms of the pair production cross sections $d\sigma_\parallel$ and $d\sigma_\perp$ by a relation similar to (3.20), the only difference being that here $d\sigma_\parallel$ and $d\sigma_\perp$ are integrated over the angles of e^+e^- emission.

The asymmetry parameter R_c, as a function of the orientation angle θ between the photon momentum $\mathbf{k}$ and the crystal axis $\mathbf{b}_1$ = <110> in the (110) plane, is shown by a solid curve in Fig.3.17 at k = 3 GeV for symmetrical pairs. The dashed curve represents the cross section $d\sigma = d\sigma_\parallel + d\sigma_\perp$ for unpolarized photons.

This method was first applied to measuring the degree of CB polarization at the 6-GeV DESY facilities [3.26]. Figure 3.18a shows the CB intensity spectrum from a diamond crystal, where the predominant peak is associated with the contribution to the cross section from a single reciprocal-lattice point (020). Another diamond crystal fixed in the goniometer inside the pair magnetic spectrometer was used to analyze the polarization. The polarization analyzer was oriented so as to provide the maximum value of the asymmetry parameter R_c (Fig.3.17). The number of symmetric e^+e^- pairs $N_\parallel$ for the photon polarization vector directed parallel to the (110) plane and the number of pairs $N_\perp$ for the crystal analyzer rotated by 90° around the vector $\mathbf{k}$ were measured in the experiment. The photon polarization was given by

$$P_\gamma = \frac{1}{R_c} \frac{N_\parallel - N_\perp}{N_\parallel + N_\perp} \quad . \tag{3.23}$$

The experimental results shown in Fig.3.18b are in good agreement with the calculations (solid line).

The technique using a thin crystal analyzer was also applied to measuring the CB beam polarization at the Erevan electron synchrotron [3.64].

Cabibbo et al. [3.65,66] suggested another method for measuring the photon beam polarization, based on selective photon absorption in a crystal with a thickness

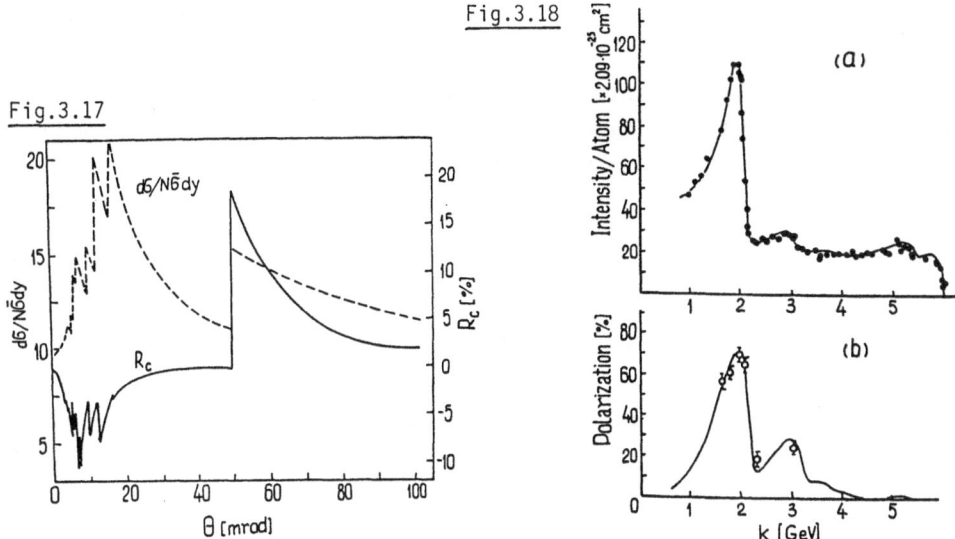

Fig.3.18

Fig.3.17

(a)

(b)

Fig.3.17. Orientation dependence of the asymmetry parameter (———, *right-hand scale*) and the cross section for pair production by unpolarized 3-GeV photons (----, *left-hand scale*) for a diamond crystal at room temperature. N is the number of atoms in the crystal and $\bar{\sigma} = z^2(e^2/mc^2)^2/137$

Fig.3.18a,b. CB intensity (**a**) and polarization (**b**) produced by 6-GeV electrons from the DESY facilities in a diamond crystal; $\theta = 50.5$ mrad, $\alpha = 49.6$ mrad

of several radiation lengths. Passing through a thickness t, a polarized photon beam undergoes different amounts of absorption depending on the angle between the polarization vector P_γ and the lattice vector **g**. By measuring the photon intensity reduction I(t)/I(0) for different φ values, one may determine the polarization from the relation

$$[I(t)/I(0)]_{P_\gamma} = [I(t)/I(0)]_{P_\gamma=0} \, [1 + |P_\gamma|P(t) \cos 2\varphi] \quad , \qquad (3.24)$$

where P(t) is the polarization acquired by an initially unpolarized photon beam after passing through the thickness t.

This method of polarization measurement was applied to a 16-GeV photon beam from SLAC [3.67], where an array of pyrolytic carbon crystals with a total thickness of 61 cm was used as an analyzer. Thick carbon and corundum crystals were used as polarizers in high-energy experiments [3.68,69] where this method seems to be quite effective.

The coherent-bremsstrahlung polarization may also be estimated from the asymmetry of photoproduction yields. Using a theoretical asymmetry value $\sum$ and measuring the reaction yields $C_\perp$ and $C_\parallel$ for perpendicular and parallel directions of the photon polarization vector, respectively, one can estimate, according to (3.19), the polarization P_γ by

$$P_\gamma = \frac{1}{\Sigma} \frac{C_\perp - C_\parallel}{C_\perp + C_\parallel} \quad . \tag{3.25}$$

In polarization measurements performed at the Kharkov 2-GeV electron linac [3.70], in order to determine the degree of polarization, we used the relation between the magnitude of the coherent effect β and the photon polarization, which may be expressed in the case of an isolated reciprocal point [3.71] as

$$P_\gamma = \frac{2(1 - x)}{1 + (1 - x)^2} \frac{\beta - 1}{\beta} \quad . \tag{3.26}$$

In the experiment, β was determined from the orientation dependence of the reaction yields exemplified in Fig.3.14a,c. This method proves to be advantageous, since no additional measurements are required here to determine the photon beam polarization, and the magnitude of polarization is naturally averaged over the energy acceptance of the detecting apparatus.

3.4.4 Effect of Secondary Collimation on CB Parameters

To produce monochromatic radiation, *Mozley* and *DeWire* [3.28] have proposed using collimation of the photon beam, and have made estimates of this effect. The essence of the method consists in the fact that CB is produced in a narrower angular cone than the incoherent radiation. Therefore, collimation of the bremsstrahlung reduces appreciably only the noncoherent part of the photon spectrum. The width of the coherent maximum Δx is related to the collimation angle θ_c [3.26] as

$$\Delta x = \frac{x\theta_c^2(1 - x)}{1 + \theta_c(1 - x)} \quad . \tag{3.27}$$

Detailed calculations of the effect of photon beam collimation on the CB intensity and polarization may be found in [3.29]. The main difficulties in radiation monochromatization by collimation are due to the initial beam divergence and multiple electron scattering in the crystal.

Experimentally, the collimation effect was studied for a silicon crystal and is reported in [3.23,26]. This effect showed up in an increase of the ratio of the coherent to the noncoherent parts of the spectrum. Because of a relatively large initial beam divergence and multiple scattering, no monochromatization was actually observed. It was in the work reported in [3.72] that coherent-bremsstrahlung monochromatization due to collimation was first obtained. A silicon crystal, 0.05 mm thick, was used as a target. Incident electrons entered the crystal at an angle θ = 9.19 mrad to the <110> axis, parallel to the (001) plane. The use of three scrapers provided a reduction of the initial beam divergence down to $(1 \pm 0.5) \times 10^{-4}$ rad. The CB spectrum (Fig.3.19) was obtained for an initial electron energy $E_0 = 1$ GeV and a collimation angle of 1.196×10^{-4} rad. The spectrum has a characteristic maxi-

54

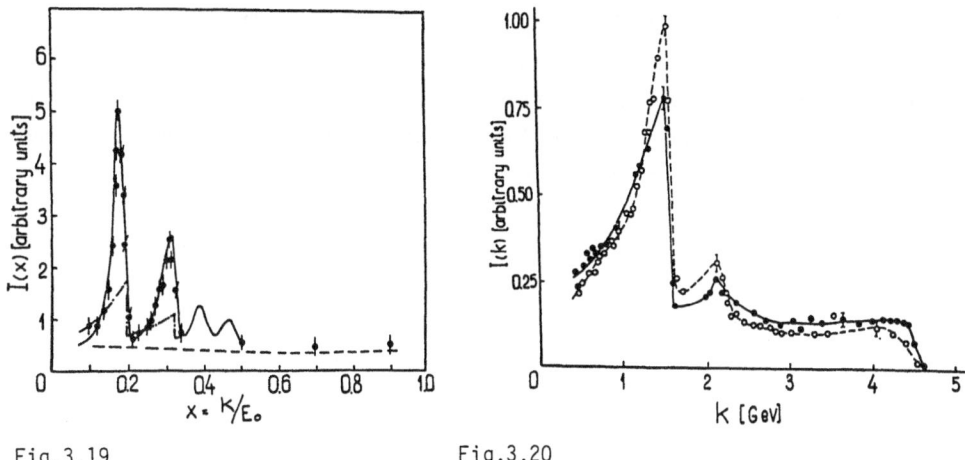

Fig.3.19   Fig.3.20

Fig.3.19. CB intensity spectrum in a silicon crystal, 0.05 mm thick, for $E_0 = 1$ GeV, $\theta = 9.19 \times 10^{-3}$rad, $\alpha = 0°$, $\theta_c = 1.196 \times 10^{-4}$rad. (-•-•-) is the CB calculation with no collimation considered and (---) represents the incoherent part of the bremsstrahlung

Fig.3.20. CB intensity spectrum in diamond crystals of 2-mm (•••, ——) and 0.08-mm (ooo, ---) thickness at electron energies of 4.5 GeV, $\theta = 50$ mrad, $\alpha = 1°$, $\theta_c = 1.6 \times 10^{-4}$ rad

mum at the photon energy $k = 185$ MeV, its FWHM being $\Delta k = 15$ MeV. The solid curve represents the calculation taking into account multiple particle scattering in the crystal, the initial beam divergence, and the energy resolution of the detecting apparatus. The agreement between experimental and calculated results is good. For comparison, the same figure gives the calculated coherent-bremsstrahlung spectrum under the same experimental conditions but without collimation. The dashed curve shows the incoherent part of the spectrum.

The collimation effect on the CB from a diamond crystal when the main contribution comes from an isolated reciprocal-lattice point has been studied by *Avakyan* et al. [3.32]. The results obtained for the initial electron energy $E_0 = 4.5$ GeV and the collimation angle $\theta_c = 0.16$ mrad are shown in Fig.3.20. The curves are drawn through the experimental points which are normalized to the incoherent part of the spectrum. For a 0.08 mm-thick crystal, the ratio of the intensity at the maximum ($k = 1.5$ GeV) to the intensity of 4-GeV photons is equal to 8.5, whereas for a 2 mm-thick crystal, it is only 5.5. The polarization of 1.5-GeV photons for diamond crystals of 2-mm and 0.08-mm thickness has been found to be, respectively, 75% and 90%.

Good progress in the production of monochromatic polarized photon beams using the CB from thin crystal targets has been made at electron linear accelerators which are capable of providing intense electron beams of small dimensions and small divergence. Here we mention the experiments performed at Kharkov [3.60] with diamond crystals 0.3 and 0.08 mm thick and at Stanford [3.73] with diamond crystals 0.05 and 0.08 mm thick.

3.4.5 Particular Properties of CB at Low Initial Electron Energies

The condition for the coherent maximum at $\alpha = 0$ may be derived from (3.13). On account of the sharp decrease of the CB cross section at $\theta \neq 0$ as δ increases, one can restrict oneself to the case of $g_1 = 0$. The decrease of the initial particle energy at a fixed θ value leads to a decrease of the energy at which the coherent maximum appears and of the magnitude of the maximum. This was shown by *Walker* et al. [3.47].

When the direction of the initial particle coincides with a crystal axis direction ($\theta = 0$), the condition for a coherent maximum may be written as $g_1 = \delta$. For ultra-relativistic initial particles, CB arises at photon energies $x \sim 1$, and its intensity is considerably lower than at a fixed crystal orientation angle $\theta \neq 0$. The decrease of primary energies results in a shift of the coherent maxima towards lower photon energies, their amplitudes remaining, however, the same. By using electron beams with a small divergence, thin crystals, and photon beam collimation, it is possible to resolve the coherent maxima against an incoherent background, when the initial energy of the electrons is low and they are moving parallel to the crystal axis. Experimentally, this was realized by *Komar* et al. [3.48]. Electrons entered a silicon crystal, 0.007-0.011 mm thick, parallel to the <110> axis. A set of collimators reduced the initial beam divergence down to 4.2×10^{-3} rad. The collimation angle of the photon beam was set at $\theta_c = 3.5 \times 10^{-3}$ rad. The bremsstrahlung spectra shown in Fig.3.21a,b were measured with a NaI(Tl) spectrometer at primary electron energies of 9.6 MeV and 7.4 MeV. The spectra distinctly show coherent maxima at photon energies of 2.02 MeV and 1.29 MeV with relative widths $\Delta = 0.09$ and 0.21, respectively. The widths of the maxima were calculated taking into account the energy resolution

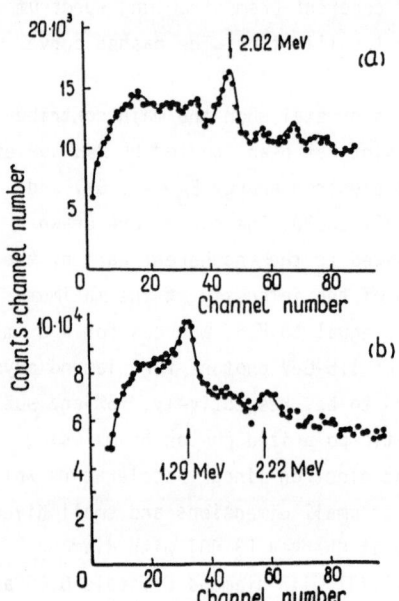

Fig.3.21a,b. CB intensity spectra in silicon crystals at electron energies of 9.6 MeV (**a**) and 7.4 MeV (**b**), for $\theta = 0$

Table 3.2. Parameters of linearly polarized photon beams produced using coherent bremsstrahlung in crystals

	Frascati [3.17]	Tokyo [3.23]	Hamburg [3.25]	Kharkov [3.34]	Cambridge [3.39]	Tomsk [3.38]	Erevan [3.32]	Cornell [3.40]	Stanford [3.41]	Daresbury [3.42]	Bonn [3.43]
Accelerator type	synchr.	synchr.	synchr.	linear	synchr.	synchr.	synchr.	synchr.	linear	synchr.	synchr.
Electron energy [GeV]	1.1	0.8	6.0	1.8	6.0	0.8	6.0	10.8	16.0	5.0	2.5
Year of beam production	1962	1965	1966	1968	1968	1969	1970	1970	1970	1973	1979
Total photon intensity N [eg.q/s]	10^8	10^8	10^9	10^{10} (10^8)[a]	10^8	10^7	10^9	10^9	10^{12}	1.5×10^{10}	
Polarization at x = 0.3 [%]	30	30 (70)	70	50 (92)	70	30	70 (90)	70	80 (90)	70 (40)	40
Monochromaticity $\Delta x/x$ [%]	30	30 (8)	25	30 (15)	30	30	30 (15)	30	20 (10)	30	30
Crystal target [mm]	^{12}C 2.0	^{28}Si 1.0 (0.05)	^{12}C 1.1 (0.04)	^{12}C 2.0 (0.08)	^{12}C 0.5	^{12}C 2.0	^{12}C 2.0 (0.1)	^{12}C 2.0	^{12}C 1.0 (0.08)	^{12}C 1.0	^{12}C 1.0
Electron beam divergence [rad]	10^{-3}	10^{-3}	5×10^{-4}	10^{-3} (10^{-4})	10^{-3}	10^{-3}	10^{-3}	10^{-4}	10^{-3}	10^{-5}	
Photon collimation angle [rad]	1.5×10^{-3}	10^{-3} (10^{-4})	10^{-4}	5×10^{-4} (10^{-4})		10^{-3}	1.6×10^{-4}	2×10^{-4}	2×10^{-4} (5×10^{-5})		5×10^{-4}
Accuracy of goniometer angle reading [rad]	10^{-4}	10^{-4}	10^{-4}	5×10^{-5}	5×10^{-5}	10^{-4}	10^{-4}	5×10^{-5}	5×10^{-5}		

[a] We give the results for thin crystals in parentheses.

of the spectrometer. The positions of the maxima are in good agreement with the values given by (2.13) when θ = 0. In this case however, unlike the CB spectrum when θ ≠ 0, the photons at the maximum are unpolarized due to the symmetry of the case.

3.5 Conclusion

Intense beams of quasi-monochromatic, linearly polarized photons have been produced using coherent bremsstrahlung at practically all electron accelerators. Table 3.2 lists the main parameters of polarized photon beams availabe at accelerators for the period from 1962 to 1979.

The use of quasi-monochromatic, polarized photon beams in studies of electromagnetic interactions has created new trends in nuclear and elementary-particle physics, and during the same period, extensive studies of meson photoproduction on nucleons and nuclei [3.74] have been performed using polarized photon beams.

Acknowledgement. It is a pleasure to thank Prof. Yu.S. Korobochko and Prof. P.V. Sorokin for useful discussions, and Dr. Yu.N. Telegin for critically reading the manuscript. Thanks are also extended to L.F. Prokopchuk for translating the text from the Russian, and preparing the manuscript for publication.

References

3.1 E.J. Williams: Phys. Rev. **45**, 729 (1934)
3.2 B. Ferretti: Nuovo Cimento **7**, 118 (1950)
3.3 L. Landau, I. Pomeranchuk! Dokl. Akad. Nauk SSSR **92**, 535 (1953)
3.4 M.L. Ter-Mikaelian: Zh. Eksp. Teor. Fiz. **25**, 296 (1953)
3.5 M.L. Ter-Mikaelian: Zh. Eksp. Teor. Fiz. **25**, 289 (1953)
3.6 F. Dyson, H. Überall: Phys. Rev. **99**, 604 (1955)
3.7 H. Überall: Phys. Rev. **103**, 1055 (1956)
3.8 H. Überall: Z. Naturforsch. **17a**, 332 (1962)
3.9 G. Fronsdal, H. Überall: Nuovo Cimento **8**, 163 (1958); Phys. Rev. **111**, 580 (1958)
3.10 E.V. Sekhposian, M.L. Ter-Mikaelian: Izv. Akad. Nauk Arm. SSR, Ser. Fiz.-Mat. Nauk **14**, No.4, 143 (1961)
3.11 W.K.H. Panofsky, A.N.S. Saxena: Phys. Rev. Lett. **2**, 219 (1959)
3.12 O.R. Frisch, D.N. Olsen: Phys. Rev. Lett. **3**, 141 (1959)
3.13 A.N.S. Saxena: Phys. Rev. Lett. **4**, 311 (1960)
3.14 L.I. Schiff: Phys. Rev. **117**, 1394 (1960)
3.15 G. Bologna, G. Diambrini, G.P. Murtas: Phys. Rev. Lett. **4**, 134 (1960)
3.16 G. Bologna, C. Diambrini, G.P. Murtas: Phys. Rev. Lett. **4**, 572 (1960)
3.17 G. Barbiellini, G. Bologna, G. Diambrini, G.P. Murtas: Phys. Rev. Lett. **18**, 112 (1962)
3.18 G. Barbiellini, G. Bologna, G. Diambrini, G.P. Murtas: Phys. Rev. Lett. **8**, 454 (1962)
3.19 G. Barbiellini, G. Bologna, G. Diambrini, G.P. Murtas: Phys. Rev. Lett. **9**, 46 (1962)
3.20 G. Barbiellini, G. Bologna, G. Diambrini, G.P. Murtas: Phys. Rev. Lett. **9**, 396 (1962)

3.21 G. Barbiellini, G. Bologna, G. Diambrini, G.P. Murtas: Frascati Report No. LNF-62/114 (1962)
3.22 G. Barbiellini, G. Bologna, G. Diambrini, G.P. Murtas: Nuovo Cimento **28**A, 435 (1963)
3.23 S. Kato, T. Kifune, Y. Kimura, M. Kobayashi, K. Kondo, T. Nishikawa, H. Sasaki, K. Kikuta, K. Kohva: J. Phys. Soc. Japan **20**, 303 (1965)
3.24 T. Kifune, Y. Kimura, M. Kobayashi, K. Kondo, T. Nishikawa: J. Phys. Soc. Japan **21**, 1905 (1966)
3.25 G. Bologna, G. Lutz, H.D. Schulz, U. Timm, W. Zimmermann: Nuovo Cimento **42**A, 844 (1966)
3.26 L. Criegee, G. Lutz, H.D. Schulz, U. Timm, W. Zimmermann: Phys. Rev. Lett. **16**, 1031 (1966)
3.27 G. Lutz, U. Timm: Z. Naturforsch. **21**A, 1976 (1966)
3.28 R.F. Mozley, J. DeWire: Nuovo Cimento **27**, 1281 (1963)
3.29 G. Bologna: Nuovo Cimento **49**A, 756 (1967)
3.30 G. Lutz: Nuovo Cimento **53**A, 242 (1968)
3.31 L. Criegee, M. Garrell, H. Sadrosinski, U. Timm, W. Zimmermann: Phys. Lett. **28**B, 140 (1986)
3.32 R.O. Avakyan, A.A. Armaganyan, L.G. Arutyunyan, G.A. Vartapetyan, L.Ya. Kolesnikov, R.M. Mirzoyan: in *Trudy Mezhdunar. konf. po apparature v fizike vysokikh ehnergij*, Dubna, USSR, 1971, Vol.2 (JINR, Dubna 1971) p.746
3.33 T. Tsuki, S. Kurokawa, T. Nishikawa, S. Suzuki, T. Katayama, M. Kobayashi, K. Kondo: Phys. Rev. Lett. **27**, 609 (1971)
3.34 Yu.V. Zhebrovskii, L.Ya. Kolesnikov, I.I. Miroshnichenko, S.I. Naisteter, A.L. Rubashkin, P.V. Sorokin, R.O. Avakyan, L.G. Arutyunyan: Kharkov preprint KFTI No.329 (1968)
3.35 Yu.V. Zhebrovskii, L.Ya. Kolesnikov, I.I. Miroshnichenko, S.I. Naisteter, A.L. Rubashkin, P.V. Sorokin, R.O. Avakyan, L.G. Arutyunyan: Prib. Tekh. Eksp. **5**, 203 (1969) [English transl.: Instrum. Exp. Tech. USSR **5**, 1327 (1969)]
3.36 Yu.V. Zhebrovskii, L.Ya. Kolesnikov, I.I. Miroshnichenko, L.M. Romas'ko, A.L. Rubashkin, P.V. Sorokin, V.G. Gorbenko, R.O. Avakyan, L.G. Arutyunyan: *Trudy Mezhdunar. konf. po uskoritelyam zaryazh. chastits vysokoj ehnergii*. Yerevan, USSR, 1969, Vol.1 (Academy Sciences of Armenian SSR, Yerevan 1970) p.500
3.37 V.G. Gorbenko, Yu.V. Zhebrovskii, L.Ya. Kolesnikov, A.L. Rubashkin, I.I. Miroshnichenko, P.V. Sorokin, V.F. Chechetenko: *Trudy Mezhdunar. konf. po apparature v fizike vysokikh ehnergij*, Dubna, USSR, 1970, Vol.2 (JINR, Dubna 1971) p.738
3.38 V.M. Kuznetsov, O.M. Stukov, E.V. Repenko, Yu.I. Sertakov: Pis'ma Zh. Eksp. Teor. Fiz. **10**, 273 (1969) [English transl.: Sov. Phys.-JETP Lett. **10**, 174 (1969)]
3.39 D. Bellenger, R. Bordelon, K. Cohen, S.B. Deutsch, W. Lobar, D. Luckey, L.S. Osborne, E. Pothier, R. Schwitters: Phys. Rev. Lett. **23**, 540 (1969)
3.40 G. Diambrini-Palazzi, G. McClellan, N. Mistry, P. Mostek, H. Odren, J. Swartz, R. Talman: Phys. Rev. Lett. **25**, 478 (1970)
3.41 R. Schwitters: SLAC report No. SLAC-TN-70-32 (1970) (unpublished); SLAC Users Handbook, p.51 (1971)
3.42 A. Jackson: Nucl. Instrum. Methods **129** , 73 (1975)
3.43 R. Brockmann, E. Dahl, H.W. Dannhausen, E. Durwen, H.M. Fischer, P. Hampe, T. Müller, U. Schäfer, W. Theisges, D. Wolf: Reports BONN-IR-79-25, Bonn University (1979); BONN-IR-79-27, Bonn University (1979) (unpublished); P. Hampe, Ph.D. thesis (BONN-IR-80-1), Bonn University, 1980; E.A. Dahl, Ph.D. thesis (BONN IR-82-26), Bonn University, 1982
3.44 V.A. Maisheev, A.M. Frolov, R.O. Avakyan, E.A. Arakelyan, A.A. Armaganyan, G.L. Bayatyan, G.S. Vartapetyan, G.A. Vartapetyan, N.K. Grigoryan, A.O. Kechiyan, S.G. Knyazyan, A.T. Margaryan, E.M. Matevosyan, R.M. Mirzoyan, S.S. Stepanyan, L.Ya. Kolesnikov, A.L. Rubashkin, P.V. Sorokin: Zh. Eskp. Teor. Fiz. **77**, 1708-1719 (1979) [English trans.: Sov. Phys.-JETP **50**, 856 (1979)]; Nucl. Instrum. Methods **178**, 319 (1980)
3.45 G. Diambrini, A. Santroni: Proc. CERN/ECFA/72/4, Geneva **1**, 232 (1972)
3.46 C.A. Heusch: Univ. of California, Santa Cruz, Report No. UCSC 76-053 (1976)
3.47 R.L. Walker, B.L. Berman, S.D. Bloom: Phys. Rev. **11**A, 736 (1975)

3.48 A.P. Komar, Yu.S. Korobochko, V.I. Mineev, A.F. Petronenko: Zh. Tekh. Fiz. **41**, 807 (1971) [English trasl.: Sov. Phys.-Tech. Phys. **16**, 631 (1971)]

3.49 G. Diambrini: Rev. Mod. Phys. **40**, 611 (1968)

3.50 U. Timm. Fortschr. Phys. **17**, 765 (1969)

3.51 M.L. Ter-Mikaelian: *Vliyanie sredy na ehlektromagnitnye protsessy pri vysokikh ehnergiyakh* (Akad. Nauk. Arm. SSR, Erevan 1969); *High-Energy Electromagnetic Processes in Condensed Media* (Wiley-Interscience, New York 1972)

3.52 V.G. Gorbenko, Yu.V. Zhebrovskii, A.S. Zelencher, L.Ya. Kolesnikov, A.L. Rubashkin, P.V. Sorokin, V.F. Chechetenko: Kharkov preprint No. KFTI 78-16 (1978)

3.53 N.A. Agarkov, V.G. Gorbenko, Yu.v. Zhebrovskii, A.S. Zelencher, L.Ya. Kolesnikov, A.L. Rubashkin, V.F. Chechetenko: Prib. Tekh. Eksp. **2**, 57 (1979) [English transl.: Exp Tech. (USSR) **22**, 350 (1979)]

3.54 I.A. Grishaev, G.D. Kovalenko, B.I. Shramenko: Zh. Eksp. Teor. Fiz. **72**, 437 (1977) [English transl.: Sov. Phys.-JETP **45**, 229 (1977)]

3.55 V.G. Gorbenko, Yu.V. Zhebrovskii, L.Ya. Kolesnikov, A.L. Rubashkin: Ukr. Fiz. Zh. **17**, 757 (1972)

3.56 I.A. Grishaev, G.D. Kovalenko, V.I. Kulibaba, V.L. Morokhovskii, B.I. Shramenko: Ukr. Fiz. Zh. **24**, 1188 (1979)

3.57 G.D. Bochek, V.I. Vit'ko, I.A. Grishaev, G.D. Kovalenko, V.I. Kulibaba, V.L. Morokhovskii, B.I. Shramenko: *Trudy VI Vsesoyuznogo soveshchaniya po fizike vzaimodejstviya zaryazhennykh chastits s monokristallami*, Moscow, 3-5 June, 1974 (Moscow State University, Moscow 1975) p.252

3.58 G. Barbiellini, G. Bologna, G. Capon, G. DeZorzi, F.L. Fasbri, G.P. Murtas, G. Diambrini, G. Sette, J. DeWire: Phys. Rev. **184**, 1402 (1969)

3.59 V.G. Gorbenko, Yu.V. Zhebrovskii, L.Ya. Kolesnikov, I.I. Miroshnichenko, L.M. Romas'ko, A.L. Rubashkin, P.V. Sorokin: Yad. Fiz. **11**, 1944 (1970 [English transl.: Sov. J. Nucl. Phys. **11**, 580 (1970)]

3.60 V.G. Gorbenko, Yu.V. Zhebrovskii, N.A. Kovalenko, L.Ya. Kolesnikov, I.I. Miroshnichenko, A.L. Rubashkin, V.M. Sanin, P.V. Sorokin: Yad. Fiz. **24**, 961 (1976) [English transl.: Sov. J. Nucl. Phys. **24**, 503 (1976)]

3.61 V.B. Ganenko, V.G. Gorbenko, L.M. Derkach, Yu.V. Zhebrovskii, L.Ya. Kolesnikov, I.I. Miroshnichenko, V.I. Nikiforov, A.L. Rubashkin, V.M. Sanin, P.V. Sorokin, S.V. Shalatskii: Yad. Fiz. **23**, 310 (1976) [English transl.: Sov. J. Nucl. Phys. **23**, 162 (1976)]

3.62 H. Olsen, L.C. Maximon: Phys. Rev. **114**, 887 (1959)

3.63 M. Kobayashi, K. Kondo: Nucl. Instrum. Methods **104**, 101 (1972)

3.64 R.O. Avakyan, A.A. Armaganyan, L.G. Arutyunyan, G.A. Vartapetyan, A.G. Iskandaryan, R.M. Mirzoyan, G.M. Ehlbakyan: Izv. Akad. Nauk Arm. SSR, Fiz. **9**, 252 (1974)

3.65 N. Cabibbo, G. Da Prato, G. De Franceschi, U. Mosco: Phys. Rev. Lett. **9**, 270 (1962)

3.66 N. Cabibbo, G. De Prato, G. De Franceschi, U. Mosco: Nuovo Cimento **27**, 979 (1963)

3.67 R.L. Eisele, D.J. Sherden, R.H. Siemann, C.K. Sinclair, D.J. Quinn, J.P. Rutherford, M.A. Shupe: Nucl. Instrum. Methods **113**, 489 (1973)

3.68 C. Berger, G. McClellan, N. Mistry, H. Orgen, B. Sandler, J. Swartz, P. Walstrom, R.L. Anderson, D. Gustavson, J. Johnson, I. Overman, R. Ralman, B.H. Wiik, D. Worcester, A. Moore: Phys. Rev. Lett. **25**, 1366 (1970)

3.69 R.O. Avakyan, A.A. Armaganyan, L.G. Arutyunyan, S.S. Danagulyan, P.Ts. Sarkisyan, G.M. Ehlbakyan, S.M. Darbinyan, R.M. Mirzoyan: Izv. Akad. Nauk Arm. SSR, Fiz. **10**, 423 (1975)

3.70 L. Ya. Kolesnikov, I.I. Miroshnichenko, P.V. Sorokin, Yu.I. Titov: Ukr. Fiz. Zh. **23**, 529 (1978)

3.71 V.G. Gorbenko, L.M. Derkach, Yu.V. Zhebrovskii, L.Ya. Kolesnikov, A.L. Rubashkin: Yad. Fiz. **17**, 793 (1973) [English transl.: Sov. J. Nucl. Phys. **17**, 413 (1973)]

3.72 T. Tsuki, S. Kurokawa, T. Nishikawa, S. Suzuki, T. Katoyama, M. Kobayashi, K. Kondo: Phys. Rev. Lett. **27**, 609 (1971)

3.73 W. Kaune, G. Miller, W. Oliver, R.W. Williams, K.K. Young: Phys. Rev. **D11**, 478 (1975)

3.74 D. Menze, W. Pfeil, R. Wilcke: Compilation of pion photoproduction data. Physics Data 7-1 (Physikalisches Institut der Universität Bonn, Bonn 1977)

4. Bent Crystal Channeling

R. A. Carrigan, Jr.[1] and W. M. Gibson[2]

With 18 Figures

The theory of channeling in bent crystals is reviewed as well as the experimental
results at high energy. Factors affecting bent crystal channeling are discussed.
Finally, some of the possible applications of the deflection of charged particles
with bent crystals are summarized.

4.1 Channeling at High Energy

During the past few years, particle channeling studies [4.1] have been extended up
to the highest energies for which beams are experimentally available [4.2,3]. The
results of these studies have led to active interest in the possibility of using
the channeling effect as a tool in high-energy physics research. For example,
Temmer and others have suggested [4.4,5] that blocking techniques be used for particle lifetime studies in the same spirit that they have been used to study compound nuclear lifetimes at MeV energies [4.6]. *Carrigan* [4.7] has suggested that
oriented single crystals could be used to investigate a process called "production
channeling", in which the nuclear interactions of short-lived particles generated
with well-collimated production processes might be studied. Following the suggestion
made by *Kumakhov* [4.8], a number of studies of radiation from high-energy channeled
electrons and positrons has been carried out [4.9-12], and extensions of this effect for particle identification at very high energies [4.13] or to give greatly enhanced yields of processes such as pair production [4.14] have been proposed. Among
the most interesting suggestions is *Tsyganov*'s idea [4.15] to use elastically bent
crystals to deflect particle beams. This opens up a whole host of possibilities including beam optics [4.16], charm-particle decays studies [4.16], and measurement
of the magnetic moment of short-lived particles [4.17]. This chapter will summarize
the current status of beam bending studies and possible applications. Before doing

1 Operated by Universities Research Association, Inc., under contract with the US
  Department of Energy.

2 Supported in part by the US Department of Energy under contract No. ER-78-5-02-
  5001.

this, however, it is useful to discuss briefly ion channeling in crystals with emphasis on aspects of particular interest at high energies and for beam bending.

Ion channeling at keV and MeV energies has been actively pursued since about 1963 [4.1] and has resulted in the development of a comprehensive classical continuum theory [4.18] which successfully describes the effect and forms the basis for numerous applications in nuclear physics and solid-state physics and technology. This same theory also forms the basis for channeling studies at high energy. Channeling is the steered motion of charged particles in the traversing of single crystals at small angles to the directions of atomic rows or planes in the crystal. The steering is a result of correlated scattering of the projectile by successive atoms in the row or plane, and is described in the continuum model in terms of electrostatic deflection by a rod or sheet of charge in the respective cases of axial or planar channeling. This geometrical channeling effect is characterized by a critical angle of the particle motion relative to the row or plane above which the steered motion breaks down and the particles no longer move in well-defined channeling trajectories. When channeled between atomic planes in a crystal, particles execute regular oscillatory motion, as shown in Fig.4.1a. When viewed along the particle path, this motion is described accurately in terms of a charged mass oscillating in a potential well defined by the screened potential of the atoms in the planes, averaged over each plane. Indeed, measurements of the transverse energy of the particle as a function of the period of the oscillatory motion allow the potential to be accurately determined [4.19]. The motion in the geometric plane perpendicular to the forward particle motion is not, in general, sinusoidal, since the potential usually has significant components of higher order than quadratic. The important feature of planar channeling from the standpoint of beam bending is that the motion of a particular channeled particle is restricted between two physical atomic planes in the crystal.

In this discussion we have, of course, referred only to positively charged incident particles. Negative particles will also experience planar potential minima when moving at small enough angles to atomic planes, and can experience similar restricted motion. There are, however, two important differences. First, the minimum in the potential is located directly on the atomic planes, not between them, so the particles are confined to a region of very high atomic density which leads to greater probability of scattering and, therefore, lower stability of the steered oscillatory trajectories. Second, the potential minima experienced by negative particles are narrow because of screening. As a result, a smaller fraction of particles in a beam are incident on the crystal in a region of low potential where they can be effectively captured. Consequently, although channeling of such particles has been seen in a variety of ways [4.20,21], no clear evidence has yet been found for bending of negative particles. As we will discuss later, it may be possible at very high particle energies for negative particles to be efficiently deflected in bent

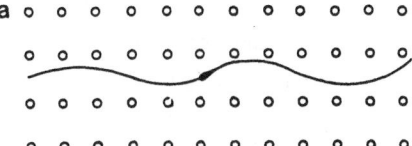

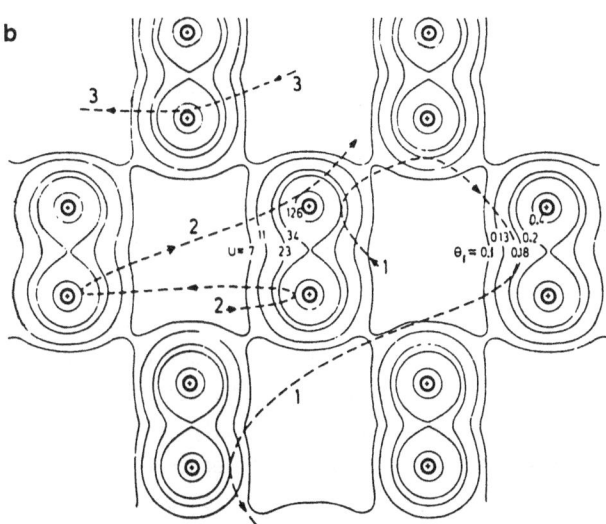

Fig.4.1. (a) Positive particle trajectory for planar channeling. (b) Positive particles within the critical angle for axial channeling wander to and fro in the transverse direction of motion as they reflect off the repulsive potential of the nuclear strings. Equipotentials are for 7,11,23,34, and 126 eV. Trajectory 1 corresponds to an angle of 21 microradians, 2 corresponds to 48 microradians, and 3 corresponds to 65 microradians at 100 GeV. Trajectories 1 and 2 are channeled. Scaled from [4.3b]

crystals, especially if the particles are produced by reactions between incident particles and atomic nuclei in crystal planes.

Positive particles incident parallel to atomic rows (or axes) in a crystal also undergo steered motion. Indeed, the effect is somewhat stronger than for planar incidence because of the higher atomic density and consequently the higher electrostatic potential along the rows. In this case, however, the motion of a channeled particle in the geometric plane perpendicular to the forward motion of the particle is somewhat more complex than in the planar case. This is shown in Fig.4.1b where the transverse motion of three channeled particles passing through a thin crystal is shown. In general, such axially channeled particles are not confined to a particular axial "channel" but can wander somewhat freely in the transverse space, with their precise trajectory being sensitively influenced by the details of the position and angle of their initial impact point on the crystal relative to the location of the atomic rows, indicated as points in Fig.4.1b. As we will show, this difference between axial and planar motion has a profound effect on the bending of incident positive particle beams. It should be noted, parenthetically, that the situation is expected to be different for negative particles, which should be captured in the deep minima centered on the atomic row positions.

For both axial and planar incidence, the probability that a particle will be channeled depends on its energy (or momentum) transverse to the atomic plane or axis of interest. The transverse energy is given by ½ pv $\sin^2\psi$ (or ½ pvψ^2 for the small angles involved in the present discussion), where ½ pv is the kinetic energy of the particle at non-relativistic energies and ψ is its angle relative to the plane or axis. For a given kinetic energy there is, therefore, an incident angle below which particles can be channeled and above which they do not undergo steered motion in the crystal. This angle is called the channeling critical angle, and in appropriate relativistic notation it is given by classical continuum theory for axially channeled high-energy particles as:

$$\psi_c = \alpha \left(\frac{4Ze}{pvd}\right)^{\frac{1}{2}} \quad , \tag{4.1}$$

where Z is the atomic charge of the target nucleus, d is the atomic spacing along the lattice string, p is the momentum, and α is a constant close to unity whose precise value depends on the thermal vibration amplitude of the crystal atoms. At high energy, this critical angle is small, e.g., it is ~40 μrad at 250 GeV/c for positive particles axially channeled along the <110> axis in germanium. Positive particles channeled along low-index crystal axes or planes show substantially lower energy loss than for random incidence because such particles are confined to a region of space where the electron density is lower than the average electron density in the crystal, and they therefore experience a reduction in the probability for close ionizing collisions. Angular distributions upon emergence from the crystal for particles incident at an angle in the vicinity of or less than the critical angle with respect to the axes or planes can also be substantially modified.

Particles in channeled trajectories are dechanneled by a variety of effects. These determine an upper limit on the thickness of the crystal that will give strong channeling at a particular energy. The theory of the dechanneling process has been studied for the axial case by *Bonderup* et al. [4.22] in terms of a diffusion mechanism in which on-axis trajectories are first scattered by electron collisions, and then increasingly by nuclear collisions, as particle trajectories approach the critical angle. Both a nuclear and an electronic diffusion length are required to model axial channeling. The electron and nuclear axial dechanneling lengths are:

$$z_e = \frac{pv}{2\pi e^2 L_e\, Nd} \quad , \tag{4.2}$$

$$z_n = \frac{3}{8\pi\rho^2}\, \frac{(0.885a_0)^2 pv}{Nd\, Z^{5/3}e^2} \quad , \tag{4.3}$$

where p and v are the momentum and velocity of the particle (the correct relativistic formula must be used at high energy), d the lattice spacing along the string, $L_e = \log(2m_e v^2 \gamma^2/I)$, I is the ionization potential, $\gamma = 1/\sqrt{1-v^2}$, m_e = mass of the electron, N is the atomic density of the material, a_0 the Bohr radius, ρ^2 the average square amplitude of the lattice vibrations, and Z the atomic number of the material. The dechanneling length is defined as the thickness of crystal in which one half of the channeled particles will be dechanneled. The dechanneling length is proportional to the energy of the particle. Table 4.1 gives the calculated dechanneling lengths for positive particles channeled along the <110> axes in silicon, germanium, and tungsten at room temperature.

Typical channeling experiments in the GeV regime are based on the technique pioneered by the CERN-Aarhus group [4.3]. High-energy particles are directed at a crystal mounted in a precise goniometer. The incident and exit angles of each particle relative to the crystal, as well as the position of impact on the crystal, are precisely measured by a set of high-resolution drift chambers. For semiconducting crystals, the energy loss can be conveniently determined by making the crystal itself a semiconducting detector. Because of the small critical angles for high-energy particles, the crystal must be essentially free of mosaic structure, strain, or other large-scale imperfections. For this reason, as well as for the convenience of having an energy-loss detector intrinsic to the target, all high-energy measurements to date have been carried out with silicon and germanium crystals, which are probably the most structurally perfect crystals in existence. A significant increase in both the critical angle and the dechanneling length could be achieved in principle by using tungsten (and possibly other) crystals, because of their high atomic number (Z) and low vibrational amplitude (ρ). Typically, the mosaic spread ($\geqslant 0.1$ degrees) in such crystals precludes their use, although in principle, highly perfect, dislocation-free tungsten crystals should be obtainable, and attempts are being made to fabricate them.

Figure 4.2 shows a schematic diagram of a typical experimental apparatus, the Albany-Chalk River-Dubna-Fermilab bending apparatus. The nearly perfect crystal is mounted in a goniometer with projected angle and azimuthal angular degrees of freedom relative to the beam. For the recent Fermilab experiments, the goniometer is remotely controlled from a computer and can be stepped in increments as small as 8 μrad. The energy deposited in the crystal is measured with the built-in diode. The diode voltage must be set high enough so that the crystal is fully depleted. Typically, this is several hundred volts.

The system is triggered by a coincidence of scintillation counters S1 and L shown in Fig.4.2, sometimes in coincidence with the crystal detector. An anticoincidence counter (A) with a hole in it smaller than the crystal is placed so that the hole partially covers the crystal. This assures that particles not triggering A have gone through the crystal. The signal from the crystal is fed into a conventional linear electronics amplifier system operating with a rise time of several microseconds. The typical energy resolution for the detector is 40 keV, which is better than required for channeling measurements.

Table 4.1. Some channeling properties for 100-GeV particles

	Silicon	Germanium	Tungsten
Z	14	32	74
A	28.09	72.59	183.85
ρ [gm/cm^3]	2.35	5.33	19.3
Axial critical angle <110> [μrad]	46	68	138
Planar critical angle (110) [μrad]	16	20	34
Minimum elastic bending radius for 1-mm bar, energy independent [cm]	76	76 cm	94 cm[a]
Critical field (110) [V/cm]	0.61×10^{10}	1.28×10^{10}	4.73×10^{10}
Tsyganov radius, planar, (110) plane	16.3	7.8	2.1
Equivalent magnetic field (110), plane [MG]	20	43	160
Nuclear dechanneling length, axial, <110> (room temperature) [cm]	10.3	2.2	1.7
Electronic dechanneling length, axial, <110> [cm]	6.44	7.7	10.8

[a]This number is an estimate. Tungsten could probably be bent to a smaller radius if dislocations were introduced.

E660 APPARATUS

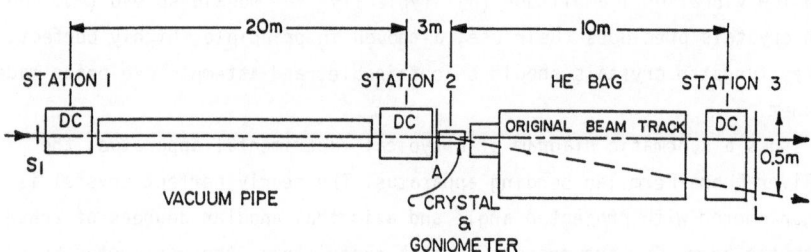

Fig.4.2. Typical channeling bending apparatus. Here S1 and L are trigger counters while A is an anticoincidence counter. The crystal is at station 2. Three sets of drift chambers (DC) define the particle direction entering and leaving the crystal

Since the directions of particles passing through the crystal are measured with a calibrated drift-chamber system, care must be taken to reduce the sources of multiple scattering. At 250 GeV, the root-mean-square (RMS) spatial resolution per module is typically 100 microns. The RMS distribution of the difference between the angles of emergence and incidence, with the crystal removed, was ~10 μrad. This distribution gives a measure of the angular resolution of the system. Much of the width of that distribution at 250 GeV was due to the spatial resolution of the drift planes, while at 35 GeV/c most of the width was due to muliple scattering in

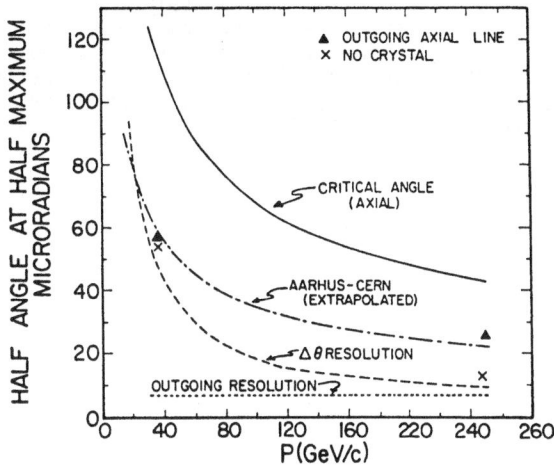

Fig.4.3. Momentum dependence of the critical angle, resolutions, and the measured half-width of the axial transmission line for a selection of small scattering angles and low energy loss in a Fermilab experiment. The Aarhus-CERN axial width is extrapolated from 1.35 GeV/c using $1/\sqrt{p}$. The triangles are the representative measurements for the Fermilab experiment. Both sets are raw, that is, resolution effects have not been removed. $\Delta\theta$ resolution: expected resolution for measuring a scattering angle. Outgoing resolution: resolution for measurement of outgoing angles. x: experimental data with the crystal removed. All angles except the critical angle are given in terms of half-width at half maximum

the system. Figure 4.3 shows the predicted and observed experimental resolution as a function of momentum for the system used in the first channeling experiments at Fermilab. It should be noted that the system resolution is different for each of three angles: the incident angle, the exit angle, and the scattering angle. This is because different sets of quantities enter the measurement of each angle, and because the angular measurements of the incoming and outgoing particles are not sensitive to multiple scattering. The experimental resolution is sufficient to clearly determine the angular spread of axial channeling effects. Planar channeling widths are typically about one third as large as axial widths, so that the magnitudes of planar effects are more affected by resolution. The system performance is also evident in Fig.4.3 from the measured width of the outgoing <110> axial transmission channeling peak, at a selection of small scattering angles and small energy losses. The Aarhus-CERN measurements [4.3] were extrapolated from 1.35 GeV/c. The effects of resolution have not been removed for either the Fermilab or CERN experiments. The agreement between the measurements is good. In addition, it is clear that the apparatus has achieved the high resolution required for experimental measurements at high energies.

The crystal is aligned by looking at the ratio of small-energy-loss events relative to all events as the crystal orientation is varied. Near planes and axes, this ratio increases, as shown by the solid line in Fig.4.4. If only particles incident and emergent within the critical angle are selected, the distribution

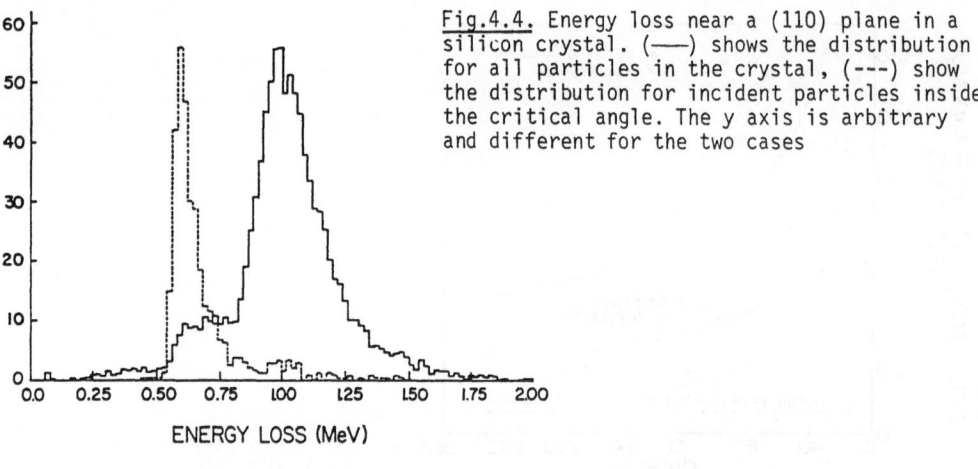

Fig.4.4. Energy loss near a (110) plane in a silicon crystal. (——) shows the distribution for all particles in the crystal, (---) show the distribution for incident particles inside the critical angle. The y axis is arbitrary and different for the two cases

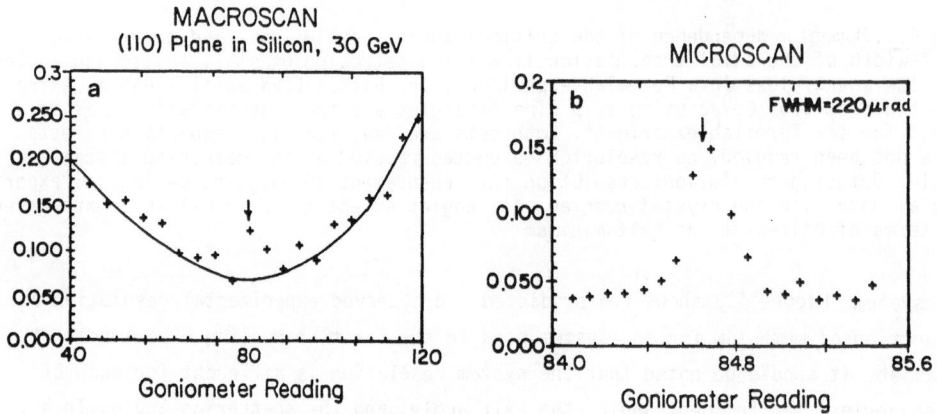

Fig.4.5a,b. Goniometer scan to find a (110) plane in silicon using +30 GeV particles; (a) macroscan, (b) microscan near the plane. The goniometer angle is in milliradians. The ordinate is the ratio of anomalously low-energy particles to particles in the main peak

sharpens to that shown by the dotted line in Fig.4.4. Figure 4.5 illustrates a typical scanning pass for a bending-alignment run. The crystal is first rotated through a macroscan of 80 milliradians or about 5°, in steps of 4 milliradians, at a low energy (30 GeV), as shown in Fig.4.5a. The ratio of low-energy events relative to events in the main peak increases near the edge of the macroscan, because for incident directions more than about two degrees away from the plane of the thin crystal slab, some particles no longer pass through the entire detector. Since the planar alignment relative to the crystal face is much better than this, the center part of the distribution can be searched for the plane. A microscan is then made in 80-microradian steps, in which one searches for the rise in the low-energy events that characterize a plane, as shown in Fig.4.5b. The width of the curve is set by the angular width of the beam, which is substantially larger than the critical angle.

4.2 Bending

What happens to a particle in a planarly channeled trajectory if the crystal through
which it moves is bent? Bending the crystal is equivalent to introducing a centrifugal potential, thereby lowering one side of the continuum potential well and rais-
ing the other. This causes the equilibrium trajectory to move away from the centerline of the midpoint between the planes toward the plane on the outside of the
curved channel. In effect, the channeling critical angle diminishes, leading to
"bending" dechanneling so that more dechanneling occurs. *Tsyganov* [4.15] has calculated the bending radius at which no high-energy particles would remain in the well
to be

$$R_T = E/eE_c \quad ,\tag{4.4}$$

where E is the total energy of the particle and E_c the interatomic field intensity
at a distance from the plane of the crystal lattice where the trajectory of the
particle no longer remains stable, due to its interactions with individual atoms.

The "Tsyganov radius" for a 100-GeV particle moving in a (110) plane in silicon
is 16 cm, so that a 1-cm arc of silicon with constant curvature could deflect particles up to 60 milliradians. More complete analyses such as those of *Ellison* [4.23,
24], *Kudo* [4.25], and *Vorobiev* and co-workers [4.26] suggest that most particles
would be lost at radii of curvature somewhat larger than the Tsyganov result.

This critical radius can be related to an equivalent magnetic field for a relativistic particle, by observing that the radius of curvature of a particle in a
magnetic field is R = p/0.03 B (where p ≅ E/c is in GeV/c, B is in kG, and R in m).
Alternatively, a proper Lorentz transformation can be used to derive the same result for the equivalent field in the particle's rest frame. The equivalent magnetic
field for the (110) plane in tungsten is 160 megagauss.

Deflection of a beam of charged particles using channeling in a bent crystal is
a distinctly different process to bending with a magnet. Up to some momentum related to the critical radius, a channeled particle near the plane direction should
be deflected independently of its momentum. The lower-momenta particles will be
lost more rapidly than the higher-momenta ones since their normal dechanneling
length will be shorter. At high momenta near the critical radius, more divergent
particles will be lost. Channeling with a bent crystal, then, is a wide-momentum
band-pass method of deflecting charged particles. This could be a distinct advantage for beam systems that require the deflection of particles with a large range
of momenta. On the other hand, the same feature ensures that such a system would
have no momentum selection capability.

Tsyganov and his collaborators, in a joint USSR-USA collaboration carried out
at Dubna [4.27], were the first to observe channeling in bent crystals. They looked
mainly at particle transmission in bent planes. The experiments were performed in
an 8.4-GeV proton beam, using silicon crystals. An experiment concentrating parti-

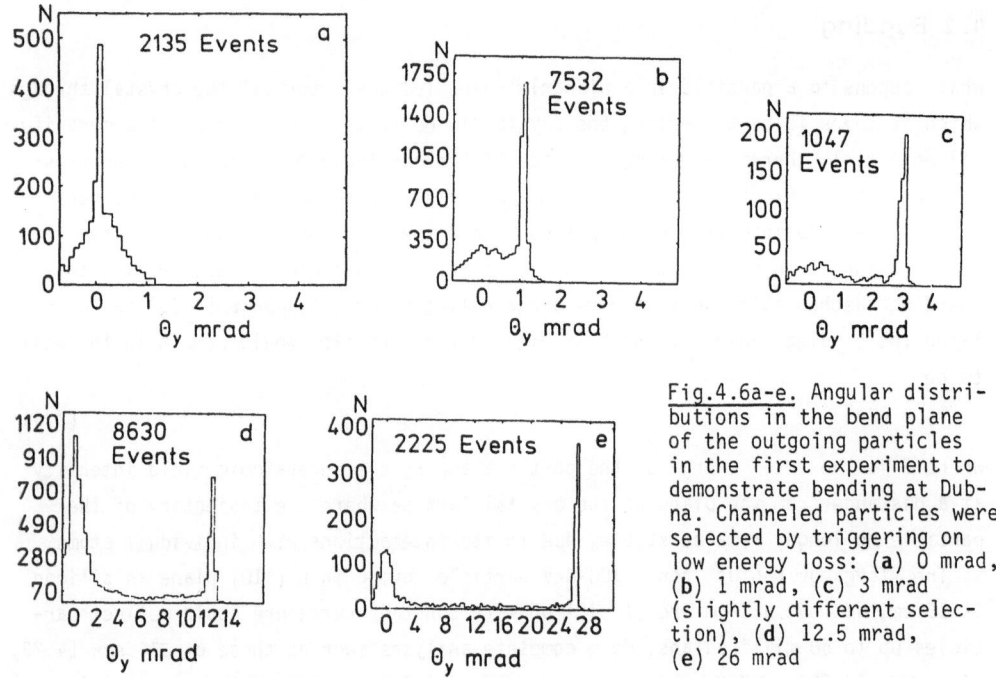

Fig.4.6a-e. Angular distributions in the bend plane of the outgoing particles in the first experiment to demonstrate bending at Dubna. Channeled particles were selected by triggering on low energy loss: (**a**) 0 mrad, (**b**) 1 mrad, (**c**) 3 mrad (slightly different selection), (**d**) 12.5 mrad, (**e**) 26 mrad

cularly on axial bending has been carried out by a CERN-Aarhus-Strasbourg group, using 12-GeV/c particles [4.28]. More recently, an Albany-Chalk River-Dubna-Fermilab group has made extensive observations of bending at energies up to 200 GeV [4.29].

In these experiments, done with bent crystals, the channeled fraction of the beam followed the direction of the downstream end of the crystal. In the Dubna experiment, the crystals were bent by up to 26 mrad. Figure 4.6 shows the distribution of the scattered channeled particles as the bending angle was increased. Eventually the bending program was stopped when a crystal broke at a bending angle slightly greater than 26 mrad. This was equivalent to a bending radius of 38 cm, far larger than the critical radius. At CERN, a bend of 52 mrad was reached with a 0.5-mm-thick crystal, corresponding to a bending radius of roughly the same amount.

In the Dubna experiment, differential dechanneling can be seen as the crystal is bent, in the form of particles spilling out of the channel as they travel around the bend. After deflections of several milliradians, the net transmission of particles decreases slowly as the bending angle increases. This is reasonable, since the critical Tsyganov radius is not approached in either experiment.

Experimental information on dechanneling will be discussed in more detail after considering the theory of bending dechanneling in the following section.

4.3 Bending Theory

As indicated in Sect.4.2, the critical radius through which a planarly channeled particle can be bent is given very simply as $R_T = E/eE_c$. This represents a lower limit on the radius of curvature within which particles will continue to be channeled as they follow the bend of the crystal, and corresponds to particles which are incident on the crystal exactly parallel to the planes of interest and exactly in the centerline between two planes prior to bending. Such particles would, in the absence of bending, be transmitted without oscillation along the centerline and are said to have zero transverse energy. If the crystal is bent, such particles will experience a centrifugal force due to bending, the magnitude of which is $pv\kappa$ where κ is the magnitude of the curvature of the bent planes, related to the radius of curvature R by $\kappa = 1/R$. The centrifugal acceleration of the particles is matched by the magnitude of the centripetal force f exerted by the plane, so $pv\kappa = f$. In this notation, the critical radius of curvature is given by $R_c = 1/\kappa_c = pv/f_c$, which is equivalent to (4.4), since $f_c = V'(x_c)$, where $V(x_c)$ is the planar potential at the distance of approach (x_c) of the particle to the plane where the particle channeling becomes unstable. For a crystal curvature less than the critical value, the trajectory for these perfectly channeled, zero-transverse-energy particles will lie at an equilibrium position x_e, which is displaced from the centerline toward the outside plane. For constant curvature, $pv\kappa = V'(x_e)$, where x_e is the distance from the centerline and $V'(x_e)$ the planar force at the equilibrium position.

If channeled particles were confined to the special class of particles with zero transverse energy, the channeling effect would be of little interest, since so few particles would meet the stringent criteria. Most channeled particles do not have zero transverse energy as they move through the crystal, since they enter the crystal at a nonzero angle to the planar direction, or they enter at a point away from the centerline. As noted in Sect.4.1, such particles oscillate about the centerline as they move through the crystal, following a periodic trajectory determined by the planar potential. The equation of motion for relativistic, planarly channeled particles is given, to a good approximation, by

$$\frac{d^2x}{dz^2} + \frac{1}{pv} V'(x) = 0 \quad , \tag{4.5}$$

where x is the distance from the centerline, z is the distance along the centerline, p and v are the momentum and velocity of the particle and $V(x)$ is the planar continuum potential. The trajectory of such a particle is commonly characterized by the wavelength λ of the periodic motion, the amplitude of the oscillation x_m, and the angle of the particle motion Ψ_m relative to the centerline. Such planarly channeled motion is shown in Fig.4.7a. Since the oscillation amplitude is related

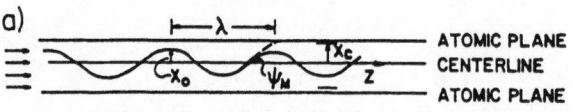

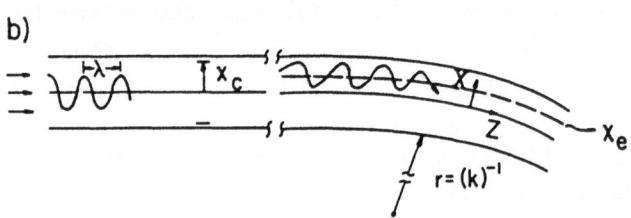

Fig.4.7a,b. Schematics of channeled planar orbits in an unbent (**a**) and bent (**b**) crystal. Note that the center of gravity of the orbit moves toward the outer plane in the bent crystal. Based in part on [4.24]

to the angle of the trajectory relative to the centerline by $\frac{1}{2} pv\psi_m^2 = V(x_m)$, it is common to discuss channeling in terms of the critical channeling angle ψ_c, where $\psi_c = [2V(x_c)/pv]^{\frac{1}{2}}$. This is simply a consequence of the conservation of transverse energy. At the peak amplitude of the oscillation, i.e., $\psi = 0$ and $x = x_m$, the energy is all potential and has the value $V(x_0)$. As the particles cross the centerline $x = 0$ and $V(x_m) = 0$, all of the energy is kinetic and has the value $E\psi_m^2$.

For a bent crystal, the oscillatory trajectories shift towards one of the atomic planes, with the center of their oscillation given by the equilibrium position discussed previously. This is shown in Fig.4.7b. It is easy to see that particles with amplitude of oscillation greater than x_c will be dechanneled. As the crystal curvature increases from zero, particles with oscillation amplitude near the critical value will begin to approach the plane almost immediately and thus be dechanneled. For a crystal bent with a constant curvature, the equation of motion for relativistic particles is well approximated by

$$\frac{d^2x}{dz^2} + \frac{1}{pv} V'(x) = \kappa \quad .$$ (4.6)

It is instructive to describe the particle motion and the effect of bending in terms of a phase space described by the position of the particle in the transverse planar space and the angle of the particle relative to the planar direction. This is shown in Fig.4.8. Particles within the open ovals will continue to be channeled in the bent crystal, whereas those within the hatched region will follow a path such as that indicated by the dashed trajectory shown. Such particles will approach the plane with amplitude larger than x_c, and hence be dechanneled. Calculation of the fraction of channeled particles that dechannel because of this centrifugal shift of the equilibrium point of their trajectory can be carried out if the form of the planar potential is known and if the critical distance of approach, x_c, is given. Such calculations have been made for a harmonic potential by *Ellison* and *Picraux* [4.23] and for more realistic Thomas-Fermi type potentials by *Ellison* [4.24] and by *Kudo* [4.25].

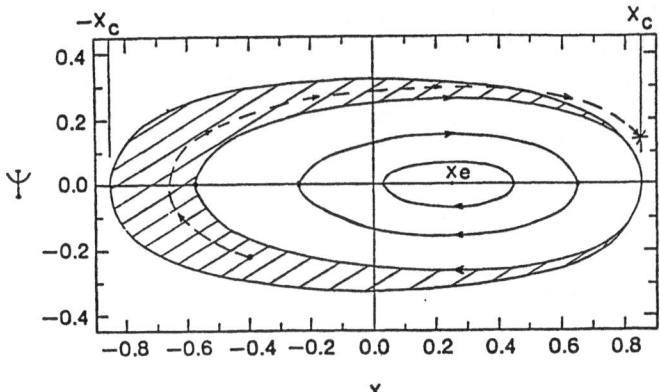

Fig.4.8. Typical phase space geometry for particles moving in a bent crystal for the wide (111) planes in silicon. x_c is the normalized distance to the charge plane, Ψ is a normalized angle relative to the local plane direction, and x_e is the center line of the displaced trajectory. Particles in the inner ovals remain channeled while those in the cross-hatched region are lost. From [4.24]

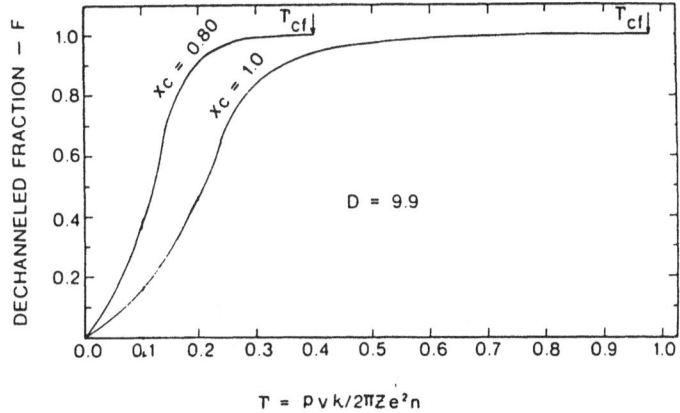

Fig.4.9. Dechanneled fraction F versus the dimensionless parameter Γ. The values of x_c correspond to dechanneling at the planes and at the Thomas-Fermi screening distance away from the planes. From [4.24]

Results for particles dechanneled upon entering a region of constant curvature are shown in Fig.4.9. This calculation uses a Molière approximation to the Thomas-Fermi potential, which has found much use in particle channeling calculations. The calculation includes the following assumptions: (1) the channeled particles have a statistical equilibrium distribution of trajectories when they reach the bend; (2) a particle is dechanneled if it penetrates too close ($x > x_c$) to a plane at any point of its trajectory; (3) the bent planes can be modeled by bent continuum planes using a planar continuum potential; and (4) the wavelength for particle oscillation in its periodic motion is short compared to the the length of the bent crystal. All of these assumptions are supported by extensive experimental studies on particles channeled over the energy range from a few keV to 250 GeV.

With respect to the last assumption, the wavelength for periodic motion increases with increasing energy, and for a given energy the maximum value is given, to a good approximation, by

$$\lambda_m = \left[\frac{\pi d_p pv}{Z e^2 nW''(0,D)}\right]^{\frac{1}{2}} \quad , \tag{4.7}$$

where the value of the function $W''(0,D)$ ranges from ~0.1 to ~0.5 for the narrowest to the broadest planes that might be considered for bending applications, d_p is the interplanar spacing, $D = d_p/a_T$, a_T is the Thomas-Fermi screening distance, and n the areal density. This gives wavelengths λ of tens of microns at the highest energies, still very small compared to the path length in the bent crystal. The two curves in Fig.4.9 correspond to $x_c = (d_p/2)-a_T$, the smallest value expected from normal de-channeling studies, and $x_c = d_p/2$, the largest possible excursion the particles can experience and still be deflected by the planar potential. Since multiple scattering is strongly reduced at very high particle energy, it might be argued that the latter is the most applicable. Note that there is a rather substantial difference between the two curves. The abscissa of Fig.4.9 is expressed in terms of a dimensionless parameter Γ,

$$\Gamma = pv\kappa/2\pi Z e^2 n \quad . \tag{4.8}$$

The value of Γ can easily be calculated for any appropriate combination of target and projectile, projectile energy, and curvature Γ, Γ_{cf} corresponds to the Tsyganov centrifugal force model. From this figure one can see that the Tsyganov formula (4.4) overestimates the effective bending by a factor of about five.

To put the prediction of particle bending into experimental terms, Fig.4.10 gives several predictions for protons channeled along (111) planes in silicon crystal, with the abscissa expressed in terms of pv/R. This figure shows *Kudo*'s calculation [4.25] for the wide and narrow (111) planes in silicon (with $a_T > 0$, to give some indication of charge smearing), as well as corresponding curves from calculations by *Ellison* [4.24], using a weighted average of these wide and narrow planes in the case of the latter curves. Finally, another curve, depicting *Kudo*'s calculations [4.25] for a parallel beam distribution incident on the wide planes, is shown. Recall that a curve lying further to the right indicates less bending dechanneling.

Note, first, that the effects of the wide planes and the narrow planes are reversed in Ellison's and Kudo's calculations. This may be due to the fact that Ellison used fewer planes away from the channel in the potential calculation. The two sets are also not exactly comparable because different screening distances were used. When this is taken into account they are reasonably comparable. Notice, however, that the Kudo parallel incidence case is drastically different. Thus differ-

74

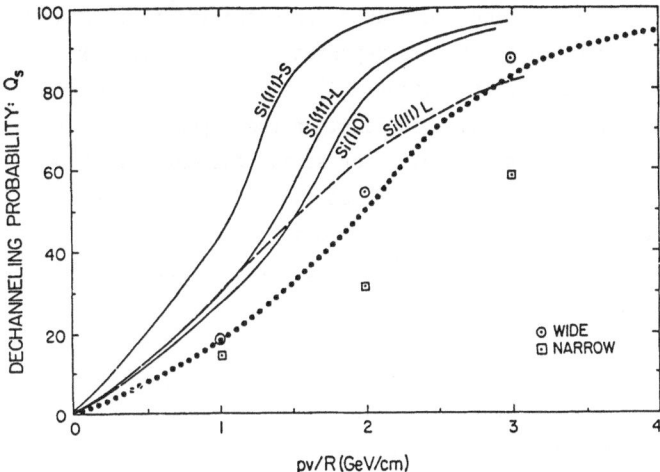

Fig.4.10. Different calculations for the wide (L) and narrow (S) (111) planes in silicon. (——) Si(111)-S,L shown are *Kudo*'s results [4.25] with a finite a_T value and an incident flux in statistical equilibrium; (---) is a similar case for the wide planes with a parallel incident flux. Kudo's calculation for Si (110) (——) in statistical equilibrium is also shown. (●●●) is *Ellison*'s result [4.24] with no charge smearing and a weighted distribution of planes. (○○○, □□□) are Ellison's calculations for wide and narrow planes, respectively. The dechanneling probability is in percent

ent choices of the potential, screening distances, and incident distributions can cause changes of almost a factor of two in the dechanneling probability.

Although silicon has been the only material used so far to test these ideas for particle deflection, as noted in Sect.4.1, it is not the most desirable material because of its relatively low atomic number (Z) and large thermal vibrational amplitude. Tungsten would probably be a far more desirable material to use. Calculations for (110) planes in tungsten are shown and compared to silicon in Fig.4.11.

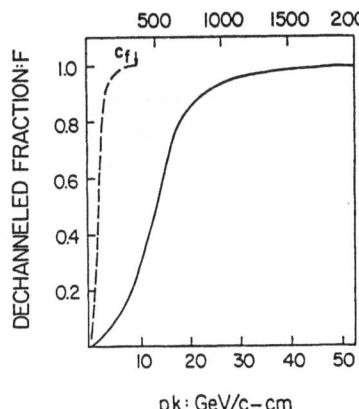

Fig.4.11. Comparison of weighted (111) plane in silicon (---) with (110) plane in tungsten (——). Both cases are for no charge smearing. The top scale is proton momentum in GeV/c for $k^{-1} = 38$ cm. c_f correspond to the Tsyganov centrifugal force model. From [4.24]

Experimental verification of calculations such as those shown in Figs.4.10 and 11 is difficult, and for the experimental studies reported to date, only approximate. A number of factors give rise to difficulties: (1) selection of channeled particles from the total ensemble of particles incident on the crystal, many of which are not within the critical channeling angle, or strike the crystal close to the atomic plane or at the crystal edges and are not captured in channeling trajectories; (2) dechanneling by processes not included in the calculation such as normal multiple scattering, scattering from crystal defects, or scattering by distortion of the crystal by the mounting apparatus; (3) the fact that the bending curvature is usually not constant. The crystal shape imposed by a simple bending apparatus such as the three-pressure-point apparatus used in most experiments to date is, e.g., more nearly parabolic than circular, giving a curvature that increases to a maximum in a nearly linear fashion and then decreases again to zero; (4) local distortion of the crystal at points of high force applied by the bending apparatus. The most complete study to date [4.29] shows the importance of these factors, especially local distortion of the crystal. In spite of these complications, and in fact by making use of nonuniform curvature and local distortion effects to provide discrimination against the effects of (1) and (2), it is shown in the next section that the curvature effects predicted by the simple theory give a reasonable estimate of the probability of a planarly channeled particle experiencing deflection in an elastically bent single crystal.

The situation for positive particles in axial channels, or negative particles in either planar or axial channels is more complex. Since axially channeled positive particles are, in general, not confined to specific "channels" in the crystal but wander somewhat freely between axial rows or "strings" of atoms, it is expected that they would not be steered by bent string potentials in the simple way that we have discussed for planes. This is verified by the experiments reported to date for the axial case [4.28,29]. It appears that the principal effect of bending is deflection of particles into planar channels, which intersect azimuthally at major crystal axial directions, with subsequent deflection by the planes, as we have discussed. Clearly, crystal planes perpendicular to the plane of the bend will have no bending effect, while crystal planes in the plane of the bend will have maximum bending effect. The effective force bending the particle is $F_1 = F \sin\eta$, where η is the angle between the crystal plane and the direction of bend (that is, $90°$ relative to the plane of the bend), and F is the bending force (that is, $pv\kappa$). thus, planes where η is not $\pi/2$ will deflect the particles less. On the other hand, a perpendicular component of the force will deflect the particles out of the plane of the bend. For a given incident particle beam collimated to projected angles of the order of the axial critical angle, the fraction of particles involved in such deflections can be substantially increased over the simple planar case, because of the multiplicity of planes involved and the larger critical angle for axial channeling. In

76

principle, axial bending might be used to separate a particle beam into components emerging in different directions — the utility of this has yet to be established.

Negative-particle deflection suffers from two obstacles: (1) the trajectories of channeled negative particles confine them to regions of the crystal where the atomic density is highest, so they experience increased multiple scattering; and (2) for an external negative-particle beam incident on a crystal, the fraction captured in stable channeling trajectories is much smaller than for the positive case, because the positive potential around the atomic rows and planes is very narrow, and the channeled particles must be bound within this potential. In any case, no significant deflection results for negative particles have yet been reported. It is possible that the small phase space for the channeling of negative particles could be overcome by production of particles at the atomically dense regions of the crystal. Until further experiments or calculations are carried out to confirm this possibility, it must be considered speculative.

4.4 Bending Experiments

As noted earlier, several sets of bending experiments have now been performed. The first objectives were to demonstrate planar [4.27] and axial [4.28] bending. More recently [4.29], these experiments have concentrated on exploring the details of the bending loss mechanisms.

The experiments follow the techniques described in Sect.4.1. The crystals used were slabs of silicon between 0.5 and 1.0 mm thick in the direction of the bend, several centimeters long in the direction of the beam, and approximately one centimeter high. Crystals were used with either a (110) or (111) plane parallel to the major face of the slab. For the recent Fermilab experiments, the crystals were fabricated so that this plane was characteristically within 1.7 mrad of the geometrical surface of the crystal [4.30]. This was verified in the experiment, both by angular distribution measurements on low energy losses and by noting that narrow slices of incident particles, taken near a surface, did not exhibit substantially different behavior than did slices in the interior for fully deflected particles.

In the Dubna and Fermilab experiments, the silicon slabs were clamped near the upstream end of the crystal just after the detector. The bending apparatus was attached to the downstream end of the slab, and it bent the crystal by deflecting it inside a system of 3 or 4 pins. This arrangement avoided the need for substantial realignment after changing the bending angle. Figure 4.12 is a photograph of the four-point bending device. For the Fermilab experiment the crystals were mounted in an automated, computer-controlled goniometer with one azimuthal (ϕ) and two tilt (θ_X, θ_Y) degrees of freedom.

Fig.4.12. Chalk River four-point bending device. Notice the shiny reflective silicon crystal projecting out of the jig. The crystal is about 10 mm high. The third and fourth pins are visible. A special differential screw controls the motion

Several experiments have been performed with different crystals, aligned for both planar and axial channeling, for several different bend angles and energies. For each particle passing through the crystal, the drift chamber coordinates and the energy loss in the detector were recorded. The incident and emergent particle directions and point of impact relative to the crystal were determined from this information. A range of bending angles was studied for particle energies from 8 to 180 GeV.

Figure 4.13, taken from the Fermilab experiments for (111) planes, shows typical angular distributions for emergent particles from a three-point bending apparatus, taken at several energies. Selection of particles that exhibited low energy loss in the semiconductor detector ensured that the particles were channeled in the upstream (unbent) region of the crystal. In all cases, there is a peak in the forward (undeflected) direction which is due, in part, to channeled particles that dechannel before the bend. For 12 GeV and 60 GeV, there is also a prominent peak in a direction corresponding to the full deflection of the crystal. The angular width of the fully deflected peak corresponds to about twice the critical channeling angle, indicating that these particles were well channeled during their passage through the crystal. However, for the 60-GeV results, an intermediate peak can be seen at half the total bending angle. For angles somewhat beyond the middle peak but less than the full bend peak there are few particles, suggesting that the particles are very well channeled between the middle and final peak. The middle peak is clearly associated with the center point of the three-point bender. At 180 GeV both the intermediate and full deflection peaks drop substantially. This indicates that the radius of curvature is small enough in this case so that a large fraction of the particles is being lost to bending dechanneling.

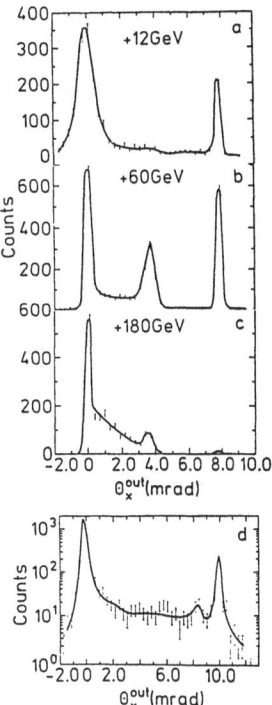

Fig.4.13a-d. Distribution of outgoing particle directions in the Fermilab experiment with small energy loss in the detector for a crystal bend of 8 milliradians and incident energies of 12, 60, and 180 GeV (**a-c**) using a three-point bender. The peak at 0 milliradians, the beam direction, is due mostly to particles dechanneled prior to the bend. Note the peak at approximately 4 milliradians that develops as the energy is increased. (**d**) Distribution of outgoing particles with small energy loss for a four-point bender and incident energy of +60 GeV. Notice that there is no half-angle loss and little loss at the two middle pins. (Note also that the y axis is logarithmic)

Figure 4.13d shows a similar distribution taken in the Fermilab experiment with the four-point bender for a total angular deflection of 10 milliradians. For a four-point bender, the deflection of an elastic beam between the two inner pins should be a circle. In addition, the force for an inner pin is forty times smaller than for the middle pin in the three-point bender. No middle peak is visible in this measurement, and only small losses occur at angular points corresponding to the two middle pins. For the four-point bender there is no prominent middle region loss, but rather a gradual loss with angle.

Understanding the origin and intensity of the observed peaks in the emergent direction spectrum, as well as the intensity of particles emerging in directions between the peaks, is important for practical applications, and also for understanding the fundamental dechanneling processes in elastically deformed crystals.

Measuring the dechanneling of particles due specifically to bending is complicated by several other effects. Ordinary dechanneling occurs in the portion of the crystal after the detector and prior to the bend. As noted earlier, dechanneling due to electron scattering is not included in the present bending dechanneling theories. Particles can also leak out of the sides of the crystal if the planes are not perfectly aligned with the geometric plane of the crystal. Crystals bent with slowly increasing curvature should produce dechanneling only between the first and second pins of the bending apparatus [4.31]. No dechanneling is expected due to the bending or centrifugal effect in regions in which the curvature is constant or decreasing. Such an effect is evident in the distributions shown in Fig.4.13, especially for the three-point bender. This appears as a larger emergent particle intensity between the first and second pins (θ_b = 1-2 mrad) compared to that between the second and third pins (θ_b = 5-6 mrad). This type of slowly varying curvature is called "global curvature". The prominent intermediate peaks are not, however, predicted by global-curvature dechanneling effects.

For both three- and four-point benders, the intermediate peaks appear at angles corresponding to the position of contact of the pins used in the bending apparatus. Dechanneling can arise from the high local pressure exerted by the bending pins, either by local elastic distortion or by defects generated in the crystal. For the Fermilab experiment, it is believed that the observed peaks are not due primarily to defects induced by the pins. Such defects should remain after the pressure is removed. Measurements in a four-point bender of a crystal previously bent in a three-point bender did not show an emergent peak from a position corresponding to the previous point of contact in the three-point bender. However, some contribution from defect generation cannot be discounted.

From the Fermilab experiment, it is believed that the principal origin of the intermediate peak is local curvature in the region of the pins. This type of distortion has been discussed in such classical treatises on elastic deformation as those of *Timoshenko* and *Goodier* [4.32], *Love* [4.33], and *Frocht* [4.34]. The distortion is greatest near the pin and decreases gradually to zero on the side of the crystal opposite the pin. Such dependence on the precise position of the particle trajectory in the crystal slab can be investigated directly, since the point of impact on the crystal can be determined by use of the drift chamber immediately in front of the crystal. The spatial resolution of the drift chambers is such that slices of the crystal transverse to the beam can be examined close to and far away from the pin. For example, the resolution of the transverse coordinate of the crystal used to obtain the Fermilab data was σ = 60 microns at 60 GeV while the crystal thickness was 1000 microns. Figure 4.14 shows, for the Fermilab experiment, the dechanneled fraction at the middle pin in a three-point bending apparatus as a function of position away from the middle pin for particles with low energy loss at 60 GeV. The dechanneled fraction was computed as the ratio of the particles dechanneled at the pin divided by the sum of the particles dechanneled plus the particles that continued on.

If the local curvature predicted by elastic deformation theory is accepted, then it is possible to utilize the effect as a "dechanneling spectrometer" to compare the observed dechanneling with that calculated from classical continuum models [4.23-26]. This is called the "local curvature" method, and has a useful feature in that it is reasonable to assume that all particles that make it through the crystal to the center peak were well channeled up to that point. This approach minimizes

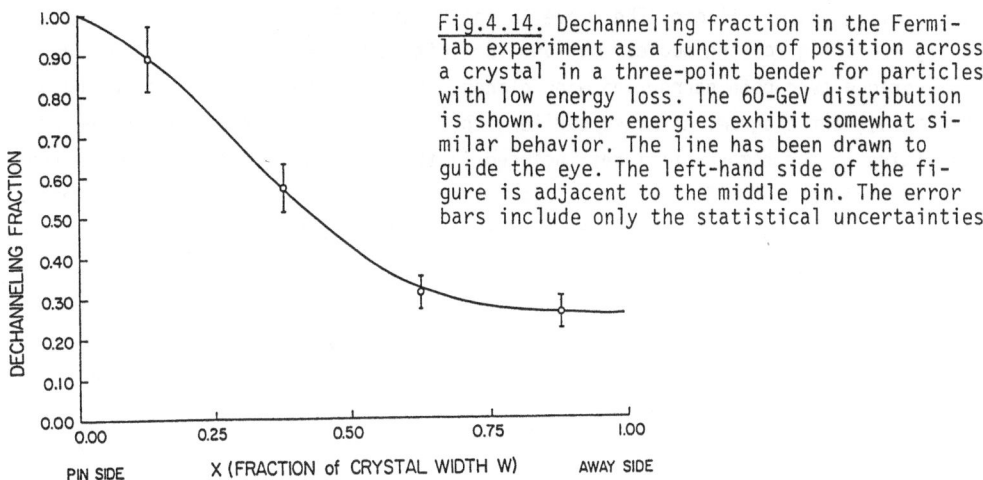

Fig.4.14. Dechanneling fraction in the Fermilab experiment as a function of position across
a crystal in a three-point bender for particles
with low energy loss. The 60-GeV distribution
is shown. Other energies exhibit somewhat similar behavior. The line has been drawn to
guide the eye. The left-hand side of the figure is adjacent to the middle pin. The error
bars include only the statistical uncertainties

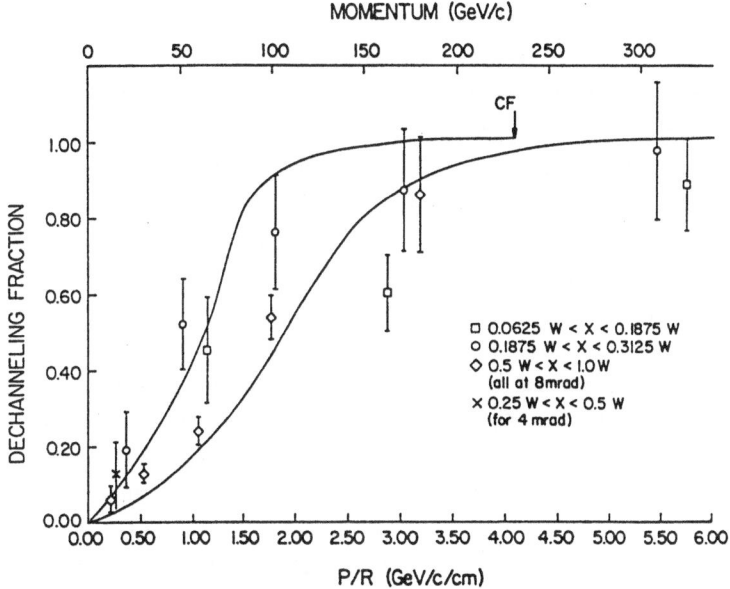

Fig.4.15. Comparison of dechanneling fraction as a function of p/R, where p is the
particle momentum and R the radius of curvature. The theoretical curves are taken
from [4.24] for the (111) plane in silicon. The right curve is for the wide planes
with no charge smearing while the left curve includes charge smearing and is averaged over planes. Experimental points are based on the losses at the middle pin in
the three-point bender. The point predicted by the Tsyganov centrifugal-force calculation for the charge-smeared case, is indicated by CF

the effects of normal dechanneling and misalignment, since the bending losses occur
in a small angular region and over a small-depth region of the crystal.

Figure 4.15 shows the dechanneling probability vs. p/R. Also shown are results
of continuum-model calculations by *Ellison* [4.24] for particles channeled between

the (111) plane of silicon. The curve to the right is for the wide planes with no charge smearing for the nuclear motion. The left curve includes charge smearing and is averaged over both narrow and wide planes. The curvature was computed using a finite-element program. As noted earlier, the theoretical estimates are somewhat sensitive to the number of planes included in the potential and the assumed potential distribution [4.25]. The local distortion situation is actually quite complex. For example, the detailed shape of the pin contact may influence the distribution. Thus the absolute value of the abscissa is uncertain but the beam energy dependence should be good. The apparent good agreement shown in Fig.4.15 between the experimental measurements and the theory must be regarded as somewhat fortuitous because of approximations and averages contained in the 'local curvature' analysis. The comparison should not, therefore, be regarded as justification for choosing a particular critical approach distance, nor can it be used to evaluate detailed differences between calculations, but it does show that the continuum calculation gives the observed functional dependence and constitutes a reasonable basis for estimating the transmission of channeled particles through bent crystals. As expected, the "Tsyganov radius" approach seriously overestimates the particle transmission probability.

Observed particle losses between the bending pins, especially for small bending radii, can be used to establish "ordinary dechanneling". When this is done, an interesting feature comes to light. Extrapolation from low energy gives lengths of about 20 mm at 10 GeV [4.35]. From measurements at Dubna, *Sun* et al. [4.36] get a dechanneling length of about 9 mm at 8 GeV for Si(110) by considering surviving particles at several points or, alternatively, differential losses between the second and third pin in a three-point bender. Similarly, from studies at CERN, *Bak* et al. [4.28] estimate 15 mm at 12 GeV for Si(110). For the Fermilab experiments for Si(111), the loss between the second and third pin gives dechanneling lengths of 10 mm (12 GeV), 40 mm (30 GeV), 140 mm (60 GeV), 100 mm (100 GeV), and 30 mm (180 GeV). The 12-GeV result is consistent with the Dubna and CERN results and starts to scale with p as one might expect. Above 60 GeV, it drops again. This occurs because the trajectories are being forced ever closer to the sides of the channels, where they are subjected to higher charge densities and more dechanneling. Alternatively, the motion of the locus of the equilibrium points for the oscillatory motion toward the outside plane can be thought of as a reduction of the effective width of the planar channel. In the Fermilab four-point bender, measurements give dechanneling lengths of 10 mm (60 GeV) and 12 mm (180 GeV). These are essentially independent of energy, suggesting that the effects of moving to the side of the channel and scaling of the ordinary dechanneling length are in this case roughly balanced.

Overall, it appears that the present theory is useful for estimating the loss to dechanneling due to bending. In a practical application, it is desirable to avoid distortion of the crystal due to the bending pins. This can be done by using more pins or a continuous clamp. As the *Ellison* [4.24] and *Kudo* [4.25] theories have al-

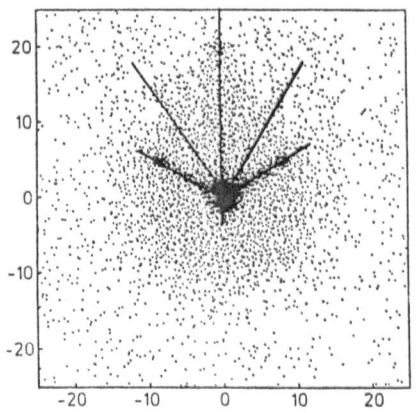

Fig.4.16. CERN experiment showing axial bending for a 12-GeV proton beam with a silicon crystal bent 20 milliradians in the upward direction. The <111> axis is originally along the beam direction which is at 0 milliradians horizontally and vertically. One (110) plane is originally horizontal. The spots showing vertical deflections of roughly 4 milliradians along with horizontal deflections are due to skew (110) planes. The horizontal and vertical scales are in milliradians

ready indicated, there will clearly be major losses prior to reaching the Tsyganov radius. In addition, the interplay of ordinary dechanneling and bending must be taken into account. Here, more sophisticated theoretical treatments must be brought to bear.

Axial distributions have been studied in both the CERN [4.28] and the Fermilab experiments. As noted earlier, crystal planes skewed relative to the plane of the bend experience less than the full deflection, but also introduce deflections perpendicular to the bend. Figure 4.16 shows axial bending in a CERN crystal for a <110> axis, with the plane of the bend containing a (110) plane. Interesting asymmetric distributions arise if the (111) plane is in the plane of the bend, because the (110) plane and the other (111) planes are asymmetrically distributed relative to the bending direction.

4.5 Factors Affecting Charged-Particle Bending with Single Crystals

4.5.1 Materials

Every facet of the channeling phenomenon is affected by the choice of single crystal. The critical angle, dechanneling length, Young's modulus, susceptibility to radiation damage, melting point, mosaic spread, and temperature effects related to lattice vibrations are all sensitive to this choice. Table 4.1 gives a summary of these parameters for silicon, germanium, and tungsten.

From many standpoints, it appears that a high-Z material is quite desirable. For example, tungsten has a high melting point, relatively small lattice vibrations, and a large critical angle. Tungsten crystals may be the ideal crystals for high-energy applications. In the last several years, a technique has been perfected (electron beam zone refining), and another is evolving (low-temperature epitaxy) for growing small, perfect tungsten crystals, i.e., crystals largely free of mosaic spread [4.37]. On the other hand, there has been no trouble otaining suitable mosaic-free silicon and germanium, while so far the tungsten crystals have been rather small. None have been tried at high energy.

4.5.2 Radiation Damage

For many applications, radiation-damage effects are important. For example, if a crystal is to be used as a septum to deflect or separate particles in a primary or secondary beam, it must be able to withstand the high levels of radiation encountered for a long enough time to make their use practicable. Extrapolation of radiation-damage results from low-energy bombardment is unreliable, because it is difficult to incorporate properly effects of showers produced in the crystal and surrounding materials, neutron-damage effects, and beam-heating effects. Indeed, it was anticipated that the residual defect density, type, and distribution might be significantly different for radiation damage in an elastically bent crystal relative to an unbent crystal. This is because of the possible motion of defects in the strain field, or even suppression of recombination of primary interstitial-vacancy pairs (Frenkel pairs) due to their separation in the strain field. To evaluate radiation-damage effects, an experiment was carried out in which elastically bent silicon crystals, as well as unbent crystals, were exposed to an intense 400-GeV proton beam for an accumulated particle fluence up to $10^{17} \mathrm{cm}^{-2}$ [4.38]. The ability of the crystals to deflect particles was only marginally affected by the radiation (an apparent reduction in dechanneling length by about 20% was observed, but this is considered to be of the same order as the uncertainty in the dechanneling length measurement). Although the net residual defect concentration was not increased in the bent crystal, an apparent migration of defects in the strain field was observed [4.38].

4.5.3 Angular Acceptance

The angular acceptance of particles into a planar channel in the direction perpendicular to the plane is determined by the critical planar channeling angle. The angular acceptance for trajectories lying within (parallel to) a plane is infinite. For planar applications, the beam angular divergence can be arranged to be small in the plane of bending and large in the crystal planes. At Fermilab, the half-angular divergence in the main ring accelerator at extraction is 20 μrad, while the angular divergence of a typical secondary beam is 200 μrad at 400 GeV. The critical planar channeling angle in the (110) plane of tungsten is estimated to be 17 μrad at 400 GeV. As the particle energy increases, both the primary and secondary beam angular spread is expected to decrease more rapidly than the channeling critical angle, resulting in an even more favorable match.

4.5.4 Spatial Acceptance

The spatial acceptance of a crystal depends on its transverse dimensions. The amount of bending will also depend on how thick the crystal is in the bending direction. The spatial acceptance will be equal to the height times the thickness in the bending direction.

The maximum elastic deflection of a crystal length L and thickness t supported at three points is

$$y = S_m L/6E_y t \quad , \tag{4.9}$$

where S_m is the maximum stress and E_y is Young's modulus. The maximum deflection can be linked to a minimum radius of curvature by noting that $y = L/8R$. The minimum radius of curvature is, therefore,

$$R = E_y t/S_m \quad . \tag{4.10}$$

Clearly, the minimum radius of curvature is directly related to thickness.

For silicon, the minimum bending radius is 90 cm for a thickness of 1 mm, quite close to the value of 76 cm found by *Tsyganov* and his colleagues [4.27].

4.5.5 Deflection

It was shown earlier that the critical bending radius is proportional to the momentum at high energies. At 400 GeV/c for the (111) plane in silicon, the useful bending radius is about 360 cm. The effective bending power of the crystal corresponds to a 4-MG magnetic field. For a 2-cm-long crystal, this would give a deflection of 5.5 mrad. For a tungsten crystal the equivalent numbers would be 40 cm, 36 MG, and 50 mrad.

4.6 Possible Applications

4.6.1 Extraction

Bent crystals could provide essentially zero-thickness septa for deflecting particles out of accelerators. Crystal bending angles are more than sufficient. Angular acceptance, as set by the critical angle, is similar to the extraction emittance. Spatial acceptance is satisfactory. Under some circumstances, it may be possible to make dechanneling losses for such a septum comparable to the loss in an electrostatic or magnetic septum. At high energy, the loss is primarily due to mismatch of particles striking charge planes on entry into the crystal. This is typically of the order of 10% or less. Such intrinsic losses preclude multiple-entry applications such as in an accelerator ring. Radiation damage is a significant factor but may not be insurmountable.

4.6.2 Secondary Beam Bending

Near a production target or at a focus, crystal size is well matched to beam size. Laminates can also be used to increase the thickness for planar geometries. At 100 GeV, the angular acceptance is about one tenth of a typical focusing beam emittance. This could be ameliorated by using beam-parallel sections but laminates might be necessary. At higher energy (multi TeV) the match between beam emittance and crystal acceptance should be more favorable.

Use of a crystal septum at Fermilab has been discussed by *Menzione* and *Elias* [4.39] and others. The idea is to use the septum to deflect high-energy, foward-produced particles down beam lines, at relatively large angles to the production distribution. Recent experiments at Fermilab have demonstrated successful operation of such devices up to 400 GeV [4.40]!

Occasionally it is noted that the bending process will give rise to synchrotron radiation. While this is true, the characteristic energy will be lower than for channeling radiation, because the radius of curvature is much larger than the wavelength of oscillation in the channel.

4.6.3 Beam Focusing

Properly sliced single crystals as shown in Fig.4.17 could give focusing. Focusing in two dimensions may be possible by using two such elements.

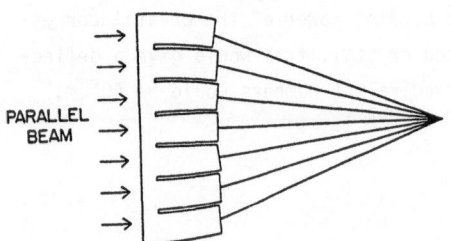

Fig.4.17. Schematic illustration of a possible focusing element using bent crystals. A parallel beam with angular dispersion less than the critical angle could be focused almost to a line in one direction. A second element in the other direction could focus to a point

4.6.4 "Separated" Beam for Short-Lived Particles

Lifetimes of charm particles are so short that they move less than 1 cm even at high energies. With crystal bending, it may be possible to deflect charm particles fifty to a hundred milliradians, well out of the forward-production cone. This could give a substantially enriched sample of charm particles if the long-lived particles in the same channel continue further around the bend (Fig.4.18). On the other hand, dechanneling of long-lived particles may constitute a serious dilution. For planar geometries, angular acceptance is of the order of 1% of the production cone. At high energies, charm-particle production is expected to be reasonably copious. The problem lies more in separating particles, that is to say, enrichment is important.

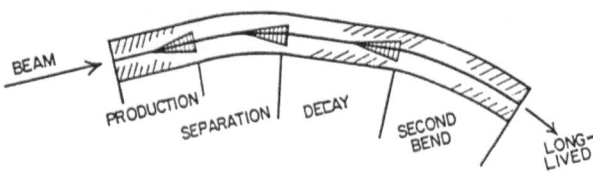

Fig.4.18. A possible "separated" charm-particle beam. All the particles within the critical angle are bent. Most of the charm particles decay in the straight section. Long-lived particles continue around the bend

This process could be enhanced if some form of rechanneling occurred. Observers at *Gatchina* [4.41] working at 1 GeV have suggested the possibility of "volume capture".

Blocking may limit the application of this process. Both the CERN and the Fermilab experiments have undertaken investigations of blocking. Unfortunately, the geometries of both sets of experiments were such that the crystal axis was never sufficiently far from the incident beam axis. As a result, the well-known "doughnut" effect becomes entangled with possible blocking effects. Further investigations of blocking at high energy, with and without bending, are needed.

4.6.5 Charm-Particle Magnetic Moment Measurement

Pondrom [4.17] has noted that the electric field of a crystal transforms into a magnetic field in the rest frame of a moving particle. The particle spin will precess around the magnetic vector. For a polarized process, it is possible to measure the precession in the same way as has been done for strange particles. Effective fields in bent crystals are sufficient to precess a magnetic moment several radians in 1 cm. *Kim* has reviewed these possibilities in detail [4.17].

4.6.6 Interstitial Site Information

Very recently, elementary particles have been used to enhance a technique used earlier in blocking measurements [4.42]. The electrons from stopping muons in crystals can be observed using multiwire proportional chambers. Bent crystals were used to effectively separate the channeled particles. Blocking effects have been observed.

In summary, there seem to be a number of potential uses of high-energy channeling and channeling of elementary particles in bent crystals. More complete information is still needed on bending dechanneling as well as on ordinary dechanneling. It is expected that over the next few years these possibilities will be investigated directly. In addition to providing information important to particle-physics experiments, such studies will establish single crystals as useful tools in high-energy physics and will lead to other applications as yet unimagined.

Acknowledgement. The authors would like to thank their colleagues for many useful conversations. We would particularly like to acknowledge contacts with C.R. Sun, E. Tsyganov, I. Kim, J. Forster, S. Baker, J. Ellison, S. Salman, and R. Wijaya-wardana.

References

4.1 D.S. Gemmell: Rev. Mod. Phys. **46**, 129 (1974); *Channeling*, ed. by D.V. Morgan (Wiley, New York 1974)

4.2 R.A. Carrigan, B.L. Chrisman, T.E. Toohig, W.M. Gibson, I.J. Kim, C.R. Sun, Z. Guzik, T.S. Nigmanov, E.N. Tsyganov, A.S. Vodopianov, M.A. Hasan, A.S. Kanofsky, R. Allen, J. Kubic, D.H. Stork, A.B. Watson: Nucl. Phys. B**163**, 1 (1980);
C.R. Sun, W.M. Gibson, I.J. Kim, G.O. Williams, M.A. Hasan, A.S. Kanofsky, R. Allen, R.A. Carrigan, B.L. Chrisman, T.E. Toohig, Z. Guzik, T.S. Nigmanov, E.N. Tsyganov, A.S. Vodopinov, A.B. Watson, J. Kubic, D.K. Stork: Nucl. Phys. B**203**, 40 (1982);
H. Hasan: Thesis, Lehigh University (1980)

4.3a O. Fich, J.A. Golovchenko, K.O. Nielsen, E. Uggerhøj, G. Charpak, F. Sauli: Phys. Lett. **57**B, 90 (1975);
H. Esbensen, O. Fich, J.A. Golovchenko, S. Madsen, H. Nielsen, H.E. Schiøtt, E. Uggenhøj, C. Vraast-Thomsen, G. Charpak, S. Majewski, G. Odyniec, G. Petersen, F. Sauli, J.P. Ponpon, P. Siffert: Phys. Rev. B**18**, 1039 (1978)

4.3b H. Esbensen, O. Fich, J.A. Golovchenko, K.O. Nielsen, E. Uggerhøj, C. Vraast-Thomsen, G. Charpak, S. Majewski, F. Sauli, J.P. Ponpon: Nucl. Phys. B**127**, 281 (1977)

4.4 G.M. Temmer: In *Proc. Kentucky Conference on Low-Medium Mass Nuclei*, ed. by J.P. Davidson, B. Korn, p.192 (1973)

4.5 S.K. Andersen, H. Esbensen, O. Fich, J.A. Golovchenko, H. Nielsen, H.E. Schiøtt, E. Uggerhøj, C. Vraast-Thomsen, G. Charpak, G. Petersen, F. Sauli, J.P. Ponpon, R. Siffert: Nucl. Phys. B**144**, 1 (1979)

4.6 W.M. Gibson: Annu. Rev. Nucl. Sci. **25**, 465 (1975)

4.7 R.A. Carrigan, Jr.: Phys. Rev. Lett. **35**, 206 (1975)

4.8 M.A. Kumakhov: Phys. Lett. **57**A, 17 (1976); Phys. Status Solidi(b) **84**, 41 (1977)

4.9 R.W. Terhune, R.H. Pantell: Appl. Phys. Lett. **30**, 265 (1977);
M.J. Alguard, R.L. Swent, R.H. Pantell, B.L. Berman, S.D. Bloom, S. Datz: Phys. Rev. Lett. **42**, 1148 (1979)

4.10 I.I. Miroshnichenko, J.J. Murray, R.O. Avakyan, T.Kh. Figut: Pis'ma Zh. Eksp. Teor. Fiz. **29**, 786 (1979) [English transl.: Sov. Phys.-JETP Lett. **29**, 722 (1979)]

4.11 N.A. Filatova, V.M. Golovatyuk, A.N. Iskakov, I.M. Ivanchenko, R.B. Kadyrov, N.N. Karpenko, T.S. Nigmanov, V.V. Palchik, V.D. Riabtsov, M.D. Shafranov, E.N. Tsyganov, I.A. Tyapkin, D.V. Uralski, A. Forycki, Z. Guzik, J. Wojtkows-ka, R.A. Carrigan, R.E. Toohig, C. Carmack, W.M. Gibson, I.J. Kim, C.R. Sun, M.D. Barizhev, N.K. Bulgakov, N.I. Zimin, I.A. Grishaev, G.D. Kovalenko, B.I. Shramenko, E.I. Denisov, V.I. Glebov, V.V. Avdeichikov: Phys. Rev. Lett. **48**, 488 (1982)

4.12 M. Atkinson, J.F. Bak, P.J. Bussey, P. Christensen, J.A. Ellison, R.J. Ellison, K.R. Eriksen, D. Giddings, R.E. Hughes-Jones, B.B. Marsh, D. Mercer, F.E. Meyer, S.P. Møller, D. Newton, P. Pavlopoulos, P.H. Sharp, R. Stensgaard, M. Suffert, E. Uggerhøj: Phys. Lett. **110**B, 162 (1982)

4.13 I.J. Kim, C.R. Sun: Private communication (to be published)

4.14 J. Kimball, N. Cue, L. Roth, B.B. Marsh: Phys. Rev. Lett. **50**, 950 (1983)

4.15 E.N. Tsyganov: Fermilab Internal Reports TM-682, TM-684 (1976)

4.16 R.A. Carrigan, Jr.: Fermilab Publications 80/45 (1980), FN-362 (1982);
R.A. Carrigan, Jr., W.M. Gibson, C.R. Sun, E.N. Tsyganov: Nucl. Instrum. Methods **194**, 205 (1982); *Silicon detectors for high energy physics*, ed. by T. Ferbel (Fermilab, Rochester 1981)

4.17 L. Pondrom: *Proc. of the 1982 DPF Summer Study on Elementary Particle Physics and Future Facilities*, ed. by R. Donaldson, R. Gustafson, F. Paige. Snowmass, CO (1982), p.98;
I.J. Kim: Nucl. Phys. B229, 251 (1983)
4.18 J. Lindhard: K. Dan. Vidensk. Selsk. Mat. Fys. Medd. **34**, No.14 (1965)
4.19 W.M. Gibson, J.A. Golovchenko: Phys. Rev. Lett. **28**, 1301 (1972)
4.20 H.J. Kreiner, F. Bell, R. Sizmann, D. Harder, W. Hüttl: Phys. Lett. **33A**, 135 (1970);
H. Kumm, F. Bell, R. Sizmann, H.J. Kreiner, D. Harder: Radiat. Eff. **12**, 53 (1972)
4.21 A.O. Aganiants, I.A. Vartanov, G.A. Vatapetian, M.A. Kumakhov, C. Trikalonos, V.I. Iaralov: Pis'ma Zh. Eksp. Teor. Fiz. **29**, 1340 (1979)
Data also reported by V.V. Beloshitsky, F.F. Komarov: Phys. Rept. **93**, 119 (1982), and in [4.12]
4.22 E. Bonderup, H. Esbensen, J.U. Andersen, H.E. Schiøtt: Radiat. Eff. **12**, 261 (1972);
H.E. Schiøtt, E. Bonderup, J.U. Andersen, H. Esbensen, M.J. Pedersen, D.J. Elliott, E. Laegsgaard: In *Atomic Collisions in Solids*, ed. by S. Datz, B.R. Appleton, C. Moak (Plenum, New York 1975), p.843
4.23 J.A. Ellison, S.T. Picraux: Phys. Lett. **83A**, 271 (1981)
4.24 J.A. Ellison: Nucl. Phys. B**206**, 205 (1982)
4.25 H. Kudo: Nucl. Instrum. Methods **189**, 609 (1981)
4.26 V.V. Kaplan, S.A. Vorobiev: Phys. Lett. **67A**, 135 (1978);
A.M. Taratin, Yu.M. Filimonov, E.G. Vyatkin, S.A. Vorobiev: Phys. Status Solidi (b) **100**, 273 (1980);
A.M. Taratin, S.A. Vorobiev: Phys. Status Solidi (b) **107**, 521 (1981)
4.27 A.F. Elishev, N.A. Filatova, V.M. Golovatyuk, I.M. Ivanchenko, R.B. Kadyrov, N.N. Karpenko, V.V. Korenkov, T.S. Nigmanov, V.D. Riabtsov, M.D. Shafranov, B. Sitar, A.E. Senner, B.M. Starchenko, V.A. Sutulin, I.A. Tyapkin, E.N. Tsyganov, D.V. Uralsky, A.S. Vodopianov, A. Forycki, Z. Guzik, J. Wojtkowska, R. Zelazny, I.A. Grishaev, G.D. Kovalenko, B.I. Shramenko, M.D. Bavizhev, N.K. Bulgakov, U.U. Avdeichikov, R.A. Carrigan, T.E. Toohig, W.M. Gibson, I.J. Kim, J. Phelps, C.R. Sun: Phys. Lett. **88**B, 387 (1979)
4.28 J. Bak, G. Melchart, E. Uggerhøj, J.S. Foster, P.R. Jensen, H. Madsbøll, S.P. Møller, H. Nielsen, G. Petersen, H. Schiøtt, J.J. Grob, P. Siffert: Phys. Lett. **93**B, 505 (1980)
4.29 W.M. Gibson, I.J. Kim, M. Pisharody, S.M. Salman, C.R. Sun, G.H. Wang, R. Wijayawardana, J.S. Forster, I.V. Mitchell, T.S. Nigmanov, E.N. Tsyganov, S.I. Baker, R.A. Carrigan, T.E. Toohig, V.V. Avdeichikov, J.A. Ellison, P. Siffert: Nucl. Instrum. Methods B**2**, 54 (1984); presented at *Int. Conf. Atomic Collisions in Solids*, Bad Iburg, July 1983;
S.I. Baker, R.A. Carrigan, C. Crawford, T.E. Toohig, W.M. Gibson, H. Jin, I.J. Kim, M. Pisharody, S. Salman, C.R. Sun, G.H. Wang, R. Wijayawardana, J.S. Forster, H. Hatton, I.V. Mitchell, Z. Guzik, T.S. Nigmanov, E.N. Tsyganov, V.V. Avdeichikov, J.A. Ellison, P. Siffert: Phys. Lett. **137**B, 129 (1984);
S.M. Salman: Thesis, State University of New York (1982)
4.30 Fabrication techniques for such detectors have been developed by V.V. Avdeichikov (Leningrad) and P. Siffert (Strasbourg)
4.31 J.A. Ellison, S.I. Baker, R.A. Carrigan, Jr., J.S. Forster, I.V. Mitchell, W.M. Gibson, I.J. Kim, M. Pisharody, S. Salman, C.R. Sun, R. Wijayawardana: Nucl. Instrum. Methods B**2**, 9 (1984); presented at *Int. Conf. Atomic Collisions in Solids*, Bad Iburg, July 1983)
4.32 S. Timoshenko, J. Goodier: *Theory of Elasticity*, 2nd ed. (McGraw-Hill, New York 1951)
4.33 A.E.H. Love: *A Treatise on the Mathematical Theory of Elasticity* (Dover, London 1927)
4.34 M.M. Frocht: *Photo Elasticity*, Vol.2 (Wiley, New York 1984), p.104
4.35 Y.H. Ohtsuki: *Charged Beam Interaction with Solids* (Taylor Francis, New York 1983)
4.36 C.R. Sun, W.M. Gibson, I.J. Kim, G.H. Wang, N.K. Bulgakov, N.A. Filatova, A. Forycki, V.M. Golovatyuk, Z. Guzik, R.B. Kadyrov, T.S. Nigmanov, V.D. Riabtsov,

A.B. Sadovsky, M.D. Shafranov, I.A. Tyapkin, E.N. Tsyganov, A.S. Vodopianov, J. Wojtkowska, N.I. Zimin, R.A. Carrigan, T.E. Toohig, M.D. Bavizhev: Nucl. Instrum. Methods B2, 60 (1984); presented at *Int. Conf. Atomic Collisions in Solids*, Bad Iburg, July 1983

4.37 R. DesLattes: Private communication

4.38 G.H. Wang, P.J. Cong, W.M. Gibson, C.R. Sun, I.J. Kim, S. Salman, M. Pisharody, S.I. Backer, R.A. Carrigan, J.S. Forster, I.V. Mitchell: Nucl Instrum. Methods 218, 669 (1983)

4.39 A. Menzione, J. Elias: Private communication

4.40 S.I. Baker, R.A. Carrigan, Jr., T.E. Toohig, W.M. Gibson, I.-J. Kim, F. Sun, C.R. Sun, J.S. Forster, H. Hatton, I.V. Mitchell, E.N. Tsyganov, T.S. Nigmanov, V.V. Avdeichikov, J.A. Ellison, P. Siffert: Nucl. Instrum. Methods A234, 602 (1985)

4.41 V.A. Andreev, V.V. Baublis, E.A. Damaskinskii, A.G. Krivshich, L.G. Kudin, V.V. Marchenkov, V.F. Morozov, V.V. Nelyubin, E.M. Orishchin, G.E. Petrov, G.A. Ryabov, V.M. Samsonov, L.E. Samsonov, E.M. Spiridenkov, V.V. Sulimov, O.I. Sumbaev, V.A. Shchegel'skii: Pis'ma Zh. Eksp. Teor. Fiz. 36, 340 (1982) [English transl.: Sov. Phys.-JETP Lett. 36, 415 (1982)]

4.42 B.D. Petterson, A. Bosshard, U. Straumann, P. Truöl, A. Wüst, T. Wichert: Phys. Rev. Lett. 52, 938 (1984)

5. Classical Theory of the Radiation from Relativistic Channeled Particles

V. V. Beloshitsky and M. A. Kumakhov

With 11 Figures

The classical theory of the radiation from channeled electrons and positrons traversing single crystals at relativistic speed is presented here. The radiation's energy and polarization spectra, and its dependence on crystal characteristics and particle dynamics at high energies are discussed.

5.1 Historical Introduction

During the last twenty years, there have been many experimental and theoretical investigations on the channeling of charged particles in crystals; see, for example, a review by *Gemmel* [5.1]. The basic physical concepts underlying particle channeling were established in the fundamental paper of *Lindhard* [5.2], which was instrumental in promoting the development of investigations on channeling. Until 1975, most studies in this field had been primarily devoted to channeling of heavy particles: protons, α particles, etc. In the last seven years, channeling of light relativistic particles, i.e., electrons and positrons, has been intensively investigated.

To a large extent, this was stimulated by *Kumakhov*'s papers [5.3-6], in which intense spontaneous radiation accompanying channeling of light relativistic particles was theoretically predicted. In these papers, the fundamental classical and quantum theory of this effect was developed, and it was shown that the spontaneous radiation which occurs during channeling has high spectral intensity and directivity (substantially higher than the intensity of bremsstrahlung), and that it is polarized and largeley monochromatic.

These characteristics attracted the attention of large nuclear research centers with electron accelerators. In 1978, the Soviet-American experiment at SLAC [5.7] was carried out using positrons with energies from 2 to 10 GeV; further, experiments were also carried out at Yerevan (U.S.S.R.) using electrons with an energy of about 5 GeV [5.8], at Tomsk (U.S.S.R.) [5.9] and Kharkov (U.S.S.R.) [5.10] at E = 1 GeV, and at Livermore (U.S.A.) at intermediate energies [5.11] (E = 56 and 28 MeV); an international experiment was carried out at Saclay (France) at E = 54 MeV [5.12], and low-energy experiments were made at Aarhus (Denmark) [5.13] and Albany, N.Y. [5.14].

In 1979-1981, a series of large experiments on electrons and positrons at $E = 10$ GeV were performed at Serpukhov (U.S.S.R.) by a U.S.S.R.-U.S.A.-Poland collaboration [5.15], and an extensive experimental investigation at CERN [5.16] using 5-55-GeV electrons and positrons was carried out. In these studies, the basic predictions of Kumakhov's theory were confirmed.

Intensive experimental and theoretical investigations of channeling radiation are continuing. In 1980, the first U.S.S.R. national conference devoted to this effect was held in the North Caucasus, with the participation of experts from various countries; the proceedings were published [5.17]. The development of field up to 1979 is covered in several review articles [5.18-20].

5.2 Classical Theory of Channeling

Channeling is the collective scattering of a particle by the set of atoms belonging to an atomic string or plane. According to Lindhard, this scattering can be considered as a deflection of the particle by the field of the atomic string (plane). In such a motion the transverse energy of the particle $E_\perp$ and its longitudinal momentum $p_\parallel$ are conserved. The conservation of the transverse energy plays a fundamental role in channeling theory. This conservation is susceptible to direct experimental verification by measuring the angular distribution of channeled particles and of the radiation [5.11,14] emitted during spontaneous transitions between quantized levels of $E_\perp$.

The conservation of $E_\perp$ for positively charged particles results in a very important restriction on impact parameters imposed by the classical condition $U(r_\perp) < E_\perp$ where $U(r_\perp)$ is the continuum potential of the atomic row or plane. It follows that at low $E_\perp$ processes involving collisions with small impact parameters (nuclear reactions, backward Rutherford scattering) are suppressed ("blocking effect"). The criterion for this blocking is the Lindhard critical angle. Since the lattice period is much larger than the screening constant a, then $U(r_\perp) \ll E_\perp$ in most of the transverse plane, and it can be assumed for most particles that $E_\perp \cong p^2 \psi_{in}^2 / 2m$, where ψ_{in} is the angle of incidence of the particle with respect to the relevant crystallographic axis (plane). We emphasize here that the critical angle is not a criterion that insists the particle motion in the transverse plane be finite. This finiteness does not at all play an important role in channeling theory. It is obviously determined by the form of the resultant potential well due to the atomic rows (planes). Following Lindhard, we will call the bound motion the channeling proper.

For negative particles the conservation of $E_\perp$ does not impose any restrictions on impact parameters. Here we will measure the transverse energy of negative and positive particles from the minimum of the total average potential of the atomic axes (planes). However, because of the concentration of the negative-particle flux near atomic rows or planes (flux-peaking effect), processes taking place at collisions with small impact parameters are enhanced. The Lindhard critical angle is also the criterion here.

Thus, the Lindhard critical angle ψ_c is the criterion governing a strong redistribution of the particle flux over impact parameters, and a strong change in the probability of processes taking place at small impact parameters. The transverse energy $E_{\perp c} = p^2 \psi_c^2 / 2m$ corresponding to this angle does not depend on the particle energy and is close to the potential well depth: $E_{\perp c} = 2Ze^2/d$ for the axial case and $2\pi N d_p a Z\, e^2$ for the planar case, where d is the periodicity of the atomic chains, a is the screening constant, d_p is the periodicity of the atomic planes, and N and Z are the atomic density and the atomic number of the target, respectively.

At $E_\perp > E_{\perp c}$, a sharp change in the particle flux distribution occurs and correspondingly, a sharp change in the probability of processes taking place at collisions with small impact parameters. In the axial case, the flux distribution over impact parameters is uniform at $E_\perp > E_{\perp c}$, as well as in the beam incident on the crystal. In the planar case this distribution reverses when $E_\perp$ goes through the value of the potential barrier U_m—the particle flux is concentrated near the potential barrier. Following *Chadderton* [5.21], we will call particles with $E_\perp \gtrsim U_m$ "quasi-channeled". It follows that quasi-channeled particles, unlike channeled particles (having $E_\perp < U_m$), are characterized by an essentially different distribution over impact parameters and, correspondingly, by a different probability for incoherent processes (multiple scattering, large-angle Rutherford scattering, nuclear reactions, etc.) to take place.

Although classical considerations are inapplicable for scattering of fast particles by isolated atoms, according to Bohr's criterion $Z\, e^2/\hbar v < 1$, the particle deflection by the average field of the atomic chain or plane exceeds that caused by quantum diffraction, as was shown by *Lervig* et al. [5.22]. Therefore, when channeling occurs, the accuracy of the classical approximation improves with increasing energy.

For electrons and positrons, this question was investigated in detail [5.23] for large-angle scattering of channeled particles. It was shown that the classical approximation adequately describes such large-angle scattering for $n \geqslant 1$, where n is the number of bound states in the potential well for transverse motion. This criterion limits the energy of positrons to $\cong 1$ MeV, and that of electrons to $\cong 10$ MeV. The same conclusion was drawn on the basis of experiments on electron penetration [5.24]. For electron radiation, however, where an observation of transitions between individual bound states of the channeled particle is possible, quantum effects are important up to energies of the order of tens of MeV [5.25].

5.3 Multiple Scattering Effects

Because of scattering by electrons and thermal atomic vibrations, the transverse energy (and other integrals of motion in the average potential) is not conserved. This leads to a dependence of the radiation spectrum on the crystal thickness.

At high energies, the main effect is the change of the particle distribution $F(E_\perp,t)$ over transverse energies with the penetration depth t. The evolution of this distribution with particle penetration depth into the crystal can be determined from the Fokker-Planck equation. For planar channeling of electrons and positrons and axial channeling of positrons, this equation has the form [5.26,27]

$$\frac{\partial F(E_\perp,t)}{\partial t} = \frac{\partial}{\partial E_\perp} \left\langle \frac{\Delta E_\perp^2}{2\Delta t} \right\rangle g \frac{\partial}{\partial E_\perp} \frac{F(E_\perp,t)}{g} \quad , \tag{5.1}$$

where g is $T(E_\perp)$ and $S(E_\perp)$ in the planar and axial cases, respectively (T is the vibration period, S is the classically accessible area of the channel cross section). Equation (5.1) is also valid in the quasi-channeling region (at $E_\perp \gtrsim U_m$). For axial electron channeling, the Fokker-Planck equation is two dimensional [5.28]:

$$\frac{\partial F(E_\perp,M_z,t)}{\partial t} = \frac{\partial}{\partial E_\perp} \left[\left\langle \frac{\Delta E_\perp^2}{2\Delta t} \right\rangle T \frac{\partial}{\partial E_\perp} \frac{F(E_\perp,M_z,t)}{T} \right]$$

$$+ \frac{\partial}{\partial E_\perp} \left[\left\langle \frac{\Delta E_\perp \Delta M_z}{2\Delta t} \right\rangle T \frac{\partial}{\partial M_z} \frac{F(E_\perp,M_z,t)}{T} \right] + \frac{\partial}{\partial M_z} \left[\left\langle \frac{\Delta M_z \Delta E_\perp}{2\Delta t} \right\rangle T \frac{\partial}{\partial E_\perp} \frac{F(E_\perp,M_z,t)}{T} \right]$$

$$+ \frac{\partial}{\partial M_z} \left[\left\langle \frac{\Delta M_z^2}{2\Delta t} \right\rangle T \frac{\partial}{\partial M_z} \frac{F(E_\perp,M_z,t)}{T} \right] \quad , \tag{5.2}$$

in view of the conservation of the component M_z of angular momentum along the atomic axis. The coefficients in angle brackets in (5.1,2) are expressed in terms of a small increase in the transverse energy, which is connected by an obvious relation with the mean-square scattering angle, $\Delta E_\perp = \frac{1}{2}E\Delta\theta^2$:

$$\left\langle \frac{\Delta E_\perp^2}{\Delta t} \right\rangle = 4\langle(\Delta E_\perp/\Delta t)[E_\perp - U(r_\perp)]\rangle \quad , \quad \text{planar case} \quad ,$$

$$= 2\langle(\Delta E_\perp/\Delta T)[E_\perp - U(r_\perp)]\rangle \quad , \quad \text{axial case} \quad , \tag{5.3}$$

$$\left\langle \frac{\Delta E_\perp \Delta M_z}{\Delta t} \right\rangle = \left\langle M_z \frac{\Delta E_\perp}{\Delta t} \right\rangle \quad , \quad \left\langle \frac{\Delta M_z^2}{\Delta t} \right\rangle = \left\langle mr_\perp^2 \frac{\Delta E_\perp}{\Delta t} \right\rangle \quad .$$

The transverse energy increase due to multiple scattering can be expressed in terms of the energy loss by ionization [5.2],

$$\Delta E_\perp = \alpha_c \frac{m_e}{m} \Delta E \quad , \tag{5.4}$$

where m_e, m are the masses of the electron and the incident particle, respectively, and α_c is the fraction of the energy loss occurring in close collisions; it is approximately equal to $\frac{1}{2}$, and approximately proportional to the electron density:

$$\Delta E = \frac{4\pi N Z L_e e^4}{m_e v^2} n(r_\perp) \Delta t \quad , \tag{5.5}$$

where $n(r_\perp)$ is the electron density profile in the channel, N is the atomic density, v is the particle velocity, and L_e is the Coulomb logarithm [5.29] with relativistic corrections taken into account. For the standard Lindhard potential [5.2], we obtain the following electron density profile:

$$n(r_\perp) = \left(\frac{r_0}{a}\right)^2 \frac{3}{(3 + r_\perp^2/a^2)^2} \quad , \quad r_0^2 = \frac{1}{\pi N d} \quad . \tag{5.6}$$

For multiple scattering by thermal vibrations, the formulas have a more complicated character. As was shown for non-relativistic protons [5.30] at an impact parameter $r_\perp < u_\perp$, multiple scattering is determined by the same expression as in an amorphous substance, taking corresponding account of the density distribution of the chain atoms over the transverse coordinates, this density being determined by the thermal vibrations. At $r_\perp > u_\perp$ *Lindhard*'s formulas are valid [5.2].

For small-angle scattering, as shown by N. Bohr, the relativistic effects reduce merely to an increase of the particle mass. This is also clear from the general way the effects of relativity should enter channeling theory, as considered above. Thus, the non-relativistic formulas for multiple scattering on nuclei are easiyl generalized, and we have at small $r_\perp$:

$$\frac{\Delta E_\perp}{\Delta t} = \frac{1}{2} E \frac{\Delta \theta_R^2}{\Delta t} P(r_\perp) \quad , \tag{5.7}$$

and at large $r_\perp \gg u_\perp$:

$$\frac{\Delta E_\perp}{\Delta t} = \frac{d}{4E} u_\perp^2 [(U')^2(r_\perp) r_\perp^{-2} + (U'')^2(r_\perp)] \quad , \tag{5.8}$$

where $\Delta \theta_R^2/\Delta t$ is the rate of change of the mean-square multiple scattering angle in an amorphous substance, namely,

$$\Delta \theta_R^2/\Delta t = \frac{E_s^2}{E^2 L_{rad}} \quad , \quad E_s \cong 21 \text{ MeV} \quad . \tag{5.9}$$

Here L_{rad} is the radiation length and $P(r_\perp) = (r_0^2/u_\perp^2) \exp(-r_\perp^2/u_\perp^2)$ is the distribution of the atoms in the transverse plane due to the thermal vibrations; $u_\perp$ is the mean-square amplitude of the thermal vibrations in two dimensions.

The angle brackets in (5.1-3) denote averages over the statistical equilibrium distribution of the transverse spatial coordinates. For a periodic motion (planar channeling of positrons and electrons and axial channeling of electrons) it reduces to an average over the period of oscillation:

$$<X> = \frac{(2m)^{\frac{1}{2}}}{T} \int_{r_{min}}^{r_{max}} \frac{X(r_\perp)dr_\perp}{[E_\perp - U_e(r_\perp)]} \qquad (5.10)$$

where $U_e = U(r_\perp)$ is the planar potential, or the effective potential

$$U_e(r_\perp) = U(r_\perp) + M_z^2/2mr_\perp^2$$

in the axial case.

For an aperiodic motion (axial channeling of positrons and axial quasi-channeling) this averaging reduces to an integral over the classically accessible region of transverse motion of the area $S(E_\perp)$.

The initial distributions $F_0(E_\perp, M_z)$ and $F_0(E_\perp)$ are determined from the condition that the transverse momentum be conserved during the particle transition through the crystal surface. We obtain from this condition:

$$F_0(E_\perp, M_z) = \int \frac{dN}{d\mathbf{p}_\perp} \delta \left[E_\perp - \frac{p_\perp^2}{2m} - U(r_\perp)\right] \delta(xp_y - yp_x - M_z) d^2r_\perp d^2p_\perp \quad ,$$

$$\qquad (5.11)$$

$$F_0(E_\perp) = \int \frac{dN}{d\mathbf{p}_\perp} \delta \left[E_\perp - \frac{p_\perp^2}{2m} - U(r_\perp)\right] d\mathbf{r}_\perp d\mathbf{p}_\perp \quad ,$$

where $dN/d\mathbf{p}_\perp$ is the incident beam distribution over the transverse momenta. For an ideal collimation of the beam, incident at angle ψ_{in} with respect to a string of atoms, we obtain

$$F_0(E_\perp, M_z) = 4\left[\pi r_0^2 \left(p_{in} - \frac{M_z^2}{r_{in}^2}\right)^{\frac{1}{2}} \left|\frac{dU(r_\perp = r_{in})}{dr_\perp}\right|\right]^{-1} \quad ; \qquad (5.12)$$

F_0 is normalized to one particle being incident on one atomic string, $p_{in} = p\psi_{in}$, r_{in} is the point at which the particle must hit the crystal in order to acquire the transverse energy

$$E_\perp = U(r_{in}) + p_{in}^2/2m \quad , \qquad (5.13)$$

and πr_0^2 is the transverse plane area per chain. Correspondingly, we obtain from (5.11):

$$F_0(E_\perp) = \frac{1}{\sigma_0} \frac{d\sigma}{dU} \quad , \quad \text{with} \quad U = E_\perp - \frac{p_{in}^2}{2m} \quad , \qquad (5.14)$$

where $d\sigma$ is a differential of area or length, and $\sigma_0 = \pi r_0^2$, or $\sigma_0 = d_p$ for the axial and planar cases, respectively.

96

Results of a numerical solution of the above equations are shown in Fig.5.1. It is seen from this figure that although the channeled electron fraction in the beam decreases very rapidly with depth, the quasi-channeled electron fraction decreases much more slowly. It is important also that up to 30% of the electrons are found at this stage of the motion. For positrons, the fraction of channeled and quasi-channeled particles decreases at the same rate.

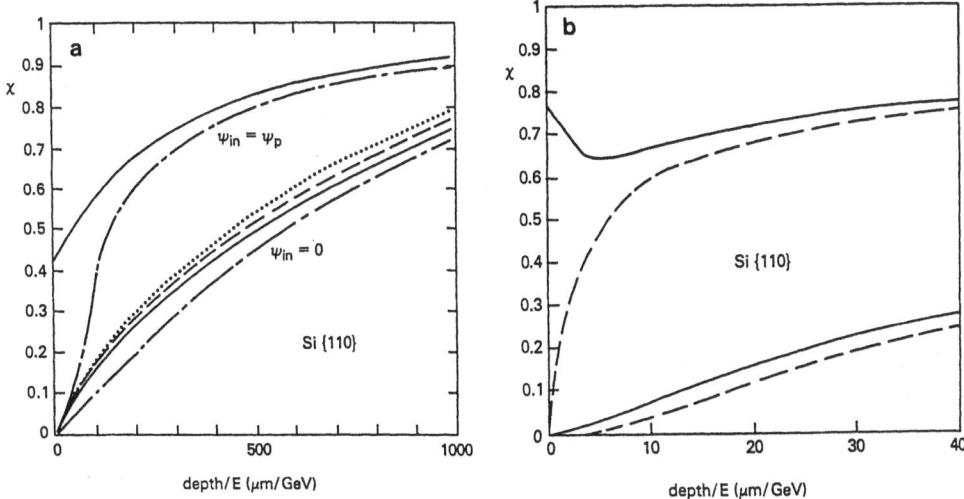

<u>Fig.5.1a,b.</u> Dechanneling function for channeled (χ_{ch}, number of electrons with $E_\perp \geqslant U_m$) and quasi-channeled (χ_{qc}, number of electrons with $E_\perp \geqslant 4U_m$) particles at different angles of crystal inclination and electron energies as a function of penetration depth. (a) Positrons in Si{110}; (———): χ_{ch}, (–·–): χ_{qc} at E = 100 MeV; (– – –): χ_{ch} at E = 1 GeV; (···): χ_{ch} at E = 10 GeV. (b) Electrons in Si{110}; (———): $\psi_{in} = \psi_p$; (– – –): $\psi_{in} = 0$; *upper pair of curves:* χ_{ch}; *lower pair:* χ_{qc} [5.27]

If the photons are collimated, the particle distribution not only over the transverse energy but also over the transverse momentum should be taken into account. In the case of axial channeling of positrons, and axial quasi-channeling, it is necessary to consider scattering by the atomic chains, which results in the establishment of a stationary axially symmetric distribution [5.2] over distances of the order 10-100 μm in the energy range 100 MeV-1 GeV. At such depths the distribution of the emitted photons will also be axially symmetric about the atomic chain. In the case of planar channeling, it is necessary to take into account the distribution over the transverse momentum component parallel to the atomic plane. This leads to a two-dimensional diffusion problem.

5.4 Classical Theory of Radiation

5.4.1 Semiquantitative Treatment of Radiation

Before presenting the exact theory of the radiation of channeled particles, we will give a simple treatment on the basis of *Kumakhov*'s work [5.3,4], which will make clear many important properties of this radiation. As mentioned above, the number of levels in the potential well formed by the atomic chains and planes at high energies is very large, so that it is possible to use classical concepts regarding the particle trajectory.

Let us consider as an example a relativistic positron moving in a planar channel. To a first approximation the potential in the channel can be taken to be harmonic, $U(x) = U_0 x^2$. Here, $U_0 = 0.35 \times 4 \, N Z_1 Z \, e^2 \ell b \, \exp(-\ell b)$, ℓ is the channel half-width, $b = 0.3/a$, a is the screening parameter, and $Z_1 e$ is the charge of the incident particle.

Neglecting changes in the longitudinal velocity, the equation of motion in the transverse plane has the form:

$$\frac{d}{dt} \left(\frac{m_0 v_x}{[1 - (v_x^2 + v_z^2)/c^2]^{\frac{1}{2}}} \right) = -2U_0 x \quad , \tag{5.15}$$

where in the following we will assume that $x_x \ll v_z$, v_z denoting the longitudinal velocity. Equation (5.15) is easily solved:

$$x(t) = x_m \cos \bar{\omega} t \quad , \tag{5.16}$$

where $\bar{\omega}$ is the frequency of the particle oscillations in the laboratory frame:

$$\bar{\omega}^2 = \Omega_0^2 (1 - \beta^2)^{\frac{1}{2}} \quad , \qquad \beta = \frac{v}{c} \cong \frac{v_z}{c} \quad . \tag{5.17}$$

Here, $\Omega_0^2 = 2v/m_0$, m_0 is the rest mass, and x_m is the amplitude of the particle oscillation in the channel. One sees that $\bar{\omega}^2 \sim 1/\gamma$. This is connected with the fact that the longitudinal motion of the particle is relativistic; this also affects the transverse motion, since the particle mass increases. The quantity Ω_0 is the particle oscillation frequency without taking the relativistic factor into account.

It is well known that the radiation power of a relativistic particle ($v_z \cong c$) moving in a circle of radius R is given by

$$I = \frac{2}{3} \frac{e^2 c}{R^2} \left(\frac{E}{mc^2} \right)^4 = \frac{2}{3} \frac{e^2 c}{R^2} \gamma^4 \quad , \tag{5.18}$$

where E is the energy of the particle and e its charge. For accelerated motion, the instantaneous value of the curvature radius is

$$R = \frac{v^2}{\dot{v}_\perp} \cong \frac{c^2}{\dot{v}_\perp} \quad , \tag{5.19}$$

where $\dot{v}_\perp$ is the time derivative of the transverse velocity. In our case,

$$\dot{v}_\perp = \dot{v}_x = -\bar{\omega}^2 x_m \cos\bar{\omega}t \quad , \tag{5.20}$$

so that one has for the inverse mean-square radius of curvature

$$1/R^2 = (2c^4/\bar{\omega}^4 x_m^2)^{-1} \quad . $$

Thus the power radiated by the channeled particle is

$$I = \frac{x_m^2 e^2 \bar{\omega}^4 \gamma^4}{3c^3} = \frac{x_m^2 e^2 \Omega_0^4 \gamma^2}{3c^3} \propto \gamma^2 \quad , \tag{5.21}$$

whence the radiate power is proportional to the square of the particle energy.

The total loss of particle energy per unit length, dI/dx, is given by

$$\frac{dI}{dx} = \frac{I}{c} \quad . \tag{5.22}$$

Equation (5.21) shows that the intensity of radiation $I \sim \Omega_0^4 \sim (\partial U/\partial x)^2$, i.e, the intensity is proportional to the potential gradient squared. Usually, the potential $V \propto Z$, i.e., the intensity increases quadratically with the atomic number of the target. The dependence of the radiation intensity on the lattice parameters also enters (5.21) through Ω_0^4.

To calculate the radiation frequency, it is necessary to take into account the Doppler effect, as well as the relativistic increase of mass in the transverse motion. The potential U' in the rest system of the positron increases γ-fold: $U' = U\gamma$. The radiation frequency in this system is

$$\omega' = \Omega_0 \gamma^{\frac{1}{2}} \quad . \tag{5.23}$$

The frequency in the laboratory system is given by the well-known Doppler formula

$$\omega(\theta) = \frac{\omega'(1 - v^2/c^2)^{\frac{1}{2}}}{1 - (v/c)\cos\theta} \quad , \tag{5.24}$$

whence we find the frequency of the radiation emitted by the channeled particle:

$$\omega(\theta) = \frac{\bar{\omega}}{1 - \beta_\parallel \cos\theta} \quad . \tag{5.25}$$

Hence, at a fixed angle, a well-defined frequency is radiated. The maximum frequency of radiation ω_m is radiated in the forward direction ($\theta = 0$):

$$\omega_m = (1 + \beta_\parallel)\bar{\omega}\gamma^2 \cong 2\Omega_0 \gamma^{3/2} \quad , \tag{5.26}$$

which increases nonlinearly with particle energy. In deriving these formulas, it was implicitly assumed that the dipole radiation condition was fulfilled. In our case, this condition can be approximately written as $\psi\gamma < 1$, where ψ is the critical channeling angle.

99

It is necessary to notice a number of interesting peculiarities of the formulas obtained.

The level spacings of the transverse energy are in the several -eV (i.e., optical) region. As the energy increases, these spacings decrease even more and they are shifted to the infrared region. Therefore, at first glance, one does not expect hard quanta to be radiated by channeled particles.

However, taking relativistic effects into account correctly, it turns out that the channeled particles can radiate in the X-ray and γ-ray regions. Moreover, the radiation intensity increases very quickly with the particle energy ($\sim E^2$), as does the maximum frequency of radiation ($\sim E^{3/2}$). The radiation energy is supplied by the total energy of the relativistic channeled particle.

5.4.2 General Relations

The classical theory of spontaneous radiation of relativistic channeled particles in the dipole approximation was developed in [5.3-6]. At very high energies, such that $\psi_c \gamma > 1$, radiation multipolarities higher than dipole have also to be considered. A calculation of radiation properties taking this into account was carried out in [5.31-38].

Channeling trajectories can be divided into two classes: periodic and non-periodic. Planar channeling and quasi-channeling of electrons and positrons and axial channeling of electrons pertain to the first class of trajectories; axial channeling of positrons and the axial quasi-channeling of positrons and electrons pertain to the second class.

We will begin by considering the first case. According to classical electrodynamics, the intensity of radiation is given by [5.39,40]

$$\frac{d^2 I}{d\omega d\Omega} = \frac{e^2 \omega^2}{4\pi^2 c\tau} |A_\omega|^2 \quad , \tag{5.27}$$

where A_ω is a vector proportional to a Fourier component of the electric field, and τ is the transit time of the particle through the crystal. It is convenient to introduce the translational motion with average velocity $\bar{\beta}$ and to express the vector potential A_ω (in analogy with the procedure in undulator theory [5.41,42] in the form

$$A_\omega = \int_{-\infty}^{\infty} \alpha(t) \exp[i\omega(1 - n \cdot \bar{\beta})t] \, dt \quad , \tag{5.28}$$

where n is the unit vector in the direction of radiation and

$$\alpha(t) = n \times [n \times \beta(t)] \exp[-i(\omega/c)n \times \delta r(t)] \quad , \tag{5.29}$$

where $\delta r(t)$ the oscillating part of the motion.

Expanding $\boldsymbol{\alpha}(t)$ in a Fourier series and letting τ go to infinity, we have

$$\boldsymbol{\alpha}(t) = \sum_{k=-\infty}^{\infty} \mathbf{a}_k \, e^{-ik\omega t} \quad , \quad \text{where} \tag{5.30}$$

$$\mathbf{a}_k = \frac{\bar{\omega}}{2\pi} \int_0^{2\pi/\omega} \boldsymbol{\alpha}(t) \, e^{-ik\bar{\omega}t} \, dt \quad , \tag{5.31}$$

and $\bar{\omega}$ is the oscillation frequency. Then the spectral-angular distribution has the form

$$\frac{d^2I}{d\omega d\Omega} = \sum_{k=1}^{\infty} \frac{e^2\omega^2}{2\pi c} \frac{\omega_k^2}{k\bar{\omega}} |\mathbf{a}_k|^2 \delta(\omega - \omega_k) \quad ,$$

$$|\mathbf{a}_k|^2 = |\mathbf{b}_k|^2 - |\mathbf{n} \cdot \mathbf{b}_k|^2 \quad , \tag{5.32}$$

$$\mathbf{b}_k = \frac{1}{T} \int_0^T \boldsymbol{\beta} \, \exp[ik\bar{\omega}t - i(\omega/c)\mathbf{n} \cdot \delta r(t)] \, dt \quad ,$$

$$\omega_k = k\bar{\omega}/(1 - \mathbf{n} \cdot \bar{\boldsymbol{\beta}}) \quad .$$

After integration over angles, one obtains the spectral distribution

$$\frac{dI}{d\omega} = \frac{e^2}{2\pi c\beta} \sum_{k=1+k_0}^{\infty} \int_0^{2\pi} |\mathbf{a}_k(\omega, \theta_k, \varphi)|^2 \, d\varphi \quad , \tag{5.33}$$

where

$$\theta_k = \cos^{-1} \frac{1 - k\bar{\omega}/\omega}{\bar{\beta}} \quad ,$$

$$k_0 = \tilde{E}[\omega(1 - \boldsymbol{\beta} \cdot \mathbf{n})/\bar{\omega}] \quad ,$$

and $\tilde{E}$ denotes the entire function (the next integer below). Therefore, for a given trajectory, the radiation spectrum in a given direction is discrete.

Although the longitudinal momentum $p_\parallel$ is conserved for motion in a continuum potential, in the relativistic case, because the mass depends on the velocity, the longitudinal velocity $v_\parallel = \beta_\parallel c$ changes. This change is not large, but it affects the radiation spectrum in the ultra-relativistic limit. From conservation of the total energy in the field $U(r_\perp)$, it follows that the particle velocity is constant to within terms of order ψ^2/γ^2, where ψ is the angle of deflection of the particle velocity from the plane or axis (channeling angle). Thus,

$$\beta_\parallel = (\beta^2 - \beta_\perp^2)^{\frac{1}{2}} \cong \beta - \beta_\perp^2/2\beta \quad ,$$

$$1 - \beta_\parallel = 1/2\gamma^2 - \beta_\perp^2/2\beta \quad . \tag{5.34}$$

The Fourier components of the current, $\mathbf{b}_k$, can be represented by integrals:

$$\mathbf{b}_{k_\perp} = \frac{1}{T} \int_{-T/2}^{T/2} \boldsymbol{\beta}_\perp(t) \, \exp[ik\bar{\omega}t - i(\omega/c)\mathbf{n} \cdot \delta\mathbf{r}(t)]dt \quad , \tag{5.35}$$

$$\mathbf{b}_{k_\parallel} = \frac{1}{T} \int_{-T/2}^{T/2} (\beta - \beta_\perp^2/2\beta) \, \exp[ik\bar{\omega}t - i(\omega/c)\mathbf{n} \cdot \delta\mathbf{r}(t)]dt \quad , \tag{5.36}$$

where

$$\delta\mathbf{r}(t) = \int_0^t (\boldsymbol{\beta} - \bar{\boldsymbol{\beta}})dt$$

and $\boldsymbol{\beta}_\perp(t)$ is determined from

$$\dot{\boldsymbol{\beta}}_\perp(t) = -(1/m_e\gamma)\nabla U \quad .$$

For the plane trajectory (planar case) there are two components $\delta\mathbf{r}$ and $\mathbf{b}_k$. In this case, one can choose $\delta r_\perp(t)$ as an even function of the time coordinate, and $\beta_\perp(t)$ as an odd function (the instant $t = 0$ must correspond to the maximum sideways deflection of the particle). In the case of a symmetrical potential well, $\delta r_\perp(t)$ can also be chosen as odd. Then the integrals in (5.35 and 36) will be real and the polarization will be linear. However, in general, the potential well is nonsymmetric (for example, for electrons channeling in diamond along {111} and the radiation will be elliptically polarized. Therefore, the polarization for a plane trajectory is not necessarily linear. In general, it is linear only in the dipole approximation.

When the dipole approximation is valid ($\gamma\psi \ll 1$), the second exponential in (5.35, 36) can be expanded in a series, retaining only the lowest-order expansion terms. Then the spectral-angular density of the radiation is

$$\frac{d^2I}{d\omega d\Omega} = \frac{e^2}{2\pi c} \sum_{k=-\infty}^{\infty} \frac{\delta(\bar{\omega}_k - \omega + \mathbf{k} \cdot \bar{\mathbf{v}})}{(1 - \mathbf{n} \cdot \bar{\boldsymbol{\beta}})^4}$$

$$\times \left[|\dot{\boldsymbol{\beta}}_{k_\perp}|^2(1 - \bar{\boldsymbol{\beta}} \cdot \mathbf{n})^2 - (1 - \bar{\beta}^2)|\dot{\boldsymbol{\beta}}_{k_\perp} \cdot \mathbf{n}_\perp|^2 \right] \quad , \tag{5.37}$$

where $\dot{\boldsymbol{\beta}}_k$ is the Fourier component of acceleration and $\bar{\omega}_k = k\bar{\omega}$, and the spectral intensity has the form:

$$\frac{dI}{d\omega} = \frac{e^2}{c} \sum_{k=1}^{\infty} \frac{|\dot{\boldsymbol{\beta}}_{k_\perp}|^2}{\bar{\omega}_k^2} \omega \left[1 - 2\frac{\omega}{\omega_k} + 2\left(\frac{\omega}{\omega_k}\right)^2 \right] \eta(\omega_k - \omega) \quad , \tag{5.38}$$

where $\omega_k = 2\gamma^2\bar{\omega}_k$ and η is the Heaviside function.

5.4.3 Planar Channeling

The average potential of the atomic planes for positrons is well approximated by the parabola $U(x) = U_m(2x/d_p)^2$ where x is measured from the median plane, d_p is the distance between planes, and U_m is the height of the potential barrier.

102

The trajectories of positrons and electrons in this field (for electrons the potential in this approximation is an inverted parabola) can be expressed analytically. For example, for positrons,

$$z(t) = ct\left(1 - \frac{1}{2\gamma^2} - \frac{\beta_m^2}{4}\right) - \frac{c\beta_m^2}{8\Omega_0}\sin 2\Omega_0 t \quad ,$$

$$x(t) = \frac{c\beta_m}{\Omega_0}\sin\Omega_0 t \quad , \qquad -\pi/\Omega \leqslant t \leqslant \pi/\Omega \quad ,$$

$$\beta_m = (2E_\perp/E)^{\frac{1}{2}} \quad , \qquad \Omega = (\pi/\alpha)\Omega_0 \quad , \qquad \Omega_0 = \frac{2c}{d_p}\left(\frac{2U_m}{E}\right)^{\frac{1}{2}} \quad , \qquad (5.39)$$

$$\alpha = \sin^{-1}(U_m/E_\perp)^{\frac{1}{2}} \quad , \qquad E_\perp \geqslant U_m \quad ,$$

$$\alpha = \pi \quad , \qquad\qquad E_\perp < U_m \quad .$$

The frequency Ω_0 corresponds to channeled particles and does not depend on the transverse energy (because of the harmonicity of the potential), and Ω corresponds to quasi-channeled particles. Naturally, the frequency Ω depends on the transverse energy $E_\perp$, and at large $E_\perp$ tends to the limit $2^{3/2}\pi c E_\perp^{1/2}/d_p E^{1/2}$, corresponding to the frequency of plane traversal in uniform rectilinear motion.

For electrons (x is measured from the plane, $\pm$ corresponds to $t \gtrless 0$):

$$x = \left[\left(\frac{E_\perp}{U_m} - 1\right)^{\frac{1}{2}}\sinh\left(\Omega_0 t \mp \frac{\pi}{\Omega}\Omega_0\right) \pm 1\right]\frac{d_p}{2} \quad , \qquad -\frac{\pi}{\Omega} \leqslant t \leqslant \frac{\pi}{\Omega} \quad , \qquad E_\perp \geqslant U_m \quad ,$$

$$= \pm\left[\left(1 - \frac{E_\perp}{U_m}\right)^{\frac{1}{2}}\cosh\Omega_0\left(\frac{\pi}{2\Omega} - |t|\right) - 1\right]\frac{d_p}{2} \quad , \qquad\qquad E_\perp < U_m \quad ,$$

$$z = c\left(1 - \frac{1}{2\gamma^2}\right)t - \frac{1}{4\beta c}\left(\frac{\Omega_0 d_p}{2}\right)^2\left(\frac{E_\perp}{U_m} - 1\right)\left[t + \frac{1}{2\Omega_0}\sinh 2\Omega_0\left(t \mp \frac{T}{2}\right) \pm \frac{1}{\Omega_0}\sinh\Omega_0 T\right] \quad ,$$

$$E_\perp \geqslant U_m \quad ,$$

$$(5.40)$$

$$= c\left(1 - \frac{1}{2\gamma^2}\right)t - \frac{1}{4\beta c}\left(\frac{\Omega_0 d_p}{2}\right)^2\left(1 - \frac{E_\perp}{U_m}\right)\left[-t + \frac{1}{2\Omega_0}\sinh 2\Omega_0\left(t \mp \frac{T}{4}\right) \pm \frac{1}{\Omega_0}\sinh\Omega_0\frac{T}{2}\right] \quad ,$$

$$E_\perp < U_m \quad ,$$

whereas for positrons,

$$\Omega = \pi\Omega_0/\Lambda \quad , \qquad T = 2\pi/\Omega \quad , \qquad \Lambda = \Lambda_0 \quad , \qquad E_\perp < U_m \quad , \qquad \Lambda = \frac{1}{2}\Lambda_0 \quad , \qquad E_\perp \geqslant U_m \quad ,$$

$$\Lambda_0 = \ln\left|(E_\perp^{\frac{1}{2}} + U_m^{\frac{1}{2}})/(E_\perp^{\frac{1}{2}} - U_m^{\frac{1}{2}})\right| \quad .$$

It should be noted that (5.40) is suitable for simple directions (one plane situated in one period). For complex directions (for example, {111} of diamond, with two planes per period) the well is formed by a combination of two parabolas with different interplanar spacings d_p, and correspondingly, in (5.40) one has to use different values of d_p for $t > 0$ and $t < 0$. In this case, the trajectory is not antisymmetric and the radiation is elliptically polarized. For positrons, the integrals in (5.35,36) can be calculated analytically. For channeled positrons, when $E_\perp \leq U_m$ [5.33-35,42]:

$$b_{kx} = \frac{\beta_\perp}{i k \kappa_x} (k S_1 + 2 S_2) \quad ,$$

$$b_{kz} = -i\left(\bar{\beta} S_1 - \frac{\beta_\perp^2}{8\beta k \kappa_z} S_2 \right) \quad ,$$

$$S_1 = \sum_{p=-\infty}^{\infty} J_p(k\kappa_z) J_{2p+k}(k\kappa_x) \quad , \tag{5.41}$$

$$S_2 = \sum_{p=-\infty}^{\infty} p J_p(k\kappa_z) J_{2p+k}(k\kappa_x) \quad ,$$

where

$$\kappa_x = \frac{\beta_\perp \sin\theta \cos\varphi}{1 - \bar{\beta} \cdot n} \quad , \quad \kappa_z = \frac{\beta_\perp^2 \cos\theta}{8\beta(1 - \bar{\beta} \cdot n)} \quad ,$$

$$\bar{\beta} = \beta(1 - \beta_\perp^2/4\beta^2) \quad ;$$

and when $E_\perp > U_0$ [5.43,44]:

$$b_{kx} = \frac{1}{i} \frac{\beta_\perp}{k\kappa_x} \frac{\alpha}{\pi} S_4 \quad , \qquad b_{kz} = -i\beta S_3 \quad ,$$

$$S_4 = \sum_{p,p'=-\infty}^{\infty} \frac{\sin D}{D} p J_p(\frac{\pi}{\alpha} k\kappa_x) J_{p'}(\frac{\pi}{\alpha} k\kappa_z) \quad ,$$

$$S_3 = \sum_{p,p'=-\infty}^{\infty} \frac{\sin D}{D} J_p(\frac{\pi}{\alpha} k\kappa_x) J_{p'}(\frac{\pi}{\alpha} k\kappa_z) \quad , \tag{5.42}$$

$$D = k\pi + (2p' - p)\alpha + \frac{k\pi}{\alpha} \kappa_x \left(\frac{U_m}{E_\perp}\right)^{\frac{1}{2}} - \frac{k\pi}{\alpha} \kappa_z \left(\frac{U_m}{E}\right)^{\frac{1}{2}} 2 \cos\alpha \quad ,$$

$$\bar{\beta}_x = d_p \Omega/2\pi c \quad , \quad \bar{\beta}_z = \beta\left(1 - \frac{E_\perp}{2E} - \frac{[U_m(E_\perp - U_m)]^{\frac{1}{2}}}{2E\alpha}\right) \quad ,$$

$$\bar{\beta} = \beta\left[1 - \frac{\beta_\perp^2}{4\beta^2}\left(1 + \frac{\sin 2\alpha}{2\alpha} - \frac{2 \sin^2\alpha}{\alpha^2}\right)\right] \quad ,$$

104

$$\alpha = \sin^{-1}(U_m/E_\perp)^{\frac{1}{2}} \quad , \qquad \beta_\perp = (2E_\perp/E)^{\frac{1}{2}} \quad ,$$

$$\Omega = (\pi/\alpha)\Omega_0 \quad , \qquad \Omega_0 = (2c/d_p)(2U_m/E)^{\frac{1}{2}} \quad .$$

The same remarks as those made after (5.39) apply here.

Notice that because of the dependence of Ω on $E_\perp$, the radiation spectrum of quasi-channeled particles in a given direction has a finite width.

In the dipole approximation, the velocity Fourier components for a trajectory in a harmonic potential have the following forms:
for positrons,

$$\dot{\beta}_{\perp k} = \frac{\Omega_0}{2i} \left(\frac{2E_\perp}{E}\right)^{\frac{1}{2}} \delta_{k,1} \quad , \qquad\qquad E_\perp \leqslant U_m$$

$$\dot{\beta}_{\perp k} = i(-1)^k \left(\frac{2U_m}{E}\right)^{\frac{1}{2}} \frac{\pi k \Omega_0}{\pi^2 k^2 - \alpha^2} \quad , \qquad E_\perp > U_m \quad ,$$

(5.43)

and for electrons:

$$\dot{\beta}_{\perp k} = -\Omega_0 \begin{cases} 0 , & k \text{ even} , \\[2mm] \left(\frac{2U_m}{E}\right)^{\frac{1}{2}} \dfrac{4\kappa\pi}{|\Lambda - i\kappa\pi|^2} , & k \text{ odd} , \end{cases} \qquad E_\perp \leqslant U_m \quad ,$$

$$\dot{\beta}_{\perp k} = i \left(\frac{2U_m}{E}\right)^{\frac{1}{2}} \frac{2\pi k \Omega_0}{\pi^2 k^2 + \Lambda^2} \quad , \qquad E_\perp \geqslant U_m \quad .$$

(5.44)

Substituting these expressions into (5.38), one can calculate the spectral distribution of the radiation. From the results obtained, it is seen that the maximum intensity (at $\omega = \omega_k$) of the kth harmonic decreases as k^{-3}. Therefore, not only for the channeled positrons, but also for the quasi-channeled positrons, and for the channeled and quasi-channeled electrons, the radiation in the fundamental frequency considerably exceeds the radiation in higher harmonics.

In Fig.5.2, we show the radiation spectrum of positrons channeled along the diamond plane (110) for angles of incidence $\psi_{in} = 0$ and $\psi_{in} = 0.7\psi_c$ [$\psi_c = (2U_m/E)^{\frac{1}{2}}$] at $E = 10$ GeV. The calculations were performed using (5.32,33,41,42).

One notices the following properties of these spectra:

1) The radiation intensity is chiefly concentrated in the fundamental harmonic.
2) A change of the angle of incidence of the beam essentially does not affect the peak location in the spectrum.
3) The relative contribution of higher harmonics increases with increasing angle of incidence.

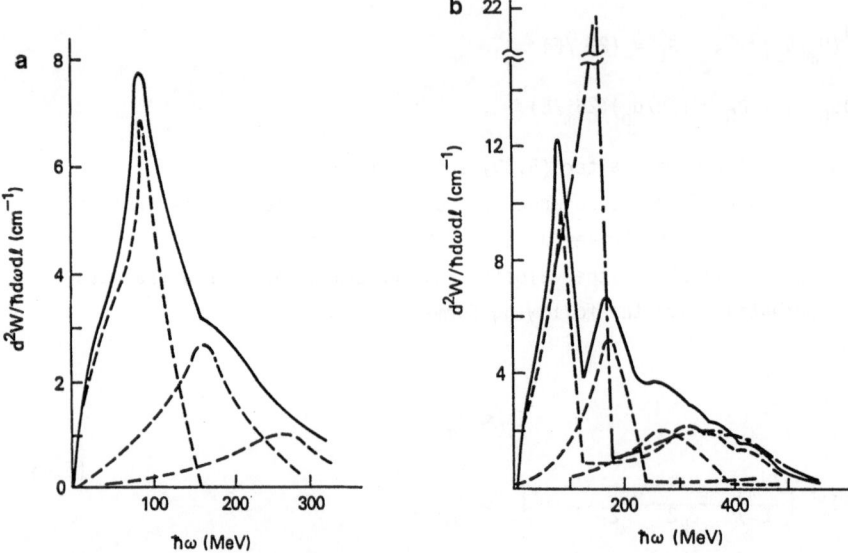

Fig.5.2a,b. Intensity spectrum of radiation by positrons of energy 10 GeV channneld in diamond along the {110} plane for different angles of incidence ψ_{in}: (a) $\psi_{in} = 0$, (b) $\psi_{in} = 0.7\psi_c = 46$ μrad [5.43]. (---) represent contributions from individual harmonics, (——) are the sums of all harmonics, (-··-) correspond to using the dipole approximation

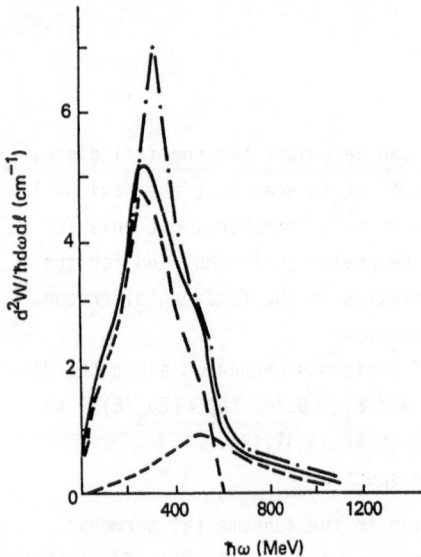

Fig.5.3. Intensity spectrum of radiation by positrons of energy 10 GeV quasi-channeled in diamond along the {110} plane for an angle of incidence $\psi_{in} = \psi_c = 66$ μrad. As in Fig.5.2, contributions from individual harmonics and results obtained in the dipole approximation are shown [5.43]

The same figures show the spectrum, calculated using (5.43) in the dipole approximation. As is seen, when higher multipoles are taken into account the main maximum of the spectrum shifts to the left, i.e., the peak radiation becomes softer and the peak value of the intensity decreases.

Figure 5.3 represents the radiation spectrum of quasi-channeled positrons at $\psi_{in} = \psi_c$, calculated according to (5.42). Here, typically, the peak location is strongly dependent on the angle of incidence, in contrast to the channeling case. However, a dipole-approximation calculation gives a harder and higher peak, so it does for channeling. It is also seen that the second harmonic is substantially less intense than the first one. Finally, we observe that the radiation frequencies of quasi-channeled positrons are shifted, by at least a factor of 2, to the harder γ-quanta region.

The radiation of planar-channeled high-energy positrons was measured in diamond at SLAC [5.7], and in silicon at Serpukhov [5.15] and CERN [5.16].

In Fig.5.4, we show data obtained at SLAC [5.7] and our theoretical results at different positron energies for an orientation where the {110} plane is parallel to the beam direction. The theoretical curves were computed [5.34,35] using (5.32, 41), with and without taking into account dechanneling according to the formulas of Sect.5.3. Good agreement between theory and experiment is seen regarding the form of the main maximum. Since in this computation the radiation of quasi-channeled positrons was not calculated, the theory yielded values which were too low at the high-frequency end of the spectrum.

The radiation spectrum of positrons in diamond as a function of orientation is shown in Fig.5.5. In the theoretical results [5.43] shown in the figures, both the channeled and quasi-channeled particles were taken into account according to (5.41, 42), as well as dechanneling. As is seen from the figure, the theoretical and experimental results are in good agreement if one assumes that the crystal was slightly misaligned: the experimental angle values should be decreased by 23 μrad. Figure 5.6 shows similar results for Si.

We now consider the main characteristics of the radiation emitted by planar-channeled electrons. For a potential approximated by $U(x) = -U_m \exp(-b|x|)$ the radiation spectrum in the dipole approximation [5.45] has been obtained. In [5.46], the radiation spectrum for a particle with given transverse energy was calculated by direct integration of (5.35,36) for the Molière potential. The most detailed spectral calculations were carried out [5.47] for a potential of the form $U(x) = -U_0/\cosh^2\alpha a$.

Results of the latter calculations for different transverse energies are shown in Fig.5.7. Particles with $|E_\perp|$ near zero radiate primarily odd harmonics. At the bottom of the potential well, where the potential is close to parabolic, the radiation intensity monotonically decreases from harmonic to harmonic, and almost all the radiated energy pertains to the fundamental. The spectral intensity of radiation is maximum at the transverse energies $|E_\perp| \cong 0.5\, U_0$.

If an electron enters the crystal at an angle ψ_{in} with respect to the atomic plane, then

$$E_\perp = U(x_{in}) + E\psi_{in}^2/2 \quad ,$$

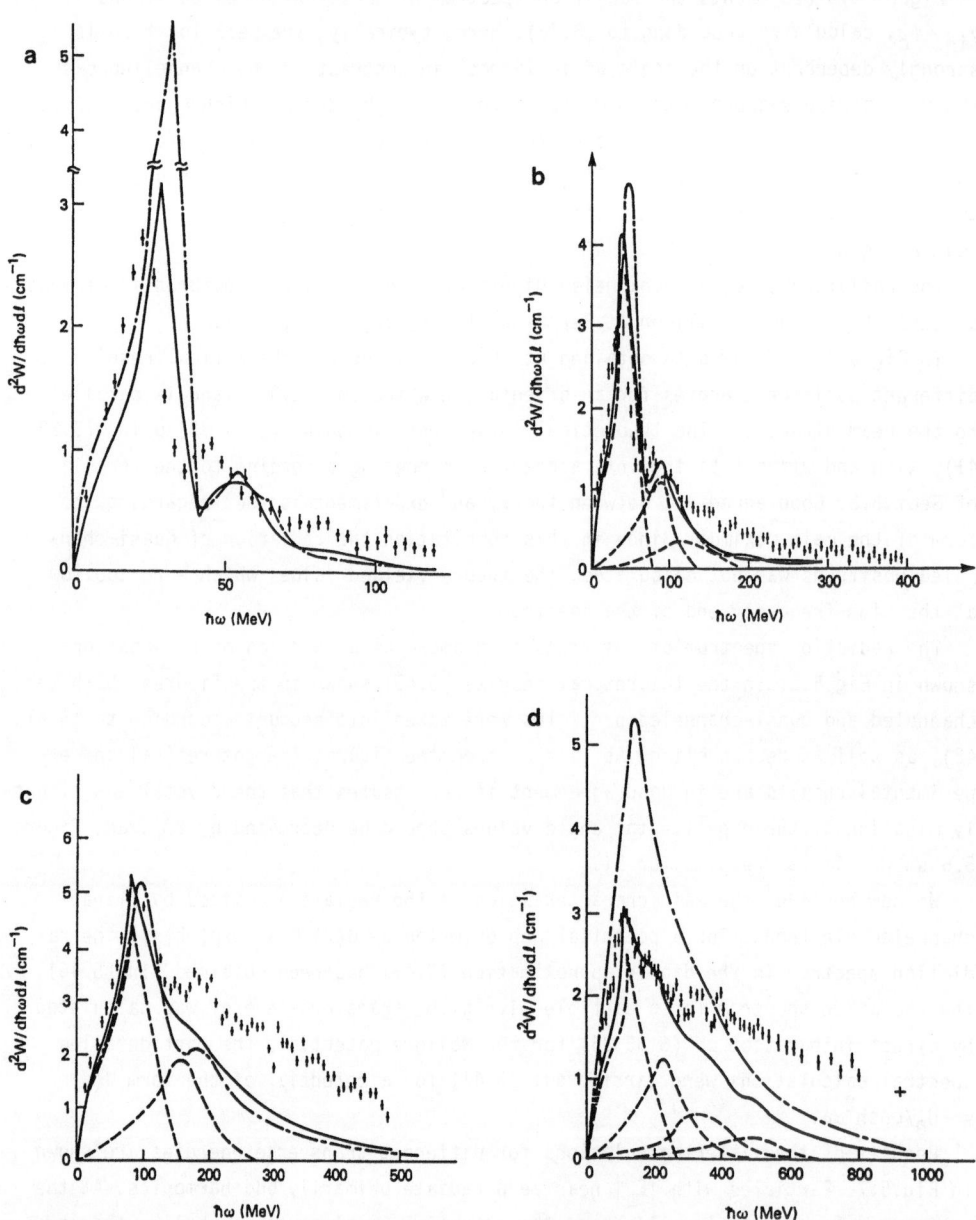

Fig.5.4a-d. Intensity spectrum of radiation by positrons channeled in diamond of thickness 80 μm along the plane {110} at different energies: (a) 4 GeV, (b) 6 GeV, (c) 10 GeV, (d) 14 GeV. (····) represent experimental data [5.7], (——) represent the theoretical calculations [5.35] without accounting for dechanneling, (—·—) correspond to theory with dechanneling taken into account. (---) show the contributions of individual harmonics without dechanneling, divided by a factor (a) 2.5, (b) 2, (c) 1.5, (d) 2.5

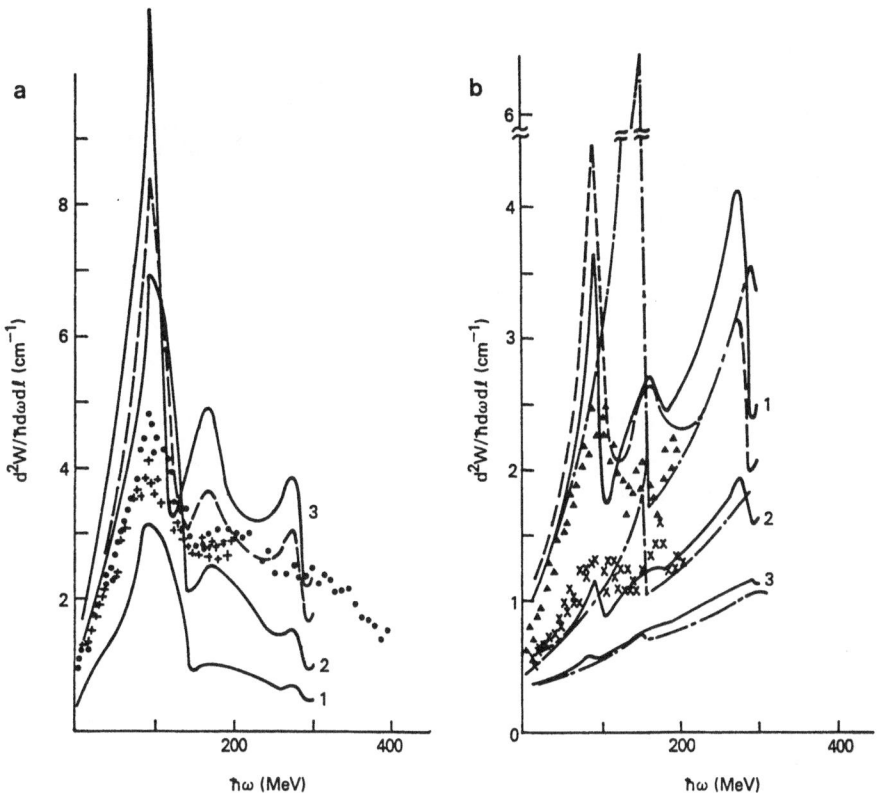

<u>Fig.5.5a,b.</u> Intensity spectrum of radiation of 10 GeV positrons channeled in diamond of thickness 80 μm along the plane {110} for an angular divergence $\Delta\psi_{in} = 10$ μrad and different angles of incidence $\bar{\psi}_{in}$: (——) 1–3 correspond to (**a**) $\bar{\psi}_{in} = 0, 23,$ 45 μrad; (**b**) $\bar{\psi}_{in} = 69, 92, 115$ μrad, respectively. (– – –) corresponds to case 2 for $\Delta\psi_{in} = 20$ μrad, (–·–) corresponds to the dipole approximation [5.43]. Experimental data were obtained for the following angles $\bar{\psi}_{in}$: (**a**) (···) at 0, (+++) at 46 μrad; (**b**) (xxx) at 115 μrad, (▲▲▲) at 92 μrad

where x_{in} is the coordinate of the point of entrance. Then for the potential $U(x) = -U_0/\cosh^2\alpha x$ we obtain for the distribution function of electrons over the transverse energies in the channel

$$f(E_\perp)dE_\perp = (\alpha d_p)^{-1}(|E_\perp| + E\psi_{in}^2/2)^{-1}[1 - (|E_\perp| + E\psi_{in}^2/2)/U_0]^{-\frac{1}{2}} dE_\perp \quad , \qquad (5.45)$$

where d_p is the interplanar spacing. Electrons with $E_\perp < 0$ are regarded as being channeled. The transverse energy of such electrons is confined to the interval

$$U_0 \cosh^{-2}(\alpha d_p/2) \leqslant |E_\perp| \leqslant U_0 - E\psi_{in}^2/2 \quad . \qquad (5.46)$$

The fraction of electrons trapped in the channeling regime for an entrance angle ψ_{in} is

109

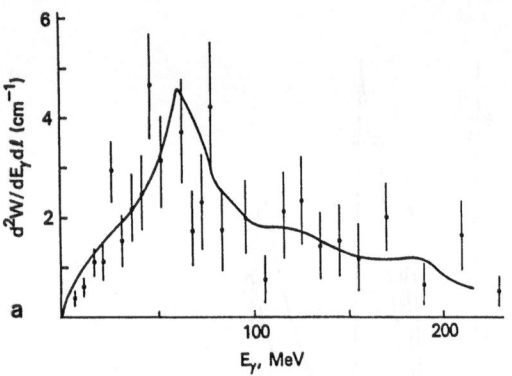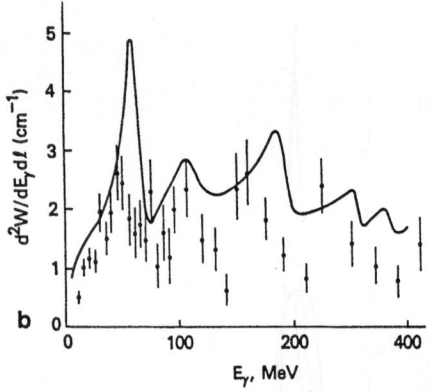

Fig.5.6a,b. Intensity spectrum of the 10-GeV positron radiation for channeling in Si of thickness 90 μm along {110}, with angles of incidence within the limits: (a) $0 \leq |\psi_{in}| \leq 20$ μrad, (b) $50 \leq |\psi_{in}| \leq 80$ μrad. ($\cdots$) represent experimental results [5.15], (——) represents the theoretical calculation [5.43]

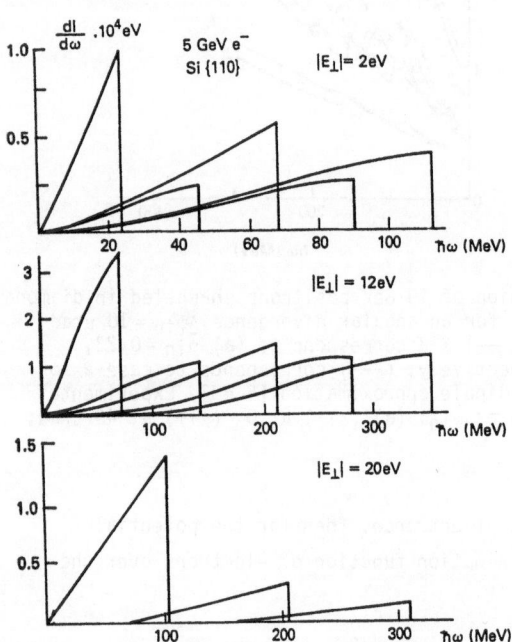

Fig.5.7. Intensity of radiation of electrons with different transverse energies for planar channeling in Si along the plane {110} at 5 GeV (individual harmonics)

$$y = \ln \left| \frac{C + 1}{C - 1} \right| / (\alpha d_p) \quad , \tag{5.47}$$

where $C = [\tanh^2(\alpha d_p/2) - (\psi_{in}/\psi_c)^2]^{1/2}$, and $\psi_c = (2U_0/E)^{1/2}$ is the critical channeling angle.

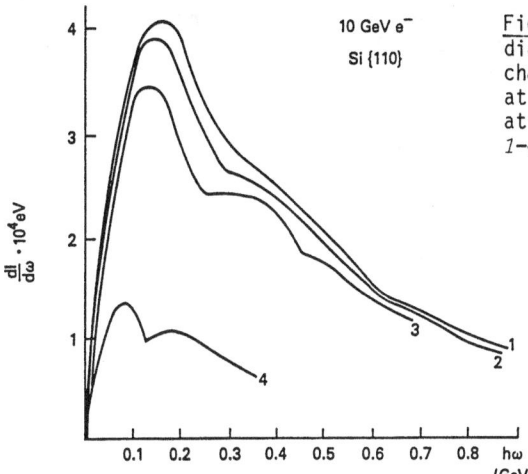

Fig.5.8. Intensity spectrum of the radiation of 10-GeV electrons for planar channeling in Si along the {110} plane at different angles of crystal inclination ψ_{in}/ψ_c =0,0.3,0.5,0.8 for curves $1\text{–}4$, respectively [$\psi_c = (2U_m/E)^{\frac{1}{2}}$]

Let us average the radiation spectrum over the distribution function $f(E_\perp)$:

$$\left\langle \frac{dI}{d\omega} \right\rangle = \int \frac{dI}{d\omega} f(E_\perp)dE_\perp \quad , \tag{5.48}$$

where the integral is taken over the energies of electrons trapped in the channel.

Figure 5.8 shows the radiation spectra of electrons with energy 10 GeV in silicon along the {110} plane at different angles of entrance to the channel. It is seen that at $\psi_{in} = 0.5\psi_c$, the radiation intensity differs very little from the intensity at normal incidence, although only ~40% of the initial beam is trapped into the channel. This is explained by the following fact. The major contribution to the radiation comes from the electrons with transverse energies from ~0.5 U_0 to ~0.75 U_0. At the angle $\psi_{in} = 0.5\psi_c$, approximately twice as many particles are in this interval than at $\psi_{in} = 0$, although the total number of electrons in the channel is less. With a further increase of the angle of incidence, the spectral intensity of the radiation rapidly decreases.

Figure 5.9 shows a comparison of the radiation spectra in silicon and germanium. Values of the intensity spectrum in the region of the maximum differ by a factor of about nine for these crystals. The squares of the potential wells for these crystals are in the same ratio. Comparison of Figs.5.8 and 9 shows that the dependence of the radiation intensity spectrum on the electron energy is somewhat weaker than $dI/d\omega \sim \gamma^{\frac{1}{2}}$.

5.4.4 Axial Channeling of Electrons

Consider the radiation in the potential field $U \sim -\alpha/r_\perp$. A bound electron trajectory in this field can be represented in the parametric form

$$x = a(\cos\delta - \varepsilon) \quad , \quad y = a(1 - \varepsilon)^{\frac{1}{2}}\sin\delta \quad ,$$

111

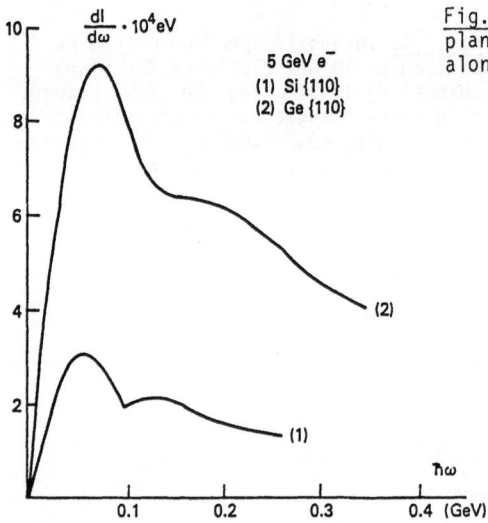

Fig.5.9. Comparison of electron radiation for planar channeling in Si (1) and in Ge (2) along the {110} plane at 5 GeV

$$t = (\delta - \varepsilon \sin\delta)/\Omega \quad ,$$

$$\bar{\beta} = \beta - (a^2\Omega^2/2c^2\beta) \quad , \quad z = c\bar{\beta}t + \delta z \quad ,$$

$$\delta z = -a^2\varepsilon\Omega\sin\delta/v \quad ,$$

(5.49)

where δ is the free parameter, a is the amplitude and ε the eccentricity of the orbit, $(a\Omega/c)^2 = 2|E_\perp|/E$, Ω is the oscillation frequency which is given by $\Omega = (2|E_\perp|/\alpha)(2E_\perp/\gamma m_e)^{\frac{1}{2}}$ and $a = \alpha/2|E_\perp|$ (here $E_\perp < 0$ when measured from the top of the potential well).

Thus we obtain [5.38,44]:

$$\frac{d^2I}{d\omega d\varphi} = \frac{e^2\omega}{2\pi c} \sum_k \frac{2E_\perp}{E} \left[\left(\frac{1}{\kappa^2} - 1\right)J_k^2(k\kappa) + J_k'^2(k\kappa)\right]$$

$$- \frac{1}{\gamma^2}\left[\left(1 - \frac{\varepsilon\cos\chi}{\kappa}\right)^2 J_k^2(k\kappa) + \varepsilon^2\sin^2\chi J_k'^2(k\kappa)\right] \quad ,$$

$$\cos\chi = K_{yz}/K \quad , \quad K = (K_x^2 + K_{yz}^2)^{\frac{1}{2}} \quad ,$$

(5.50)

$$K_x = \frac{\Omega a}{c} \frac{\sin\theta_k \cos\varphi}{1 - \bar{\beta}\cos\theta_k} \quad ,$$

$$K_{yz} = \varepsilon + \frac{\Omega a}{c} \frac{(1 - \varepsilon^2)\sin\theta_k \sin\varphi}{1 - \bar{\beta}\cos\theta_k} - \frac{\Omega^2 a^2}{c^2} \frac{\varepsilon\cos\theta_k}{1 - \bar{\beta}\cos\theta_k} \quad .$$

In the dipole approximation, where $2E_\perp\gamma^2/E \ll 1$, we arrive at the result

$$\frac{d^2I}{d\omega d\varphi} = \frac{e^2}{\pi c}\frac{|E_\perp|}{E}\omega \sum_k \left\{ J_k'^2(k\varepsilon)\left[1 - \frac{(1-\beta^2)\sin^2\theta_k \cos^2\varphi}{(1-\bar\beta\cos\theta_k)^2}\right]\right.$$

$$\left. + \left(\frac{1}{\varepsilon^2} - 1\right)J_k^2(k\varepsilon)\left[1 - \frac{(1-\beta^2)\sin^2\theta_k \sin^2\varphi}{(1-\bar\beta\cos\theta_k)^2}\right]\right\} . \tag{5.51}$$

Here the angle φ is measured from the semimajor axis of the trajectory. Therefore, one has to average over this angle, since for different particles and atomic chains the orientation of ellipses will be different. Consequently, the radiation spectrum must not depend on azimuthal angle. An averaging of (5.51) results in (5.38) with

$$|b_{\perp k}|^2 = \frac{2E_\perp}{E}|k\bar\omega|^2\left[J_k'^2(k\varepsilon) + (\varepsilon^{-2} - 1)J_k^2(k\varepsilon)\right] . \tag{5.52}$$

It is also necessary to average the intensity over the transverse energies $E_\perp$ and ε. Since the oscillation frequencies Ω strongly depend on $E_\perp$, this averaging leads to a spreading of the maxima and an overlap of harmonics, and to the formation of a continuous spectrum with one broad maximum.

Consider now the simple case of parallel incidence of a beam on a crystal along the channel axis. Then $\varepsilon = 1$, and the initial distribution over $E_\perp$ has the form

$$\frac{dN}{dE_\perp} = \frac{2\alpha^2}{r_0^2|E_\perp|^3} , \qquad -U_m \leqslant E_\perp \leqslant -U_1 , \qquad U_1 = \frac{\alpha}{r_1} ,$$

where the limits U_m and U_1 are chosen on the basis of the following considerations: $U_m \approx \alpha/u_\perp$ is the potential well depth for an isolated chain and is determined by averaging over thermal vibrations; $u_\perp^2$ is the mean-square amplitude of these vibrations in the transverse plane, and U_1 is determined by the influence of neighboring atomic chains on the potential. The quantity U_1 is approximately equal to α/r_1, where $2r_1$ is the distance between nearest chains.

We now give some results of calculations of the radiation spectrum in the dipole approximation for the potential $U = -\alpha/r_\perp$ where α is chosen to best-fit the Lindhard potential, $\alpha = 1.5Ze^2 a_F/d$, a_F being the screening parameter and d the distance between atoms in the chain. For the spectral intensity distribution, we obtain:

$$\frac{dI_k}{d\omega} = \frac{3\bar I_k}{\omega_{km}}\frac{\omega}{\omega_{km}}\left[1 - 2\frac{\omega}{\omega_{km}} + 2\left(\frac{\omega}{\omega_{km}}\right)^2\right] ,$$

where $\omega_{km} = k\Omega/(1-\bar\beta)$,

$$\bar I_k = \frac{4e^2 k^2 a^2 \tilde\omega^4 \gamma^4}{3c^3}\left[J_k'^2(k\varepsilon) + \frac{1-\varepsilon^2}{\varepsilon^2}J_k^2(k\varepsilon)\right] , \tag{5.53}$$

k is the number of the harmonic in the Fourier expansion (5.30), and the eccentricity ε is given by

$$\epsilon = \left[1 - \frac{2M_z^2 c^2}{\alpha^2 E}\left(|E_\perp| + U_2\right)\right]^{\frac{1}{2}} \quad .$$

Here U_2 is the cutoff parameter of the potential, $U_2 = \alpha/r_0$, and $\pi r_0^2 = (Nd)^{-1}$ is the area per chain in the transverse plane; finally,

$$\tilde{\omega} = (2|E_\perp| + 2U_2)^{3/2} C\alpha^{-1} E^{-\frac{1}{2}} \quad .$$

If an electron with energy E enters the channel at a point with transverse polar coordinates r_{in}, φ_{in}, making an angle ψ_{in} with the atomic axis, then its transverse energy $E_\perp$ and angular momentum M_z with respect to this axis are:

$$E_\perp = U(r_{in}) + E\psi_{in}^2/2 \quad , \qquad M_z = Er_{in}\psi_{in}\sin\varphi_{in}/c \quad , \tag{5.54}$$

where c is the speed of light in vacuum.

Let us average (5.33) over all entrance points to the channel:

$$\left\langle \frac{dI_k}{d\omega} \right\rangle = S_0^{-1} \int \frac{dI_k}{d\omega}\, r_{in}\, dr_{in}\, d\varphi_{in} \quad , \tag{5.55}$$

where S_0 is the channel area in the transverse plane. The integral in (5.55) is taken over all r_{in} and φ_{in} corresponding to the channeling regime. For the distribution function of electrons over transverse energies and angular momenta, we have from (5.54):

$$dN = S_0^{-1} r_{in}\, dr_{in}\, d\varphi_{in} = 4r_{in}^3 (\alpha S_0)^{-1}(p_\perp^2 r_{in}^2 - M_z^2)^{-\frac{1}{2}} dM_z\, dE_\perp \quad , \tag{5.56}$$

where $p_\perp = E\psi_{in}/c$ is the projection of the electron momentum onto the transverse plane and $r_{in} = \alpha(U_2 + |E_\perp| + E\psi_{in}^2/2)^{-1}$. For the fraction of electrons trapped in the channel, we obtain from (5.56):

$$N_{ch} = \left(1 + \frac{E\psi_{in}^2}{2U_1}\right)^{-1} \quad . \tag{5.57}$$

The distribution function is such that states with transverse energy values near zero are more populated. At $\psi_{in} = 0$ all electrons have zero angular momenta with respect to the axis. The entire beam is then trapped in the channel. With an increase of the angle ψ_{in}, the fraction of particles with larger angular momenta increases, and the electron orbits in the transverse plane become closer to circular ones. While integrating, we must recall that the potential has a singularity at $r_\perp = 0$. This leads to (5.53) overestimating by a large amount the contribution to radiation from electrons with small angular momenta. It makes sense to limit the range of integration in (5.55) for small angular momenta. The minimum value of angular momentum M_{min} can be found from the condition that the minimum of the effective potential U_{eff}^{min} must not be lower than the potential well depth $U_m \cong U(u_\perp)$, where $u_\perp$ is the amplitude of thermal vibration of atoms in the chain, i.e, given by the condition

$$U_{eff}^{min}(M = M_{min}) = U(u_\perp/2) \quad ,$$

whence $M_{min} = (\alpha E u_\perp/2c^2)^{1/2}$. The range of integration in (5.55) is determined by the conditions:

$$E_\perp^{min} \leqslant E_\perp \leqslant 0 \quad , \qquad M_{min} \lesssim M_z \lesssim p_\perp r_{in} \quad , \tag{5.58}$$

where $|E_\perp^{min}|$ is the minimum of the values $|E_\perp'| = |U_{min}| - E\psi_{in}^2$ and $|E_\perp''| = p_\perp \alpha/M_{min}$ $- U_2 - E\psi_{in}^2/2$, and $E_\perp''$ is the minimum possible value of the electron transverse energy at $M_z = M_{min}$.

Figure 5.10 shows the dependence of the fraction of electrons trapped in the channeling regime on the entrance angle to the crystal. Curve 1 is the result predicted by (5.57). Curve 2 is the fraction of electrons corresponding to the inequalities (5.58):

$$N_{ch} = \frac{1}{\pi} \left(\frac{4cM_{max}}{r_0 E \psi_{in}} \right)^2 \left[\cos^{-1} \eta + \eta (1 - \eta^2)^{1/2} \right] \quad , \tag{5.59}$$

where $\eta = M_{min}/M_{max}$, $M_{max} = \alpha E c^{-1} (U_1 + E\psi_{in}^2/2)^{-1}$ being the maximum possible value of the angular momentum of an electron in the channel at given ψ_{in}. Figure 5.11 shows the spectra of radiation from electrons of energy 200 MeV axially channeled in the <111> directions in tungsten [5.47]. In the framework of the present model, the frequencies of emitted photons first increase and then decrease with an increase of the entrance angle. The number of channeled electrons decreases rapidly with increasing ψ_{in}, while the spectral intensity increases at small angles. This is connected with the limitations on possible values of angular momenta.

From Fig.5.11, it follows that the radiation intensity in the region of the maximum exceeds the analogous value for bremsstrahlung in an amorphous medium by a factor of ~12. By increasing the energy up to 1 GeV, this ratio increases by approximately a factor 50.

We mention the following features distinguishing radiation of axially channeled electrons from the radiation of planarly channeled positrons:

1) The electron channeling radiation is harder than that from positron channeling.
2) The radiation intensity per electron is substantially higher than that per positron.
3) The electron radiation spectrum is broader than that of positrons, since the potential for axial channeling of electrons is essentially anharmonic.

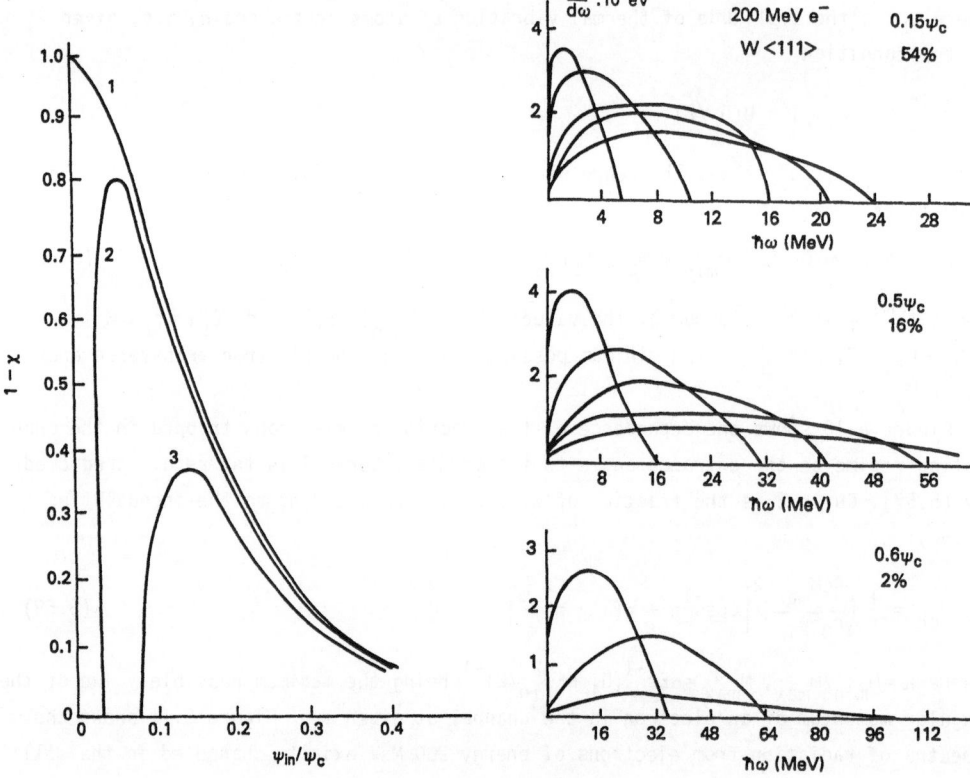

Fig.5.10. The fraction $|-\chi$ of electrons: (1) trapped into states associated with the <111> axis in tungsten, for an angle of incidence ψ_{in} relative to the axis, and with limitations on: (2) angular momentum $M \geqslant M_{min}$, and (3) distance of closest approach to the chain $\rho \geqslant u_\perp$

Fig.5.11. Radiation intensity of individual harmonics for 200-MeV electrons channeled in tungsten along the <111> axis for different angles of incidence. The channeled particle fraction is given as a percentage of the entire incident beam intensity

5.4.5 Quasi-Channeling in the Axial Case

Particles traveling on non-periodic trajectories scatter on the atomic chain. This is true for all axially channeled positrons and axially quasi-channeled electrons and positrons, except for a small fraction of the beam particles which have a transverse energy $|E_\perp| < U_S$, where U_S is the potential energy value on the separatrix of equipotential curves which divides the family of these curves into two subfamilies —equipotential curves that close themselves around the atomic chains or else around the center of symmetry of the channel. For positrons, this excludes the properly channeled particles from consideration. Since U_S amounts to ~5 eV, the fraction of such particles is usually small.

In the field $\pm\alpha/r_\perp$ (± correspond to positrons and electrons, respectively, $\alpha > 0$), these trajectories have the form:

$$x = a(\varepsilon \pm \cosh\xi) \quad , \qquad y = a(\varepsilon^2 - 1)^{\frac{1}{2}}\sinh\xi \quad ,$$

$$z = \frac{v}{\Omega}\,\varepsilon\left|1 - \frac{a^2\Omega^2}{2v^2}\right|\,\sinh\xi \pm \frac{v}{\Omega}\,\xi\left(1 + \frac{a^2\Omega^2}{2v^2}\right) \quad , \tag{5.60}$$

$$t = \frac{1}{\Omega}\,(\varepsilon\,\sinh\xi \pm \xi) \quad , \qquad \left(\frac{a\Omega}{v}\right)^2 = \frac{2|E_\perp|}{\beta^2 E} \quad ,$$

where Ω is as defined after (5.49).

The energy radiated during the scattering on the chain is given by a formula similar to (5.27):

$$\frac{d^2W}{d\omega d\theta} = \frac{c^2\omega^2}{4\pi^2 c}\,|\mathbf{n}\times\boldsymbol{\beta}_\omega|^2 \quad , \qquad \boldsymbol{\beta}_\omega = \int_{-\infty}^{\infty}\boldsymbol{\beta}\,\exp[i\omega t - i\mathbf{k}\cdot\mathbf{r}(t)]\,dt \quad . \tag{5.61}$$

Up to a constant phase factor, we obtain

$$\beta_{\omega x} = \frac{a}{c}\,2\,e^{-\pi\nu/2}\left[\frac{\eta\nu}{\xi^2}K_{i\nu}(\xi) - \frac{i\delta}{\xi}K'_{i\nu}(\xi)\right] \quad ,$$

$$\beta_{\omega y} = -\frac{a}{c}\,(\varepsilon^2 - 1)^{\frac{1}{2}}2\,e^{-\pi\nu/2}\left[\frac{\delta\nu}{\xi^2}K_{i\nu}(\xi) - \frac{i\eta}{\xi}K'_{i\nu}(\xi)\right] \quad ,$$

$$\beta_{\omega z} = -\frac{\beta\varepsilon}{\omega}\left(1 - \frac{a^2\Omega^2}{2v^2}\right)2\,e^{-\pi\nu/2}\left[\frac{\delta\nu}{\xi^2}K_{i\nu}(\xi) - \frac{i\eta}{\xi}K'_{i\nu}(\xi)\right] \tag{5.62}$$

$$+ \frac{\beta}{\Omega}\left(1 + \frac{a^2\Omega^2}{2v^2}\right)2\,e^{-\pi\nu/2}K_{i\nu}(\xi) \quad ,$$

where

$$\eta = \mp\frac{\omega}{\eta}\left(\frac{2E_\perp}{E}\right)^{\frac{1}{2}}\sin\theta\cos\varphi \quad ,$$

$$\delta = \frac{\omega}{\Omega}\left\{\varepsilon\left[1 - \beta\left(1 - \frac{E_\perp}{E\beta^2}\right)\cos\theta\right] - \left[\frac{2E_\perp}{E}(\varepsilon^2 - 1)\right]^{\frac{1}{2}}\sin\theta\sin\varphi\right\} \quad ,$$

$$\xi = (\delta^2 - \eta^2)^{\frac{1}{2}} \quad ,$$

$$\nu = \pm\frac{\omega}{\Omega}\left[1 - \beta\left(1 + \frac{E_\perp}{E\beta^2}\right)\cos\theta\right] \cong \pm\frac{\omega}{2\Omega}\left(\theta^2 + \frac{1}{\gamma^2} - \frac{2E_\perp}{E}\right) \quad .$$

Hence,

$$\frac{d^2W}{d\omega d\theta} = \frac{e^2\omega^2 2E_\perp}{\pi^2 c\Omega^2 E}\,e^{-\pi\nu}\left\{\left[1 - \frac{\nu^2}{\xi^2} - \left(1 - \frac{\varepsilon\delta\nu}{\xi}\right)^2\frac{1}{\gamma^2\psi_m^2}\right]K_{i\nu}^2(\xi)\right. \tag{5.63}$$

$$\left. + \left(1 - \frac{\varepsilon^2\eta^2}{\psi_m^2\gamma^2\xi^2}\right)K'^2_{i\nu}(\xi)\right\} \quad .$$

Here $\psi_m = \sqrt{2E_\perp}/E \approx \psi_{in}$ and $K_{i\nu}(\xi)$ is the McDonald function. This formula was also obtained in [5.48,49], and with the exception of a misprint [the term $(1 - \epsilon\delta\nu/\xi)^2$ in these papers should read $1 \mp \epsilon\delta\nu/\xi$] the formulas are identical.

In the dipole approximation, (5.62,63) reduce to the following for the energy radiated during scattering on the chain:

$$\frac{d^2W}{d\omega d\Omega} = \frac{e^2}{4\pi^2 c} \left(\frac{\omega}{\Omega'}\right)^4 \left[\dot{\boldsymbol{\beta}}_{\perp\Omega'}^2 (1 - \mathbf{n}\cdot\boldsymbol{\beta})^2 - (1 - \beta^2)(\mathbf{n}_\perp \cdot \dot{\boldsymbol{\beta}}_{\perp\Omega'})^2\right] \quad , \tag{5.64}$$

where

$$\Omega' = \omega(1 - \mathbf{n}\cdot\boldsymbol{\beta}) \quad , \quad \dot{\boldsymbol{\beta}}_{\Omega'} = \int_{-\infty}^{\infty} \dot{\boldsymbol{\beta}} \, e^{i\Omega't} \, dt \quad , \tag{5.65}$$

$$\frac{dW}{d\omega} = \frac{e^2\omega}{2\pi c} \int_{\omega/2\gamma^2}^{\infty} \frac{|\dot{\boldsymbol{\beta}}_{\Omega'}|}{\Omega'^2} \left[1 - \frac{\omega}{\Omega'\gamma^2} + \frac{\omega^2}{2\Omega'^2\gamma^4}\right] d\Omega' \quad .$$

For the potential $\pm\alpha/r$ we have, up to a constant phase,

$$\dot{\beta}_{x\Omega'} = 2i\nu\left(\frac{2E_\perp}{E}\right)^{1/2} e^{\mp\pi\nu/2} K_{i\nu}'(\nu\epsilon) \quad ,$$

$$\dot{\beta}_{y\Omega'} = 2\nu\left(\frac{2E_\perp}{E}\right)^{1/2}\left(\frac{\epsilon^2 - 1}{\epsilon^2}\right)^{1/2} e^{\mp\pi\nu/2} K_{i\nu}(\nu\epsilon) \quad , \tag{5.66}$$

where the upper sign applies to positrons and where $\nu = \Omega'/\Omega$.

Substituting (5.66) into (5.64) we obtain the spectral angular distribution of the radiated energy:

$$\frac{d^2W}{d\omega d\Omega} = \frac{e^2}{c\pi^2} \frac{\psi_m^2\omega^2}{\Omega^2} e^{\mp\pi\nu}\left\{\left[K_{i\nu}'^2(\epsilon\nu) + (1 - \epsilon^{-2})K_{i\nu}^2(\epsilon\nu)\right]\right.$$

$$\left. - \frac{(1 - \beta^2)\sin^2\theta}{(1 - \mathbf{n}\cdot\boldsymbol{\beta})^2}\left[K_{i\nu}'^2(\epsilon\nu)\cos^2\varphi + (1 - \epsilon^{-2})K_{i\nu}^2(\epsilon\nu)\sin\varphi\right]\right\} \quad . \tag{5.67}$$

The velocity can be considered as constant in the dipole approximation, and hence in (5.64-67), the velocity $\boldsymbol{\beta}$ is equal to the initial velocity of a particle incident on the chain. Note that this velocity does not coincide with that of a particle incident on the crystal, since in the crystal, scattering on the atomic chains takes place. At $\psi_{in}\gamma \ll 1$, we can assume that the angle φ is measured from the atomic chain axis, and not from the velocity direction. In this case (5.67) can be trivially averaged over the azimuthal angle   because the distribution of particles is axially symmetric in a sufficiently thick monocrystal. Then the radiation distribution is azimuthally symmetric and has the form

$$\frac{d^2W}{d\omega d\theta} = \frac{e^2}{\pi^2 c} \psi_m^2 \frac{\omega^2}{\Omega^2} e^{\mp\pi\nu}\left[K_{i\nu}'^2(\epsilon\nu) + (1 - \epsilon^{-2})K_{i\nu}^2(\epsilon\nu)\right] \times \left[1 - \frac{(1 - \beta^2)\sin^2\theta}{2(1 - \beta\cos\theta)^2}\right] \quad . \tag{5.68}$$

The radiation energy should be averaged over the transverse energy $E_\perp$ and eccentricity ε. The averaging over $\varepsilon = (1 + 4E_\perp^2\rho^2/\alpha^2)^{1/2}$, where ρ is an impact parameter for collision with a chain, reduces to averaging over the impact parameters. In calculating the intensity, it is also necessary to multiply the radiation energy by the average frequency of collisions $d\rho Nd\nu_\perp = Ndc(2E_\perp/E)^{1/2}d\rho$.

Thus, the average intensity of radiation of a beam of particles is found to be

$$\frac{dI}{d\omega} = \iint_{\varepsilon_{min}}^{\varepsilon_{max}} \frac{dW}{d\omega} \frac{\varepsilon d\varepsilon}{(\varepsilon^2 - 1)^{1/2}} \frac{Ndc\alpha}{2E_\perp} \left(\frac{2E_\perp}{E}\right)^{1/2} \frac{dN}{dE_\perp} dE_\perp \quad , \tag{5.69}$$

where

$$\varepsilon_{min} = (1 + 4E_\perp^2 u_\perp^2/\alpha^2)^{1/2} \quad ,$$

$$\varepsilon_{max} = [1 + eE_\perp^2/(\pi Nd\alpha^2)]^{1/2} \quad ,$$

and $dN/dE_\perp$ is the distribution of particles over the transverse energy, which for the $1/r$ potential has the form

$$\frac{dN}{dE} = \frac{2\alpha^2}{r_0^2(E_\perp - E\psi_{in}^2/2)^3} \quad , \quad |U_1| \leqslant E_\perp \leqslant \frac{E}{2}\psi_{in}^2 + U_b \quad ,$$

with $U_b = U_m \cong \alpha/u_\perp$ for positrons and $U_b \cong 0$ for electrons. According to (5.68), at small frequencies $\omega \ll \omega_1 = 2^{5/2}U_m^{3/2} \cdot \gamma^{3/2}/(m_e^{1/2}\alpha)$ the radiation spectrum of electrons and positrons is a slowly varying function of frequency, but at $\omega \gg \omega_1$, it falls off exponentially [5.48,49]. However, for $U(r) \sim \ln(1/r_\perp)$, one obtains a power-law decrease of the intensity $dW/d\omega \sim \omega^{-1}$ at large frequencies [5.50].

5.5 Comparison of Spectral Intensity with That of Bremsstrahlung

For channeling radiation as well as for bremsstrahlung, approximately the same angular relationships ($\Delta\theta \approx \gamma^{-1}$, where $\gamma = m_e c^2/E$, with E the particle energy) are characteristic. However, the effective thickness of a crystal for generation of these radiation types is not the same because the multiple scattering mechanisms are essentially different. As was shown by *Kumakhov* [5.51] this results in a strong increase of the spectral-angular distribution of channeling radiation in comparison with bremsstrahlung at energies of several GeV.

Let us consider first the relationship for the angular distribution of ordinary bremsstrahlung. The maximum of the angular distribution of bremsstrahlung $(dW/d)_B$ is limited by multiple scattering. The characteristic thickness x_B, from which the radiation is emitted into the angular interval γ^{-1}, is determined by the mean-square scattering angle with logarithmic precision:

$$1/\gamma^2 = (E_s/E)^2 x_B/L_{rad} \quad , \tag{5.70}$$

where L_{rad} is the radiation length, and $E_s = 21$ MeV. Then we obtain from (5.70):

$$x_B = L_{rad}/1700 \quad . \tag{5.71}$$

The values of x_B are 56 μm for Si, 1.7 μm for W, and 73 μm for diamond.

The energy emitted into the angular interval γ^{-1} is given by

$$\Delta E = \frac{dE}{dx} \Delta x = \frac{E}{R} x_B = \frac{E}{1700} \quad . \tag{5.72}$$

Approximately, one can also write

$$\Delta E = \left(\frac{dW}{d\theta}\right)_B \Delta\theta = \frac{E}{1700} \quad \text{for} \quad \Delta\theta = \pi/\gamma^2 \quad . \tag{5.73}$$

One can deal similarly with the case of bremsstrahlung in an aligned crystal. However, in this case, information about the energy loss by spontaneous radiation is needed. The energy dependence of the energy loss for spontaneous radiation has the form $dE/dx \sim E^2$. Calculations of this quantity for different channels of a crystal and different intervals of energy are reported in [5.52,53]. The characteristic length over which the detected radiation originates in an oriented crystal is the dechanneling length of the particle, x_{ch}.

The emission of the radiated energy into the angular interval γ^{-1} in the fundamental harmonic, i.e, in a narrow spectral range of frequencies, takes place up to some optimal energy determined by the radiation having dipole character. For planar channeling, this energy is close to 5-10 GeV, and for axial channeling, close to 1 GeV for light crystals such as diamond or silicon.

We introduce the ratio of the radiation angular distributions,

$$\eta_1 = \frac{(dW/d\theta)_{ch}}{(dW/d\theta)_B} \quad . \tag{5.74}$$

Since the angular intervals in which bremsstrahlung and spontaneous radiation are concentrated are equal, one has

$$\eta_1 \cong \frac{(dE/dx)_{ch}}{E/L_{rad}} \frac{x_{ch}}{x_B} \quad , \tag{5.75}$$

where $x_{ch} = \int \chi(x)dx$, and $\chi(x)$ is the fraction of particles which have remained in the channeling regime at the depth x. For example, the dechanneling lengths for positrons in (110) channels of silicon are equal to 500 μm (E = 1 GeV), or 5000 μm (E = 10 GeV), and for diamond, respectively, 300 μm and 3000 μm. Similar values are also characteristic for planar channeling in tungsten crystals (for positrons $x_{ch} \cong x_{1/2}$, where $x_{1/2}$ is the depth at which half the particles originally trapped into channels dechannel). For axial channeling of positrons, the dechanneling lengths increase by approximately one order of magnitude.

120

For axially channeled electrons, $x_{1/2} \cong 30$ µm at $E = 1$ GeV for tungsten. Similar values are obtained for Si. However, here it is characteristic that at large depths (i.e., at a depth $> x_{1/2}$) the rate of dechanneling sharply decreases. Therefore, in an experiment it makes sense to use crystals with a thickness substantially larger than $x_{1/2}$, i.e., 200-300 µm at $E = 1$ GeV and 1 mm at $E = 10$ GeV.

It follows from (5.75) that, already at $E = 1$ GeV, $\eta_1 \cong 10$ for axial channeling.

We now define the ratio of spectral-angular distributions of the radiations:

$$\eta_2 = \frac{(d^2 W/d\theta d\omega)_{ch}}{(d^2 W/d\theta d\omega)_B} \quad . \tag{5.76}$$

Let us find the thickness at which η_2 has its maximum value. One can write, approximately,

$$\eta_2 = K_1 K_2 \quad , \tag{5.77}$$

where $K_1 = (dW/d\omega)_{ch}/(dW/d\omega)_B$ is the ratio of the spectral densities, obtained in several experiments [5.8-16] and $K_2 = x_{ch}/x_B$.

The dechanneling length x_{ch} is $\sim E$. Consequently, if K_1 does not change, or only changes slightly, then η_2 increases with energy. As obtained experimentally, $K_1 \cong (60-70)$ at $1 \leqslant E \leqslant 15$ GeV for positrons in diamond [5.7-9,54-60] and silicon [5.10,15,16,61,62]. Therefore $\eta_2(E = 1 \text{ GeV}) \cong 300-400$; η_2 ($E = 10$ GeV) $\cong 3-4 \times 10^3$. Estimates of this quantity for different cases are given in Table 5.1.

Table 5.1. Estimates of the gain coefficient (η) for various particles and crystals

Particles E [GeV]	Plane Crystal, thickness, η_2	Axis Crystal[a], thickness, η_2
1	2	3
e^+, 1	Diamond, Si, W 500-1000 µm η_2 (diamond) $\cong 3 \times 10^2$ η_2 (Si) $\cong 5 \times 10^2$ η_2 (W) $\cong 1 \times 10^4$	Diamond, Si, W 5 mm-1 cm η_2 (diamond) $= 3 \times 10^3$ η_2 (Si) $\cong 5 \times 10^3$ η_2 (W) $\cong 1 \times 10^3$
e^+, 10	Si, W 5 mm-1 cm η_2(Si) $\cong 5 \times 10^3$ η_2(W) $\cong 1 \times 10^4$	Si, W 5-10 cm η_2(Si) $\cong 5 \times 10^3$ η_2(W) $\cong 2 \times 10^3$
e^-, 1	Si 25 µm η_2(Si) $\cong 1 \times 10^1$	W 200-300 µm $\eta_2 \cong 3 \times 10^2$
e^-, 10	Si 250 µm η_2(Si) $\cong 1 \times 10^2$	W 1 mm $\eta_2 \cong 5 \times 10^2$

[a]At $\gamma\psi_c \gg 1$ ($E \gg m_e^2 c^4/U_m$) it is necessary to consider that CR is emitted into the angle ψ_c. This is important at the energies discussed for tungsten.

An analysis of the data indicates that the radiation peculiarities of channeling show up most strikingly in the spectral-angular distribution. In order to obtain maximum values of the ratio η_2 one should choose: (a) the maximum particle energy at which the dipole condition is still valid; (b) a crystal thickness of the order of the dechanneling length of the particles; (c) the dimensions of the detector and its distance from the crystal as determined by the value of γ^{-1}. The values of the spectral-angular distributions η_2 indicate the possibility of constructing a powerful source of directed radiation. As an estimate, one can show that the number of photons emitted into the angle γ^{-1} is given by the simple formula

$$N_{ch}^{ph} \cong \frac{1}{137} \frac{x_{ch}}{\lambda_{ch}} \kappa \quad , \tag{5.78}$$

where $\kappa = \gamma^2 \psi_c^2$ (at $E = 1\text{-}10$ GeV, $\kappa \cong 1$), with ψ_c the Lindhard critical angle, $\lambda_{ch} \cong d_p/\psi_c$, and d_p the planar channel width. For bremsstrahlung photons,

$$N_B = \frac{x_B}{L_{rad}} \ln \frac{\omega_2}{\omega_1} \quad , \tag{5.79}$$

[if ω_2 and ω_1 do not differ by more than a factor of 10, then $\ln(\omega_2/\omega_1) \cong 1$]. Hence it is seen that at high energies when $x_{ch} \gg \lambda_{ch}$, one channeled particle can emit more than one quantum into an angular interval of the order of magnitude of several γ^{-1}.

An investigation of the spectral-angular distribution of the radiation of channeled particles in thick crystals is difficult since the usual electronic detectors are incapable of selecting a few photons emitted by the same particle. It is possible, however, to use a suitable resonance reaction for this purpose, for example (γ, n). Photonuclear reactions have widths of several MeV. At the same time, in channeling of GeV-particles, quanta of up to several tens of MeV are usually emitted. Therefore the reaction yield can be written in the form

$$\gamma \sim \int \frac{d^2 W}{d\theta d\omega} \sigma(\omega) d\omega \quad , \tag{5.80}$$

where σ is the reaction cross section. Since the reaction linewidth is small, the ratio of the yields for aligned and nonaligned crystals will essentially furnish the value of η_2. The same method can be used to determine the dechanneling lengths; for this, it is necessary to use several crystals of different thicknesses.

In the foregoing, a comparison of channeling radiation (CR) with ordinary bremsstrahlung (BS) has been carried out. Let us now proceed to a comparison with coherent bremsstrahlung (CB).

A particle radiating CB is scattered as an ordinary particle, i.e., it does not suffer a suppression of scattering; therefore, the factor x_{ch}/x_B, which gives rise to a sharp increase of the spectral angular density of CR, is absent in CB.

The radiation from channeled particles is concentrated in a narrower spectral region, its density in this region is substantially higher than for CB, and due to the suppression of multiple scattering, CR is more sharply directed than CB; further, the total fraction of energy emitted in CR into a given interval of frequencies and angles is substantially larger than in CB.

These differences are due to the different mechanisms responsible for channeling radiation and coherent bremsstrahlung. CR is due to deflection of particles by the field of the atomic rows and planes, whereas CB is due to periodic collisions of particles with the atomic rows and planes, with a negligible deviation of their trajectory from a straight line. Therefore, they have different characteristic frequencies and are usually observed at different crystal orientations. CB in the pure form is observed at angles of incidence substantially exceeding the critical angles of channeling. However, for beam motion along an atomic plane at a small angle to an axis it is possible to observe both CR and CB effects in different frequency regions [5.52,53]. In this case, combination frequencies due to the effects of channeling will appear in the CB spectrum [5.52,53,62].

5.6 Concluding Remarks

Theoretical and experimental investigations have shown that channeling radiation has many interesting properties which distinguish it from other types of radiation.
Its main distinguishing properties are:

1) In contrast to bremsstrahlung and Čerenkov radiation, it depends strongly on the sign of the charge.
2) Unlike Čerenkov radiation, it depends not only on velocity but also on the energy of the particles.
3) As distinct from bremsstrahlung, at ultra-relativistic energies its spectral density depends not only on the energy but also on the mass of the particles.
4) In contrast to ordinary bremsstrahlung, it has well-separated characteristic frequencies (peaks) which depend on the characteristics of the particle and crystal.
5) The location of the characteristic frequencies is continuously adjustable by a change of the direction of bombardment of the crystal and/or the particle energies, as well as of the angle of observation.
6) The radiation possesses a high degree of linear polarization for planar channeling. The direction of this polarization changes smoothly as the crystal is rotated.
7) In the hard X-ray and γ-ray regions, it exceeds other types of radiation by almost two orders of magnitude in spectral intensity, and by three orders in luminosity.

These unique properties of channeling radiation lead one to expect that it will find various applications:

Its applicability to investigating the channeling phenomenon itself and interactions of electrons and positrons with monocrystals is quite obvious. An elastic (coherent) interaction is determined by the interaction potential of a particle with atoms in a crystal. A detailed determination of this potential on the basis of data on photon energies and their intensity is already in progress. A study of inelastic (incoherent) interactions resulting in dechanneling allows us to determine details of the physics of small-angle scattering of particles, and of crystal imperfections. Work on this subject is being carried out at present.

Besides the atomic potentials, it is possible to investigate other properties of monocrystals, for example, the spatial anisotropy of thermal vibrations.

Channeling radiation is a source of polarized X-ray radiation with adjustable frequency which can be widely utilized for X-ray structural investigations of solids and organic compounds, and for applications in medicine and lithography.

It also seems natural to apply it to phononuclear reactions [5.63] in the study of the giant resonance structure and the excitation of Mössbauer levels, and to use it in the construction of pulsed sources of polarized neutrons [5.64]. An increase in the neutron yield as compared with the yield obtained by using ordinary bremsstrahlung has already been demonstrated experimentally [5.65].

Finally, the use of this radiation to detect ultra-high-energy particles and to calibrate photometric devices [5.66] appears promising.

Acknowledgements. This article was expertly translated from the original Russian by Dr. Basil Stoyanov.

References

5.1 D.S. Gemmel: Rev. Mod. Phys. **46**, 129 (1974)
5.2 J.L. Lindhard: K. Dan. Vidensk. Selsk. Mat. Fys. Medd. **34**, No. 14 (1965)
5.3 M.A. Kumakhov: Phys. Lett. A**57**, 17 (1976)
5.4 M.A. Kumakhov: Dokl. Akad. Nauk. SSSR **230**, 1077 (1976) [English transl.: Sov. Phys.-Dokl. **21**, 581 (1976)]
5.5 M.A. Kumakhov: Zh. Eksp. Teor. Fiz. **72**, 1489 (1977) [English transl.: Sov. Phys.-JETP **45**, 781 (1977)]
5.6 M.A. Kumakhov: Phys. Status Solidi (b) **84**, 41 (1977)
5.7 I.I. Miroshnichenko, J.J. Murray, R.O. Avakyan, T.Kh. Figut: Pis'ma Zh. Eksp. Teor. Fiz. 29, 786 (1979) [English transl.: Sov. Phys.-JETP Lett. **29**, 722 (1979)]
5.8 A.O. Aganyants, N.Z. Akopov, Yu.A. Vartanov, G.A. Vartapetyan: Preprint No. EFI-312 (37)-78, Erevan Phys. Inst., Erevan (1978)
5.9 B.N. Kalinin, V.V. Kaplin, A.P. Potilitsin, S.A. Vorobiev: Phys. Lett. A**70**, 447 (1979)
5.10 V.B. Ganenko, L.E. Gendenshtein, I.I. Miroshnichenko, V.L. Morokhovsky, E.V. Pegushin, V.M. Sanin, S.V. Shalatsky: Pis'ma Zh. Eksp. Teor. Fiz. 32, 397 (1980) [English transl.: Sov. Phys.-JETP Lett. **32**, 373 (1980)]

5.11 M.J. Alguard, R.L. Swent, R.H. Pantell, B.L. Berman, S.D. Bloom, S. Datz: Phys. Rev. Lett. **42**, 1148 (1979)
5.12 M. Gouanère, S. Sillou, M. Spighel, N. Cue, M.J. Gaillard, R.G. Kirsch, J.-C. Poizat, J. Remilleux, B.L. Berman, P. Catillon, L. Roussel, G.M. Temmer: Preprint No. LAPP-Exp-05 (1981)
5.13 J.U. Andersen, E. Laegsgaard: Phys. Rev. Lett. **44**, 1079 (1980)
5.14 N. Cue, E. Bonderup, B.B. Marsh, H. Bakhru, R.E. Benenson, R. Haight, K. Inglis, G.O. Williams: Phys. Lett. A**80**, 26 (1980)
5.15 N.A. Filatova, V.M. Golovatyuk, A.N. Isakov, I.M. Ivanchenko, R.B. Kadyrov, N.N. Karpenko, T.S. Nigmanov, V.V. Palchik, V.D. Riabtsov, M.D. Shafranov, E.N. Tsyganov, I.A. Tyapkin, D.V. Uralski, A. Forycki, Z. Guzik, J. Wojtkowska, Yu.S. Khodirev, V.A. Maisheev, R. Carrigan, T. Toohig, C. Carmack, W. Gibson, C.R. Sun, M.D. Bavizhev, N.K. Bulgakov, N.I. Zimin, I.A. Grishaev, G.D. Kovalenko, B.I. Shramenko, E.I. Denisov, V.I. Glebov, A.S. Khlebnikov, M.A. Kumakhov, V.V. Avdeichikov: Fermilab Report No. 81/34-Exp. 7850.507-11 (1981) (unpublished)
5.16 M. Atkinson, J.F. Bak, P.J. Bussey, P. Christensen, J.A. Ellison, R.J. Ellison, K.R. Eriksen, D. Gidings, R.E. Hughes-Jones, B.B. Marsh, D. Mercer, F.E. Meyer, S.P. Møller, D. Newton, P. Pavlopoulos, P.H. Sharp, R. Stensgaard, M. Suffert, E. Uggerhøj: Phys. Lett. B**110**, 162 (1982)
5.17 *Proc. First National Conf. on Rad. Charg. Part. in Crystals*, North Caucasus, USSR, May 21-23, 1980: Radiat. Eff. **56**, 2 (1981)
5.18 R. Wedell: Phys. Status Solidi (b) **99**, 11 (1980)
5.19 V.A. Bazylev, N.K. Zhevago: Radiat. Eff. **54**, 41 (1981)
5.20 G.M. Temmer: Chin. J. Nucl. Phys. **2**, 353 (1980) [English transl.: Chin. Phys. **1**, 1024 (1981)]
5.21 L.T. Chadderton: Radiat. Eff. **27**, 13 (1975)
5.22 P. Lervig, J. Lindhard, V. Nielsen: Nucl. Phys. A**96**, 481 (1967)
5.23 J.U. Andersen, S.K. Andersen, W.M. Augustyniak: K. Dan. Vidensk. Selsk. Mat. Fys. Medd. **39**, No. 10 (1977)
5.24 H.J. Kreiner, F. Bell, R. Sizmann, D. Harder, W. Huttl: Phys. Lett. A**33**, 135 (1970)
5.25 R.L. Swent, R.H. Pantell, M.J. Alguard, B.L. Berman, S.D. Bloom, S. Datz: Phys. Rev. Lett. **43**, 1723 (1979)
5.26 V.V. Beloshitsky, M.A. Kumakhov: Zh. Eksp. Teor. Fiz. **62**, 1144 (1972) [English transl.: Sov. Phys.-JETP **35**, 605 (1972)]
5.27 V.V. Beloshitsky, Ch.G. Trikalinos: Radiat. Eff. **56**, 71 (1981)
5.28 V.V. Beloshitsky, M.A. Kumakhov: Zh. Eksp. Teor. Fiz. **82**, 462 (1982) [English transl.: Sov. Phys.-JETP **55**, 265 (1982)]
5.29 R.M. Sternheimer: Phys. Rev. **145**, 247 (1966), Eq. (15); see also E.M. Lifshitz, L.P. Pitaevskii: *Physical Kinetics* (Pergamon, New York 1981)
5.30 M.K. Kitagawa, Y.H. Ohtsuki: Phys. Rev. B**8**, 3117 (1973)
5.31 M.A. Kumakhov, Ch. Trikalinos: Phys. Status Solidi (b) **99**, 449 (1980)
5.32 M.A. Kumakhov, Ch. Trikalinos: Zh. Eksp. Teor. Fiz. **78**, 1623 (1980) [English transl.: Sov. Phys.-JETP **51**, 815 (1980)]
5.33 V.A. Bazylev, V.V. Beloshitsky, V.I. Glebov, N.K. Zhevago, M.A. Kumakhov, Ch.G. Trikalinos: Dokl. Akad. Nauk SSSR **253**, 1100 (1980) [English transl.: Sov. Phys.-Dokl. **25**, 624 (1980)]
5.34 V.A. Bazylev, V.V. Beloshitsky, V.I. Glebov, N.K. Zhevago, M.A. Kumakhov, Ch. G. Trikalinos: Zh. Eksp. Teor. Fiz. **80**, 608 (1981) [English transl.: Sov. Phys.-JETP **53**, 306 (1981)]
5.35 V.A. Bazylev, V.V. Beloshitsky, V.I. Glebov, M.A. Kumakhov, Ch. G. Trikalinos, N.K. Zhevago: Radiat. Eff. **56**, 87 (1981)
5.36 A.I. Akhiezer, V.F. Boldyshev, N.F. Shul'ga: Fiz. Elem. Chastits At. Yadra **10**, 51 (1979) [English transl.: Sov. J. Part. Nucl. **10**, 19 (1979)]
5.37 A.I. Akhiezer, I.A. Akhiezer, N.F. Shul'ga: Zh. Eksp. Teor. Fiz. **76**, 1244 (1979) [English transl.: Sov. Phys.-JETP **49**, 631 (1979)]
5.38 V.N. Baier, V.M. Katkov, V.M. Strakhovenko: Zh. Eksp. Teor. Fiz. **80**, 1343 (1981) [English transl.: Sov. Phys.-JETP **53**, 688 (1981)]
5.39 L.D. Landau, E.M. Lifshitz: *The Classical Theory of Fields* (Addison-Wesley, Reading, Massachusetts 1951)

5.40 J.D. Jackson: *Classical Electrodynamics* (Wiley, New York 1962)
5.41 H. Motz: J. Appl. Phys. **22**, 527 (1951)
5.42 D.F. Alferov, Yu.A. Bashmakov, E.G. Bessonov: Trudy Fiz. Inst. Akad. Nauk SSSR **80**, 100 (1975) [English transl.: Proc. P.N. Lebedev Phys. Inst. **80** (1976)]
5.43 V.V. Beloshitsky, Yu.A. Bykovsky, M.Kh. Kumekhov: Radiat. Eff. Lett. **76**, 93 (1983)
5.44 V.A. Bazylev, V.I. Glebov, N.K. Zhevago: Zh. Eksp. Teor. Fiz. **78**, 62 (1980) [English transl.: Sov. Phys.-JETP **51**, 31 (1980)]
5.45 V.N. Baier, V.M. Katkov, V.M. Strakhovenko: Dokl. Akad. Nauk SSSR **246**, 1347 (1979) [English transl.: Sov. Phys.-Dokl. **24**, 469 (1979)]
5.46 S. Kheifets, T. Knight: J. Appl. Phys. **50**, 5937 (1979)
5.47 M.A. Khokonov: Radiat. Eff. (to be published)
5.48 A.L. Avakian, C. Yang, N.K. Zhevago: Radiat. Eff. **56**, 39 (1981)
5.49 A.L. Avakian, N.K. Zhevago, C. Yang: Zh. Eksp. Teor. Fiz. **82**, 573 (1982) [English transl.: Sov. Phys.-JETP **55**, 573 (1982)]
5.50 N.F. Shul'ga: Pis'ma Zh. Eksp. Teor. Fiz. **32**, 179 (1980) [English transl.: Sov. Phys.-JETP Lett. **32**, 166 (1980)]
5.51 M.A. Kumakhov: Usp. Fiz. Nauk **127**, 531 (1979) [English transl.: Sov. Phys.-Usp. **22**, 193 (1979)]
5.52 V.V. Beloshitsky, M.A. Kumakhov: Dokl. Akad. Nauk. SSSR **251**, 331 (1980) [English transl.: Sov. Phys.-Dokl. **25**, 196 (1980)]
5.53 V.V. Beloshitsky, M.A. Kumakhov: Radiat. Eff. **56**, 25 (1981)
5.54 N.F. Shul'ga, L.E. Gendenshtein, I.I. Miroshnichenko, E.V. Pegushin, S.P. Fomin, R.O. Avakyan: Zh. Eksp. Teor. Fiz. **82**, 50 (1982) [English transl. Sov. Phys.-JETP **55**, 30 (1982)]
5.55 A.O. Aganyants, Yu.A. Vartanov, G.A. Vartapetyan, M.A. Kumakhov, Ch. Trikalinos, V.Ya. Yaralov: Pis'ma Zh. Eksp. Teor. Fiz. **29**, 554 (1979) [English transl.: Sov. Phys.-JETP Lett. **29**, 505 (1979)]
5.56 A.O. Aganyants, G.O. Marukyan, Yu.A. Vartanov, G.A. Vartapetyan: Preprint, Erevan Phys. Inst., Erevan (1981)
5.57 A.O. Aganyants, R.O. Avakyan, S.P. Troyan, Yu.A. Vartanov, G.A. Vartapetyan: Preprint, Erevan Phys. Inst., Erevan (1981)
5.58 S.A. Vorob'ev, A.N. Didenko, V.N. Zabaev, B.N. Kalinin, V.A. Kaplin, A.A. Kurkov, A.P. Potylitsyn, V.M. Tomachakov: Pis'ma Zh. Eksp. Teor. Fiz. **32**, 261 (1980) [English transl.: Sov. Phys.-JETP Lett. **32**, 241 (1980)]
5.59 Yu.N. Adishchev, P.S. Ananyin, A.N. Didenko, B.N. Kalinin, V.A. Kaplin, E.I. Rozum, A.P. Potylitsin, S.A. Vorobiev, V.N. Zabaev: Phys. Lett. **A77**, 263 (1980)
5.60 Yu.A. Adishchev, P.S. Anan'in, S.A. Vorobiev, V.N. Zabaev, B.N. Kalinin, V.V. Kaplin, A.P. Potylitsyn, V.K. Tomchakov, E.I. Rozum: Pis'ma Zh. Eksp. Teor. Fiz. **30**, 430 (1979) [English transl.: Sov. Phys.-JETP Lett. **30**, 402 (1979)]
5.61 N.A. Filatova, V.M. Golovatyuk, A.N. Isakov, I.M. Ivanchenko, R.B. Kadyrov, N.N. Karpenko, T.S. Nigmanov, V.V. Palchik, V.D. Riabtsov, M.D. Shafranov, E.N. Tsyganov, I.A. Tyapkin, D.V. Uralski, A. Forycki, Z. Guzik, J. Woitkowska, R.A. Carrigan, Jr., T.E. Toohig, C. Carmack, W.M. Gibson, I.J. Kim, C.-R. Sun, M.D. Bavizhev, N.K. Bulgakov, N.I. Zimin, I.A. Grishaev, G.D. Kovalenko, B.I. Shramenko, E.I. Denisov, V.I. Glebov, V.V. Avdeichiko: Phys. Rev. Lett. **48**, 488 (1982)
5.62 J. Bak, F. Komarov, P.H. Meyer, J. Pedersen, R. Stensgaard, A. Sørensen, E. Uggerhøj, K. Østergard, S.P. Møller, F. Grob, P. Siffert, M. Suffert: Proposal No. CERN/PSCC/81-79 PSCC/P51 (1981)
5.63 A.P. Antipenko, I.A. Grishaev, V.I. Kasilov, N.I. Lapin, S.F. Shcherbak: Yad. Fiz. **33**, 310 (1981) English transl.: Sov. J. Nucl. Phys. **33**, 163 (1981)
5.64 I.P. Eremeev, M.A. Kumakhov: Phys. Lett. **A72**, 359 (1979)
5.65 A.P. Antipenko, I.A. Grishaev, V.I. Kasilov, N.I. Lapin, V.L. Morokhovsky, S.F. Shcherbak: Sov. J. Probl. Atom. Sci. Tech. Nucl. Constants **1**, 25 (1981)
5.66 M.A. Kumakhov: Pis'ma Zh. Tekh. Fiz. **3**, 1025 (1977) [English transl.: Sov. Phys.-Tech. Phys. Lett. **3**, 420 (1977)]

6. Channeling Radiation – Quantum Theory

J.U.Andersen, E.Bonderup, and E.Laegsgaard

With 16 Figures

A general introduction to channeling and channeling radiation is followed by a derivation of a wave equation for the transverse motion of a channeled particle. Owing to the predominance of forward scattering, the depth of penetration can be treated as a time variable in this equation, and it assumes the form of a time-dependent Schrödinger equation. It is obtained from the Klein-Gordon equation, including interaction with the radiation field. The neglect of spin simplifies the formulation considerably, and it is argued to be well justified in the description of channeling radiation. The treatment also allows a natural inclusion of interaction with the crystal degrees of freedom, which leads to instability of channeling states and line broadening of the radiation. Both thermal and electronic contributions to scattering are discussed in detail, and their influence on the frequency distribution and intensity of channeling radiation is illustrated by comparisons with experiment. This scattering is important for the potential applications considered in the last section.

6.1 Channeling and Channeling Radiation

When charged particles penetrate matter, they are deflected by atoms, and this gives rise to the emission of bremsstrahlung. In a crystal, the spectrum is modified, and we may distinguish between normal or incoherent bremsstrahlung, coherent bremsstrahlung, and channeling radiation. Classically, this distinction is related to the correlation of the deflections in time and space. Normal bremsstrahlung originates in uncorrelated deflections by atoms. Even when the projectiles are moving at a small angle to a set of crystal planes or to an axis, the deflections in close collisions with atoms are uncorrelated, owing to the thermal fluctuations in atomic positions. However, the deflections in more distant collisions are highly correlated, as illustrated in Fig.6.1. They may be described as an interaction with a one- or two-dimensional continuum potential, obtained as an average of the crystal potential over the coordinates parallel to the plane or axis, respectively [6.1,2]. The associated radiation is denoted coherent bremsstrahlung when the projectile motion is only slightly affected by the potential, i.e., when the

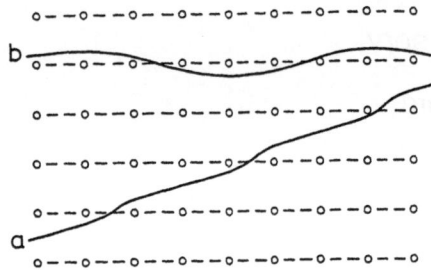

Fig.6.1. Qualitative illustration of classical electron motion, governed in effect by
a continuum planar potential. The angles of
the paths to the crystal planes are greatly
exaggerated

trajectories are nearly straight lines. The characteristic photon frequencies are
then determined by the periodicity of the perturbing potential (trajectory a in
Fig.6.1). For very small angles of incidence to a plane or axis, however, the projectile motion is governed by the correlated deflections or, in effect, by the continuum planar or axial potential (trajectory b in Fig.6.1). The particle is channeled, and the radiation associated with the transverse motion perpendicular to a
plane or an axis is called "channeling radiation" [6.3]. The intensity of ordinary
bremsstrahlung is modified due to the redistribution of particle flux; normally it
is reduced for channeled positrons and increased for channeled electrons.

Usually, quantum perturbation theory is applied to describe bremsstrahlung from
relativistic electrons and positrons [6.4,5]. Within this framework, coherent bremsstrahlung is distinguished by a quantized momentum transfer to the lattice as a
whole, equal to ℏ times a reciprocal-lattice vector, and the intensity is proportional to the square of the corresponding Fourier component of the thermally averaged crystal potential. In this picture, the normal bremsstrahlung is associated
with scattering by the remaining fluctuating part of the potential, and its intensity is somewhat reduced compared to that obtained from an amorphous target [6.5].
In a quantal picture, channeling radiation is associated with jumps between quantum states of transverse motion in the planar or axial continuum potential. The
states are characterized by a well-defined transverse energy, and the transitions
are analogous to spontaneous radiative decay of excited atomic states.

As an introduction, we consider the simple example of a positron moving in a
nearly harmonic potential between adjacent atomic planes [6.6]. For energies below
the GeV region, the transverse motion is non-relativistic in the "rest system" R,
where the translatory longitudinal motion parallel to the planes has been transformed away, and this leads to a particularly simple description. If the classical
oscillation frequency of the projectile is ω_0 in the laboratory, it is equal to
$\gamma_z \omega_0$ in the rest system, due to time dilation. The factor γ_z is given by
$\gamma_z = (1 - \beta_z^2)^{-\frac{1}{2}}$, where $\beta_z c$ is the longitudinal velocity of the positron. The assumption that the transverse motion in R is non-relativistic implies that γ_z can be replaced by $\gamma = E/mc^2$, where E is the total energy and m the rest mass of the positron.

In a classical description, the positron radiates in the rest frame at the frequency $\gamma\omega_0$, with an intensity distribution characteristic of a dipole. A Lorentz transformation to the laboratory frame gives the observed photon frequencies. Also within a quantal picture, the description is particularly simple in the rest system. The energy of the emitted photon is equal to the spacing $\Delta E_\perp^R$ of energy levels in the oscillator, and the photon frequencies in the laboratory are obtained from the Doppler relation

$$\hbar\omega = \frac{\Delta E_\perp^R}{\gamma(1 - \beta\cos\vartheta)} \simeq \frac{2\gamma\Delta E_\perp^R}{1 + \gamma^2\vartheta^2} \quad , \quad \gamma \gg 1 \quad , \tag{6.1}$$

where ϑ is the angle in the laboratory between the z direction and the direction of the photon.

From correspondence arguments, we expect to have the relation $\Delta E_\perp^R = \gamma\Delta E_\perp = \gamma\hbar\omega_0$, and this is indeed easily seen to hold: The transverse force and hence also the continuum potential governing the motion is increased by a factor of γ in R since the (transverse) momentum transfer in a collision with an atom is invariant under the transformation, while the rate of collisions is higher by a factor of γ due to Lorentz contraction of the distance between atoms. Also, the kinetic energy is increased by a factor of γ in R since the transverse momentum is invariant under the transformation and since the effective mass connecting transverse momentum and energy is the relativistic mass γm in the laboratory.

The characteristic frequency ω_0 is proportional to the square root of the force constant divided by the mass and hence scales as $\gamma^{-\frac{1}{2}}$. This leads to a scaling with $\gamma^{3/2}$ of the maximum frequency of the distribution (6.1). The Lorentz transformations involved are illustrated in Fig.6.2. As a result of these relativistic effects, transitions within a transverse potential of depth of the order of a Rydberg lead to photon energies in the keV region for MeV projectiles.

From this discussion, we may also draw conclusions concerning the applicability of a classical description of the radiation process. For a harmonic potential, the radiation energies are the same in classical and quantal descriptions, but in general, the quantal picture predicts discrete, separated lines in the rest system, while the classical electromagnetic frequency spectrum is continuous. A transition to the classical limit occurs as the number of energy levels increases. For a harmonic potential, the number of bound channeling states is proportional to $\gamma^{\frac{1}{2}}$: In the laboratory, the potential is independent of γ, and the spacing of levels decreases as $\gamma^{-\frac{1}{2}}$ owing to the relativistic increase of the effective mass, while in the rest system the potential scales as γ and the spacing as $\gamma^{\frac{1}{2}}$.

In general, the number of bound states ν may be estimated semiclassically from the magnitude of the available phase space [6.7]; for planar channeling one obtains, for electrons and positrons, respectively,

Motion and Potential:

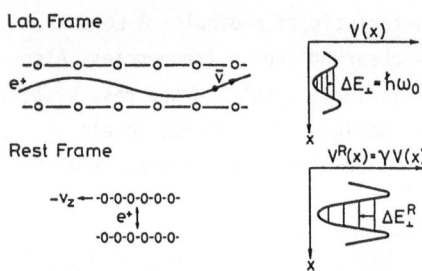

Lab. Frame

Rest Frame

Fig.6.2. Illustrations of Lorentz transformations between the laboratory and the
rest frame for a radiating positron with
a kinetic energy of 1 MeV, corresponding
to $\gamma = 3$

Radiation Intensity and Frequency:

Rest Frame

$$\omega^R = \Delta E_\perp^R / \hbar = \gamma \omega_0$$

$$\omega = \frac{\omega^R}{\gamma(1-\beta\cos\vartheta)}$$

Lab. Frame

$$\simeq \frac{2\gamma^2\omega_0}{1+\gamma^2\vartheta^2}$$

$$\nu_p^- \simeq \gamma^{\frac{1}{2}}\left(\frac{4a_0}{d_p}\right)(Nd_p^3)^{\frac{1}{2}} \quad , \tag{6.2}$$

$$\nu_p^+ \simeq \gamma^{\frac{1}{2}}Z^{1/3}(Nd_p^3)^{\frac{1}{2}} \quad , \tag{6.3}$$

where a_0 is the Bohr radius, $a_0 \simeq 0.53$ Å, N is the density of atoms with atomic number Z, and d_p is the planar spacing. For beam energies in the low MeV region corresponding to $\gamma \lesssim 10^2$, the number of quantum states is fairly small in the planar
case, being somewhat higher for positrons than for electrons, i.e.

$$\nu_p^+ / \nu_p^- \simeq Z^{1/3} \simeq 2\text{-}4 \quad .$$

For channeling of electrons along an axis with atomic spacing d, this type of
estimate gives for the number of bound states:

$$\nu_s^- \simeq \gamma\left(\frac{4a_0}{d}\right)Z^{1/3} \quad , \tag{6.4}$$

which increases more rapidly with γ. Axially channeled positrons are not localized
in a potential minimum, but the number of states per unit cell with transverse energy below the barrier for penetration into strings is large, even for small γ. Thus,
a quantal treatment of channeling is necessary only for planar channeling at beam
energies $\lesssim 100$ MeV and for axial electron channeling at energies $\lesssim 10$ MeV.

130

6.2 Transverse Wave Equation

Most detailed quantum treatments of channeling radiation are based on the Dirac equation, which describes the behavior of spin-½ particles [6.8-12]. However, the influence of spin is insignificant with respect to both the transverse energy levels and the probabilities for radiative transitions between these, and one may therefore apply the much simpler Klein-Gordon equation, valid for spinless particles, to introduce the basic separation into longitudinal and transverse motion [6.7,13,14]. The order of magnitude of the neglected spin terms may easily be estimated in the rest frame, where the transverse motion is non-relativistic. The spin magnetic moment $\boldsymbol{\mu}$ couples to a position-dependent transverse magnetic field $\mathbf{B}$, which, in the limit $\gamma \gg 1$, has the same strength as the transverse electric field. The relative magnitude of the resulting change $-\boldsymbol{\mu} \cdot \mathbf{B}$ in the Hamiltonian is of the order of the fine-structure constant α, but since the average value of $\mathbf{B}$ is zero, the perturbation does not lead to a first-order shift in transverse energy. Also, the ratio of the intensities of magnetic and electric dipole transitions is very low, typically of the order α^2.

The effect of spin on the radiative transitions between transverse states can, of course, also be estimated directly in the laboratory frame. To isolate the spin terms, one may first transform the time-independent Dirac equation into an equation for the square of the energy [6.15]. The usual expression for the last spinor components in terms of the first two [6.9] is then inserted, and the result is the Klein-Gordon equation with additional spin terms. These are easily seen to be very small, but a detailed comparison with the rest-frame estimates of relative changes in photon energy and intensity is somewhat complicated by the contribution from the Thomas precession in the laboratory [6.16].

The time-independent Klein-Gordon equation expresses the relativistic relation between momentum and energy for a particle of charge e and rest mass m:

$$\left\{c^2\left[-i\hbar\boldsymbol{\nabla} - \frac{e}{c}\mathbf{A}(\mathbf{R})\right]^2 + m^2c^4\right\}\psi(\mathbf{R},\ldots)$$

$$= [E - H_r - H_\ell - V(\mathbf{R},\ldots)]^2 \psi(\mathbf{R},\ldots) \quad . \tag{6.5}$$

Here, E is the energy of the entire system, and H_r and H_ℓ denote the Hamiltonians of the radiation field and the lattice, which interact with the projectile through the potentials $\mathbf{A}$ and V. Only the coordinates $\mathbf{R}$ of the particle are shown explicitly. In contrast to the spin, the coupling to lattice degrees of freedom is important since the lines of channeling radiation are broadened by thermal and electronic scattering, which limits the coherence length for the emission of radiation.

We separate from the wave function ψ a factor $\exp(iKz)$ corresponding to uniform motion in the channeling direction with energy E,

$$\psi(\mathbf{r},z,\ldots) = e^{iKz} w(\mathbf{r},z,\ldots) \quad , \quad \mathbf{r} = (x,y) \quad , \tag{6.6a}$$

with

$$E^2 = (\hbar K)^2 c^2 + m^2 c^4 \quad .\tag{6.6b}$$

In the equation for w, obtained by substituting (6.6) into (6.5), we neglect terms $\partial^2/\partial z^2$ and $(V + H_r + H_\ell)^2$, where the initial energy of the lattice and the radiation field is taken to be zero. This approximation is based on the assumption that the scattering is limited to small angles [6.13] and amounts to omission of terms of relative magnitude $(k_\perp/K)^2$, where $\hbar k_\perp$ is the magnitude of a typical transverse momentum. In addition, excitation energies for the lattice and the radiation field are assumed to be very small. In the Coulomb gauge, div$\mathbf{A} = 0$, we then obtain, after division by $2E = 2\gamma mc^2$,

$$i\hbar v \frac{\partial}{\partial z} w = \left[-\frac{\hbar^2}{2\gamma m} \nabla_{\mathbf{r}}^2 + V(\mathbf{r}, z, \ldots) + H_r + H_\ell + H_{e,r} \right] w \quad ,$$

$$H_{e,r} = -e\left(\beta A_z + \frac{1}{\gamma mc} \mathbf{A} \cdot \mathbf{p} \right) \quad ,\tag{6.7}$$

where $v = \beta c$ is given by the relation $\hbar K = \gamma mv$. As usual, contributions of order A^2 have been neglected in the term $H_{e,r}$. This equation has the form of a time-dependent Schrödinger equation.

The coupling of the particle to the degrees of freedom of the radiation field and of the lattice may be treated as a perturbation, and to zeroth order, we neglect the term $H_{e,r}$ and replace the potential by the thermally averaged lattice potential $V_T(\mathbf{r}, z)$, which is the expectation value of $V(\mathbf{r}, z, \ldots)$ in the ground state for the target electrons and in the lattice vibrational states weighted by Boltzmann factors. For a lattice potential approximated as a sum of potentials from atoms, the thermal average corresponds to a convolution of the atomic potential with the Gaussian distribution of the atomic displacements.

The final simplification is obtained when $V_T(\mathbf{R})$ is replaced by the continuum approximation, which, for an axis in the z direction, is given by

$$U_T(\mathbf{r}) = \frac{1}{d} \int_0^d dz \, V_T(\mathbf{r}, z) \quad .\tag{6.8}$$

The solutions of (6.7) are then product functions,

$$w = u(\mathbf{r}) | \text{lattice} \rangle | \text{radiation} \rangle \exp(-i\varepsilon z/\hbar v) \quad ,\tag{6.9}$$

where the lattice and radiation parts are eigenstates of the Hamiltonians H_ℓ and H_r. The energy ε is a sum of the eigenvalues of these two operators and the energy $E_\perp$ of the transverse motion of the projectile, which is described by a "stationary Schrödinger equation",

$$\left[-\frac{\hbar^2}{2\gamma m} \nabla_{\mathbf{r}}^2 + U_T(\mathbf{r}) \right] u(\mathbf{r}) = E_\perp u(\mathbf{r}) \quad .\tag{6.10}$$

In the planar case, the continuum potential V_T depends on one coordinate only,

$$V_T(x) = \frac{1}{A} \int_A dy\, dz\, V_T(x,y,z) \quad , \tag{6.11}$$

and the transverse wave function $u(\mathbf{r})$ separates into a product of a one-dimensional function, describing the motion perpendicular to the plane, and a plane wave in the y direction, parallel to the plane. The choice of z direction in the plane is arbitrary as long as the x and y momenta remain small (Sect.6.4). Usually, the z axis is chosen as the projection of the incident beam direction on the plane, but sometimes other choices are more convenient.

Both the electronic and thermal scattering and the difference between the continuum potential and the thermally averaged lattice potential with three-dimensional periodicity may usually be treated as perturbations. Their contribution to the linewidth of the channeling radiation is discussed later. At this point, we shall briefly consider the corresponding corrections to the transverse energy. They are of second order, and an indication of the magnitude of possible shifts of the transition frequencies is obtained from the result for a classical damped harmonic oscillator [6.17], $\delta\omega/\omega = -(5/8)(\Gamma/\omega)^2$, where Γ is the full width of the Lorentzian frequency distribution. Typically, one has $\Gamma/\omega \lesssim 10^{-1}$, and the shifts are therefore expected to be very small. Cursory estimates for thermal scattering, based on the sudden-collision expansion introduced in Sect.6.8, lead to shifts of the same order, but further work is needed on this problem before an accuracy at the 1% level can be considered as established.

The corrections due to the periodic variations of the lattice potential have been investigated in more detail. For the axial case, the accuracy of the continuum approximation was studied by *Lervig* et al. [6.13], and expressions for the lowest-order correction terms were derived. The correction decreases with increasing γ, and for electrons of a few MeV, it is already very small, $|\delta\omega/\omega| \lesssim 10^{-3}$.

Similarly, for planar channeling the variation of the potential in a direction almost parallel to the projection of the beam on the plane is not important. But when the beam becomes nearly parallel to a strong axis in the plane, the influence of 'nonsystematic reflections' increases, corresponding to reciprocal-lattice vectors perpendicular to the axis but not to the plane, and there is a gradual transition from one-dimensional planar states to two-dimensional axial states. As discussed in [6.18], the shifts of planar energy levels in this transition region can be evaluated numerically.

We also briefly mention the radiation induced by the perturbations. Thermal and electronic scattering causes a continuous spectrum of bremsstrahlung, and the periodic perturbations of the continuum potential lead to a coherent bremsstrahlung for channeled particles, which is at higher frequencies and much weaker than channeling radiation.

133

6.3 Emission of Radiation

We now consider radiative transitions between the eigenstates determined by (6.10). The transition probabilities may be obtained from a standard perturbation treatment of the coupling term $H_{e,r}$ in (6.7), and at first we neglect the lattice degrees of freedom. If, for $z = 0$, the transverse state of the particle is $u_i(\mathbf{r})$ and the radiation field is in its ground state $|0\rangle$, the amplitude for finding the particle at depth $z = L$ in a state $u_f(\mathbf{r})$ together with a photon of momentum $\hbar\kappa$ and energy $\hbar\omega$ is given by

$$\langle u_f, \kappa | w(z = L, \ldots)\rangle$$

$$\simeq \frac{1}{i\hbar v} \int_0^L dz \, \langle u_f(\mathbf{r}), \kappa | - e\left(\beta A_z + \frac{1}{\gamma mc} \mathbf{A} \cdot \mathbf{p}\right)|u_i(\mathbf{r}), 0\rangle$$

$$\times \exp[i(E_{\perp f} + \hbar\omega - E_{\perp i})z/\hbar v] \quad . \tag{6.12}$$

Consider, for simplicity, emission in the z direction. The photon polarization vector ϵ_A is then in the xy plane, and the relevant matrix element of the radiation field, quantized within a volume L_0^3, is given by [6.17]

$$\langle \kappa | \mathbf{A} | 0\rangle = \epsilon_A c\sqrt{2\pi/L_0^3} \, \sqrt{\hbar/\omega} \, e^{-i\kappa z} \quad . \tag{6.13}$$

The photon momentum and frequency are connected through the relation $\kappa = n_r \omega/c$, where n_r is the index of refraction, and for large L the integral in (6.12) approaches a delta function which determines the photon energy:

$$\hbar\omega = (E_{\perp i} - E_{\perp f})/(1 - n_r\beta) \simeq (1 + \beta)\gamma^2 \frac{E_{\perp i} - E_{\perp f}}{1 + 2\gamma^2\delta} \quad , \tag{6.14}$$

with $\delta \equiv 1 - n_r$. The correction for refraction is important only for low beam energies, since δ is proportional to ω^{-2} for not too low frequencies ω [6.19]. With $\delta = 0$, (6.14) corresponds to (6.1), which contains the generalization to other photon directions.

To obtain the number of photons dN_{ph} emitted in a crystal of thickness L within a solid angle $d\Omega$ around the forward direction, we multiply the square of the expression in (6.12) by a photon phase-space factor $L_0^3\omega^2 d\omega \, d\Omega/(2\pi c)^3$ and integrate over frequency, with the result

$$dN_{ph} = 2(1 + \beta^{-1})(1 + 2\gamma^2\delta)^{-1}\alpha\omega\left(\frac{L}{c}\right)(mc)^{-2}|\langle u_f| \mathbf{p}_\perp \cdot \epsilon_A |u_i\rangle|^2 \frac{d\Omega}{4\pi} \quad . \tag{6.15}$$

Discussions of the polarization and generalization to other angles of emission may be found in the literature [6.8-12,20].

6.4 Corrections to the Wave Equation

The properties of channeling radiation have been derived on the basis of (6.7). The main advantage of this approximate equation, describing retarded propagation of the transverse wave function, is that it allows inclusion of incoherent scattering in the description. Here we discuss the small corrections to the radiation energy, which account for the approximations made in the derivation of (6.7).

We consider states ψ of the type given by (6.6,9), but with a more accurate value of the energy ε. The basic equation, (6.5), with a continuum potential and without the term $\frac{e}{c}\mathbf{A}$, is multiplied by ψ^* and integrated over all variables. We then obtain, to first order, for the correction to the value $\varepsilon_0 = E_\perp + E_\ell + E_r$,

$$\varepsilon - \varepsilon_0 \simeq \varepsilon_0 \left(\frac{\varepsilon_0}{2\gamma^2 E} + \frac{1}{E}\left\langle \frac{p_\perp^2}{2m\gamma}\right\rangle\right) - \frac{1}{2E}\left\langle \left(\frac{p_\perp^2}{2m\gamma}\right)^2\right\rangle \ . \tag{6.16}$$

The brackets $<\ >$ denote an expectation value in the state $u(\mathbf{r})$, and the projectile velocity has been assumed to be close to c.

The last term in (6.16) gives the lowest-order correction to the transverse kinetic energy in the non-relativistic expansion of the expression $E_\perp^{kin} = [p_\perp^2 c^2 + (m\gamma)^2 c^4]^{\frac{1}{2}} - m\gamma c^2$. It is always negligible and will be omitted in the following. As we shall see, the other two terms also have a simple interpretation.

Consider a transition between the states ψ_i and ψ_f, with emission of a photon in the forward direction. According to (6.12,13) (with $n_r = 1$), the photon energy is determined by the requirement that $\beta\hbar\omega$ be equal to the change in ε, which may be evaluated by application of (6.16) to the initial and final states. All correction terms except those containing a factor $\hbar\omega$ may be ignored, and we then obtain

$$\hbar\omega \simeq 2\gamma^2 \Delta E_\perp \left(1 + \frac{2(E_{\perp f} + E_\ell) + \hbar\omega}{E} + \frac{2\gamma^2}{E}\left\langle \frac{p_\perp^2}{2m\gamma}\right\rangle_f \right)^{-1} \ . \tag{6.17}$$

Both of the correction terms in this expression may be interpreted as modifications of the Doppler-shift factor $(1+\beta)\gamma^2$ in (6.14). The first corresponds to the replacement $\gamma \rightarrow (E - E_{\perp f} - E_\ell - \frac{1}{2}\hbar\omega)/mc^2$ and therefore contains the frequency shift due to an energy loss E_ℓ, $\delta\omega/\omega \simeq -2E_\ell/E$. The second amounts to a replacement of $\gamma m = E/c^2$ by $(E - \hbar\omega)/c^2$ in the expression for the transverse kinetic energy in the final state. This term may also be written approximately as $\gamma^2 <\psi^2>_f$, where ψ is the angle the motion makes with the z direction, and it may therefore, according to (6.1), be interpreted as the Doppler correction for a finite angle between the directions of the projectile and of the photon.

The shift due to energy loss is usually very small, but that associated with the angle ψ becomes important at very high projectile energies. The maximum transverse energy of a channeled particle is equal to the potential barrier, which is independent of γ, and thus the second correction term in (6.17) is proportional to γ. It may also be expressed as twice the kinetic energy in the rest frame, divided by mc^2. Hence the condition for using (6.7) to calculate the radiation energies is just the condition for the simple derivation of (6.1), i.e., the transverse motion must be non-relativistic in the rest frame.

The correction is small for $E < 1$ GeV and is therefore not important in the energy region where a quantal treatment of channeling is required. It does, however, limit the freedom of choice of z direction in the planar case. Radiation formulas derived from (6.7) can be applied only if the z direction is chosen at an angle to the beam direction much smaller than γ^{-1}, such that the motion parallel to the plane is also non-relativistic in the rest frame.

6.5 Axial-Channeling Radiation

In the axial case, only negative particles (electrons) have localized motion and discrete transverse energy levels. Except for exceedingly low transverse energies, positrons move about freely between atomic rows and hence have a continuous transverse energy spectrum. Their motion may usually be described by classical mechanics, and we shall therefore limit our discussion of the transverse wave equation (6.10) to electrons. Methods known from atomic, molecular, and solid-state physics may be applied to solve the equation, and the analysis is much simplified by the fact that here we are dealing with a genuine one-particle problem, i.e., we do not have the complication of exchange and correlation. The coupling to lattice degrees of freedom can be treated as a perturbation, as discussed in Sect.6.8.

The axial-channeling states of low transverse energy are localized in the vicinity of atomic strings. They may therefore be determined with the potential from a single string, which resembles a two-dimensional atom [6.21,22]. The rotational symmetry of the potential leads to eigenstates which in polar coordinates (r,φ) may be expressed as

$$u(\mathbf{r}) = R_\ell(r) \frac{1}{\sqrt{2\pi}} e^{\pm i\ell\varphi} \quad , \quad \ell = 0,1,2,\ldots \quad , \tag{6.18}$$

where the radial wave function is a solution of the eigenvalue problem

$$\left\{- \left(\frac{d^2}{dr^2} + \frac{1}{r}\frac{d}{dr}\right) + \frac{\ell^2}{r^2} + \frac{2m}{\hbar^2} [\gamma U_T(r) - \gamma E_\perp]\right\} R_\ell(r) = 0 \quad . \tag{6.19}$$

The energy levels are doubly degenerate for $\ell \neq 0$. The combination of the factor γ with the potential and transverse energies may be interpreted as a transformation to the rest system. The potential γU_T in (6.19) depends on projectile energy and atomic spacing only through the ratio γ/d, and this leads to a scaling rule for radiation spectra obtained for electron incidence along different axes in a given

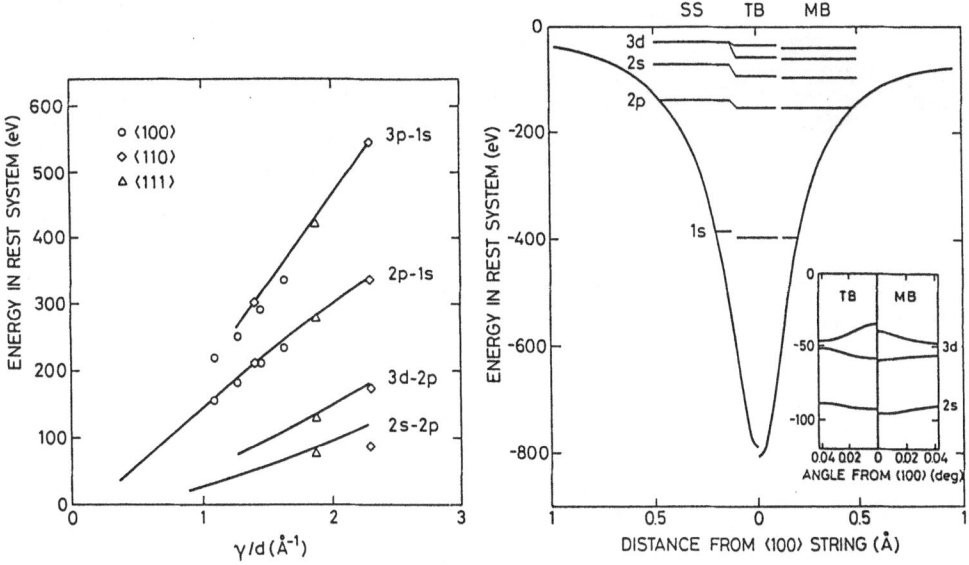

Fig.6.3   Fig.6.4

Fig.6.3. Comparison between calculated and measured photon energies in the rest frame for varying beam energy and for three different axes in Si. The curves have been calculated with the single-string approximation and a Doyle-Turner potential, (6.19, 20). From [6.20]

Fig.6.4. Continuum potentials in rest system for 4-MeV electrons channeled along a <100> direction in Si. To the left is shown the single-string potential, (6.20), and to the right, the averaged crystal potential is plotted in the direction towards a nearest-neighbor string. Transverse energy levels are from single-string (SS), tight-binding (TB), and many-beam (MB) calculations. The insert shows band structures corresponding to a tilt in the direction of a nearest-neighbor string. From [6.20]

crystal [6.23], as illustrated in Fig.6.3. The selection rule for the dipole transitions described by (6.15) is $\Delta\ell = \pm 1$, as for atomic transitions.

An example of a single-string potential $\gamma U_T(r)$ is shown in Fig.6.4 on the left-hand side, together with the energy levels. The potential is derived from calculated atomic scattering factors, and is given by

$$U_T(r) = -\frac{e^2}{a_0}\frac{2a_0^2}{d}\sum_{i=1}^{4}\frac{a_i}{B_i + \rho_2^2}\exp\left[-r^2/(B_i + \rho_2^2)\right] \quad , \tag{6.20}$$

where a_i and $b_i = (2\pi)^2 B_i$ are the coefficients tabulated by *Doyle* and *Turner* [6.24], and ρ_2^2 is the two-dimensional mean-square thermal displacement. The underlying atomic potentials are based on relativistic Hartree-Fock calculations, but since the coefficients have been adjusted to give accurate Fourier transforms only for reciprocal distances less than 8π Å^{-1}, they are not valid at small r. Corrections are necessary when the vibrational amplitude is small, $\rho_2 \lesssim (8\pi)^{-1}$ Å, and also for the evaluation of incoherent scattering (6.49).

137

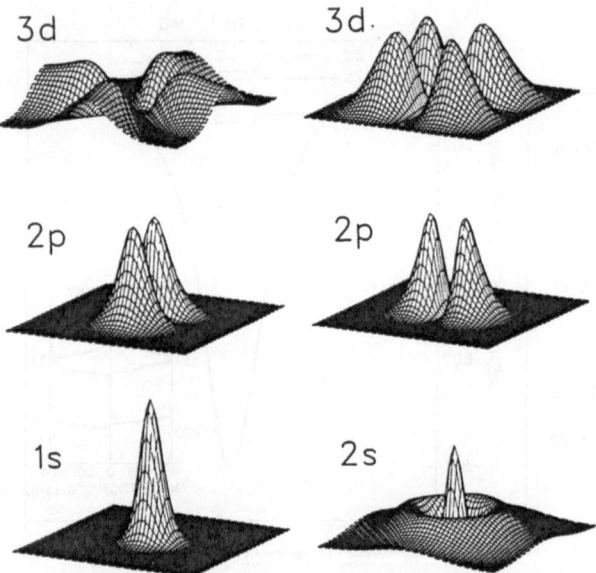

3d   3d

2p   2p

1s   2s

Fig.6.5. Squares of wave functions from many-beam calculations for transversely bound states of 4-MeV electrons, channeled along a <100> axis in Si. The edges of the unit cell are in the directions of nearest-neighbor strings, i.e., parallel to (110) planes. For the 1s and 2s states, the intensities have been multiplied by factors of 0.2 and 0.4. From [6.20]

On the right-hand side of Fig.6.4 is shown the potential obtained when the contribution from neighboring strings is included. The energy levels determined from an expansion in plane waves (many-beam calculation, discussed below) are compared to the results from a tight-binding approximation, where the difference between the two potentials is treated a perturbation which splits the upper levels. As shown in the insert, the finite overlap between states in neighboring string potentials leads to a dependence on the transverse wave vector $k_\perp$, related to the angle of incidence as in (6.26) below.

The squares of the wave functions corresponding to the energy levels indicated in Fig.6.4 are illustrated in Fig.6.5. The results are from many beam calculations, but the 1s and the 2p states show no deviation from the single-string picture. The 1s wave function has rotational symmetry, and for the two nearly degenerate 2p states the intensity distributions are proportional to $\cos^2\varphi$ and $\sin^2\varphi$. However, one of the two 3d states is clearly not localized within the unit cell. This state corresponds to the lower 3d level in the insert of Fig.6.4. The 2s distribution shows a small admixture of a single-string 3d state with four-fold symmetry.

Plots of intensity distributions are useful for understanding the sensitivity of different energy levels to changes in the potential. As an example, thermal vibrations mainly affect the potential close to a string, and hence mainly s levels are sensitive to crystal temperature. The variation of the line energies shown in Fig. 6.6 reflects changes in the 1s level, while the p levels are almost independent of temperature. Transitions between states with $\ell \neq 0$, e.g., 3d-2p, should therefore exhibit little temperature dependence, and this was confirmed by experiment [6.25]. For the most accurately determined line, corresponding to a 2p-1s transition, there

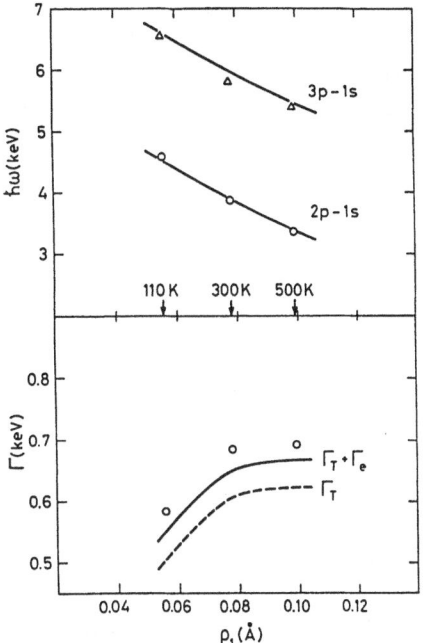

Fig.6.6. Line energies and widths for radiation from 3.5-MeV electrons channeled along a <111> direction in Si as functions of the one-dimensional vibrational amplitude. The line energies were obtained from a 169-beam calculation but agree to within 1% with a single-string calculation. The thermal widths Γ_T are for the 2p-1s transition and were calculated from a formula analogous to (6.50), without the reduction term and with an enhancement factor of 1.2-1.3, due to correlation of thermal vibrations. The electronic contribution was estimated on the basis of (6.61). From [6.25]

is agreement within 1% between experiment and calculations. With the present accuracy, such measurements determine the vibrational amplitude to within about 5%.

It is interesting to ask whether line energies can be determined with an accuracy sufficient to reveal deviations of the crystal potential from the sum of potentials of free atoms, as given in (6.20) with the Doyle-Turner representation. An example that has been studied experimentally [6.26] is 2p-1s radiation from electrons incident along a <110> direction in diamond. In this direction, the atomic strings are arranged in close-lying pairs, and the molecular-like transverse states are sensitive to charge accumulation in the region between the members of a pair, where half of the bonds from atoms in the two strings are centered.

Intensity distributions in the transverse plane, analogous to those presented in Fig.6.5, are shown in Fig.6.7 for 4-MeV electrons. In analogy to the LCAO method for molecules, the wave functions may be approximated by linear combinations of single-string states. They are classified as σ or π according to parity under reflection in the line connecting the two neighboring strings, and as gerade or ungerade according to parity under reflection in the center between the strings. There is little difference between the distributions for the two 1s states since there is little overlap between the single-string wave functions. For the π-2p states, this overlap is also fairly small, but for σ-2p it is large, and the even and odd combinations have very different intensity distributions. The σ_g-2p level is lowest. This "bonding orbital" has high intensity in the region between the two strings and is therefore particularly sensitive to bond charge. It was indeed found by *Andersen*

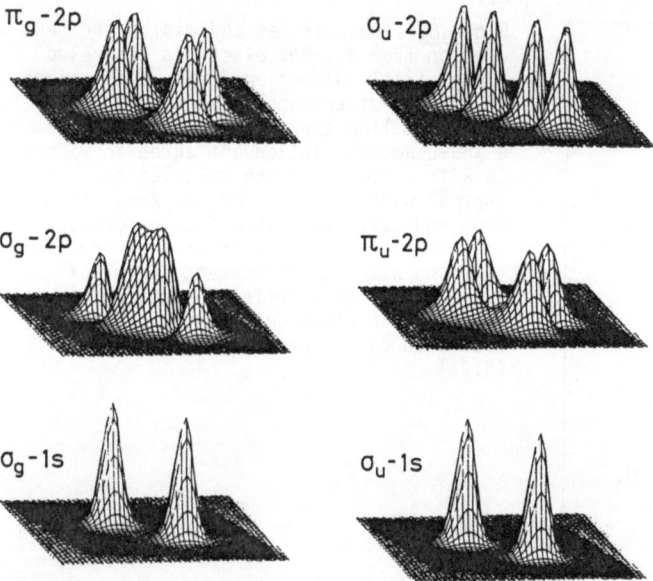

π_g-2p σ_u-2p

σ_g-2p π_u-2p

σ_g-1s σ_u-1s

Fig.6.7. Intensity distributions in the transverse plane for the lowest bound states of 4-MeV electrons channeled along the <110> direction in diamond. The peaks of the 1s distributions are at the positions of the two <110> atomic rows 0.89 Å apart. The transverse unit cell is indicated by the dark area, which has sides parallel to the (1Ī1) and (Ī11) planes, while the rows are in a (2Ī0) plane. As explained in the text, a notation borrowed from molecular physics has been used to designate the distributions, which have been obtained from a 225-beam calculation. From left to right and bottom to top, they are arranged in order of decreasing binding energy. The 1s distributions have been reduced by a factor of two relative to the others

et al. [6.26] that the total charge in a bond could be estimated from the observed splitting between the different 2p-1s transitions.

The intensity of channeling radiation also contains information on the states of channeled electrons [6.20,22,27], as we shall discuss below for planar channeling. The linewidths reflect mainly incoherent scattering and will be discussed in detail in later sections.

6.6 Planar-Channeling Radiation

In the planar case, both electrons and positrons have localized transverse motion with discrete transverse energy levels, determined from the one-dimensional Schrödinger equation,

$$\left\{ -\frac{\hbar^2}{2m}\frac{d^2}{dx^2} + [\gamma V_T(x) - \gamma E_\perp] \right\} u(x) = 0 \quad . \tag{6.21}$$

Since the planar potential is periodic with period d_p, it may be expanded in a Fourier series, and the eigenfunctions become Bloch waves [6.7,28]:

$$V_T(x) = \sum_j V_j^T \exp(ijgx) \quad , \quad j = 0,\pm1,\ldots \quad ,$$

$$u_{n,k_x}(x) = \exp(ik_x x) \sum_j C_j^{(n)}(k_x) \exp(ijgx) \quad , \quad g = 2\pi/d_p \quad . \tag{6.22}$$

For a fixed value of k_x within the first Brillouin zone, $|k_x| \leqslant g/2$, the band index n distinguishes the solutions of (6.21), which assumes the many-beam matrix form [6.29]

$$\left[\frac{\hbar^2}{2m}(k_x + jg)^2 - \gamma E_\perp^{(n)}(k_x) \right] C_j^{(n)} + \sum_\ell \gamma V_{j-\ell}^T C_\ell^{(n)} = 0 \quad . \tag{6.23}$$

For electrons in a cubic lattice, the Doyle-Turner expression for the Fourier component V_j^T is given by

$$V_j^T = -2\pi N a_0^2 \frac{e^2}{a_0} \sum_{i=1}^{4} a_i \exp\left[-\frac{1}{4}(B_i + 2\rho_1^2)(jg)^2 \right] \quad , \tag{6.24}$$

where ρ_1^2 is the one-dimensional mean square vibrational amplitude. This corresponds to a potential from a single plane:

$$V_T(x) = -2\sqrt{\pi} N d_p a_0^2 \sum_{i=1}^{4} \frac{a_i}{(B_i + 2\rho_1^2)^{\frac{1}{2}}} \exp[-x^2/(B_i + 2\rho_1^2)] \quad . \tag{6.25}$$

Figure 6.8 shows this potential, corrected for the contribution from neighboring planes, for a (111) plane in nickel. There is a significant temperature dependence of the energy levels, in particular for the ground state $n = 0$. Also shown are the squares of the wave functions for the two lowest-lying states, with even and odd parity, respectively. The alternating parity leads to the selection rule $\Delta n = $ odd for dipole transitions, where $\Delta n = n_i - n_f$, and the matrix element in (6.15) strongly favors transitions with $\Delta n = 1$. Photon energies for the $1 \rightarrow 0$ transition, measured at three different target temperatures, are compared with calculations in Fig.6.9 for two different planes. As for axes, it appears that such measurements can be used to determine vibrational amplitudes to an accuracy of about 5%.

The description is very similar for positrons, and (6.24,25) also apply, apart from a change of sign. The potential now has a minimum between planes, and for the calculation of bound states it may be approximated by the sum of the potentials (6.25) from two neighboring planes. The shape is nearly harmonic, and hence the energy levels are almost equally spaced. The spectral lines corresponding to transitions between neighboring levels are therefore very close and, experimentally, a single line broadened by anharmonicity is observed [6.3].

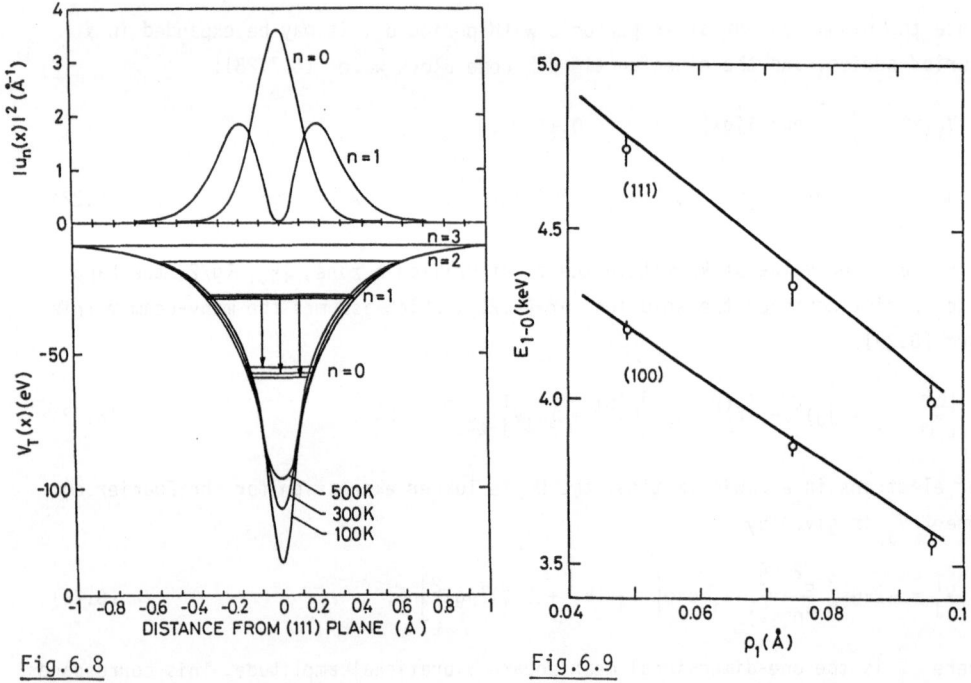

Fig.6.8

Fig.6.9

Fig.6.8. The (111) planar potential in Ni evaluated from (6.25) but with the contributions from neighboring planes included. The vibrational amplitudes for the three temperatures, as obtained from the Debye model with $T_D = 375$ K, are $\rho_1 = 0.049$ Å, 0.074 Å, and 0.095 Å. The planar spacing is $d_p = 2.03$ Å. Energy levels for bound states of 4-MeV electrons are indicated for the three temperatures, and the intensity distributions across the plane are shown for the two lowest levels at room temperature. From [6.18]

Fig.6.9. Line energies for 1→0 transitions, obtained from many-beam calculations using the Doyle-Turner potential, (6.24), are compared with measurements for 4-MeV electrons channeled along two different planes in Ni at temperatures 100, 300 and 500 K. From [6.18]

Examples of complete band-structure calculations, also including free transverse states, are given in Fig.6.10 for 4-MeV electrons incident along various planes in silicon. A beam incident at an angle ψ to a plane populates only transverse states with a reduced value of the Bloch vector k_x given by

$$K\psi = \pm k_x + jg \quad , \tag{6.26}$$

where $\hbar K = \gamma\beta mc$ is the electron momentum and j is an integer. This relation has been used to relate k_x to the angle given as abscissa in Fig.6.10 and in the insert of Fig.6.4. The transverse energy depends only weakly on angle for the low-lying bands, and in Fig.6.11 the measured energy and intensity for the 1→0 transition are compared to results of a many-beam calculation. The calculated intensity is proportional to the population of the initial state, which is given by

142

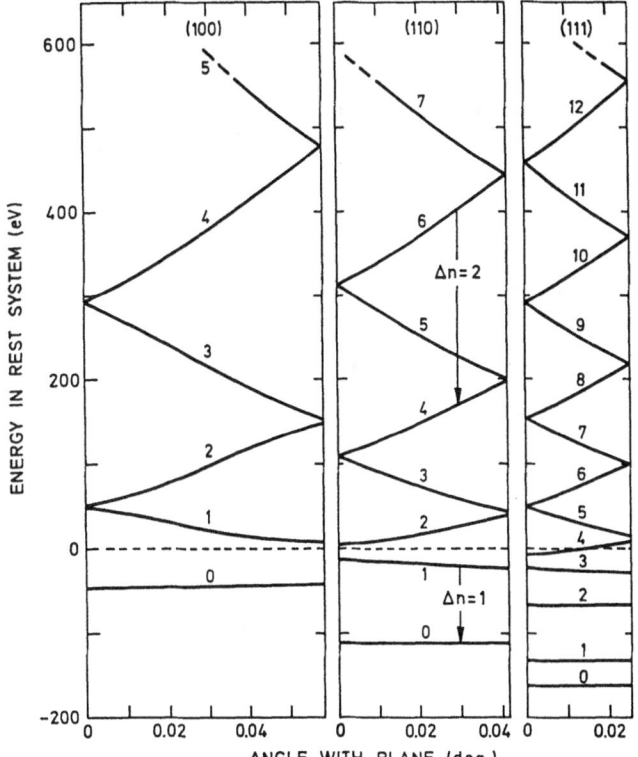

Fig.6.10. Band structure of the transverse energy $\gamma E_\perp(k_x)$ for 4-MeV electrons along three planes in Si, calculated from (6.23). The angle ψ with the plane is related to k_x through (6.26), and the angular range is from $\psi = 0$ to the Bragg angle, $\psi_B = g/2K$. From [6.28]

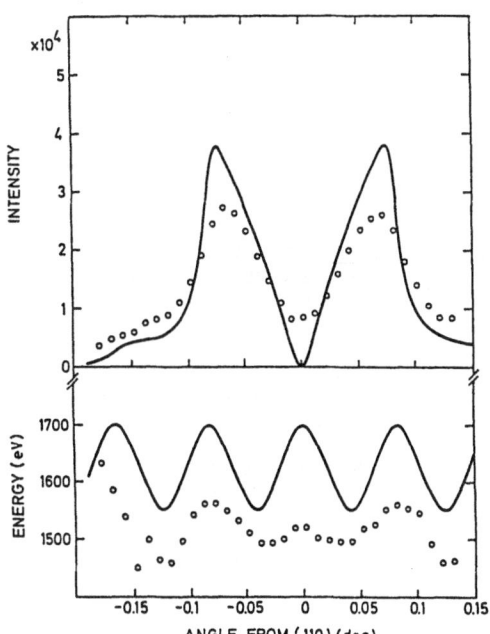

Fig.6.11. Comparison with measurements of the $1 \rightarrow 0$ line energy and intensity for 4-MeV electrons channeled along a (110) plane in Si. The influence of refraction in (6.14) (-3%) and a further small correction (-2%) for an error in beam energy have not been included in the calculation. The intensity has been calculated from (6.15) multiplied by the $n = 1$ population, as given by (6.27) without corrections for interband scattering. From [6.28]

143

$$P_1(\psi) = \frac{1}{d_p} \left| \int_{-d_p/2}^{d_p/2} dx\, u_1(x)\, \exp(iK\psi x) \right|^2 \quad , \tag{6.27}$$

where $u_1(x)$ is the $n=1$ wave function, normalized within one period d_p of the planar potential. The variation of the intensity therefore reflects the square of the $n=1$ wave function in momentum space. The deviations between measurements and calculations may be explained by beam divergence and scattering between states. Since the calculated intensity vanishes at $\psi=0$, these effects are important at small angles of incidence. Apart from this region, the measurements agree with the predicted oscillatory band structure, which is due to a small overlap of $n=1$ wave functions centered on neighboring planes.

In Fig.6.12, line energies measured over a larger angular range are compared with the transition energies calculated from the level structure shown in Fig.6.10. At the smallest angles, there is, in addition to the $1 \to 0$ line, a weak transition to the ground state from the free state $n=2$ or 3. Free-bound transitions are visible up to an angle $\psi \sim 0.4°$. At larger angles, only transitions between free states occur, with a selection rule $\Delta n = $ even, because neighboring levels correspond to opposite directions of the transverse momentum in an extended-zone picture. At a given angle of incidence, only one band is populated significantly, and the dispersion relation approaches that for a free particle, i.e.,

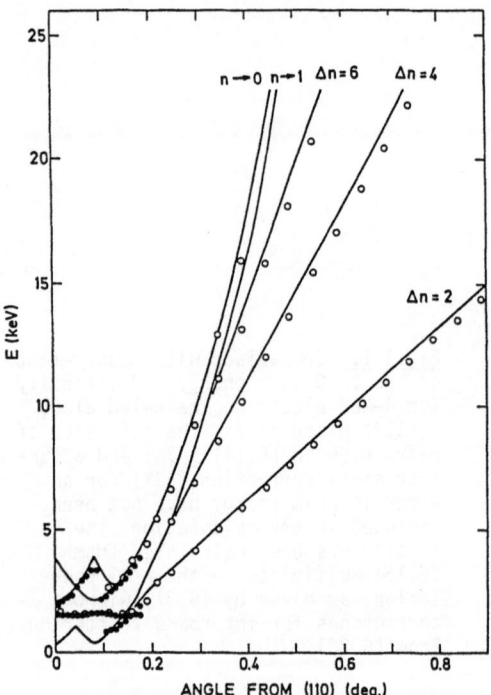

Fig.6.12. Comparison with measurements of line energies calculated from the band structure in Fig.6.10 for 4-MeV electrons incident at a small angle to a (110) plane in Si. For a given angle, only transitions with significant intensity are indicated. (●●●,ooo) correspond to two series of measurements with different energy dispersion in the spectra. The transitions indicated at the smallest angles, $\psi \lesssim 0.1°$, are $2 \to 1$, $1 \to 0$, $2 \to 0$, and $3 \to 0$ in order of increasing energy. As in Fig.6.11, small corrections for refraction and for an error in beam energy have not been included. From [6.28]

$$E_\perp(\psi) \simeq \frac{\hbar^2(K\psi)^2}{2m\gamma} + V_0^T \quad . \tag{6.28}$$

A $\Delta n = 2$ transition results from a transverse momentum transfer $\hbar g$, and the magnitude of the corresponding change in transverse energy is given by

$$\Delta E_\perp \simeq \frac{\hbar^2 Kg\dot\psi}{m\gamma} - \frac{\hbar^2 g^2}{2m\gamma} \quad . \tag{6.29}$$

The photon energy in the laboratory, as given by (6.14), is therefore linear in ψ, with a small offset at $\psi = 0$. Also the energies of transitions with $\Delta n = 4$ and $\Delta n = 6$ are straight lines as functions of ψ, with slope proportional to Δn. The free-bound transitions have a parabolic shape according to (6.28). These features are confirmed by the experiment, and when a correction is made for the small error in beam energy (-1%), there is quantitative agreement between theory and experiment.

6.7 Coherent Bremsstrahlung

Transitions between high-lying levels correspond to the well-known phenomenon of coherent bremsstrahlung [6.30]. The transverse wave functions are plane waves, slightly perturbed by the planar potential. In the conventional description of coherent bremsstrahlung, unperturbed plane waves are used as basis states, and the radiation is associated with the process of momentum transfer $q = \hbar(\Delta n/2)g$ from the projectile to the lattice. The photon energy may then be derived from the energy-momentum relation for a free particle. If E and $\mathbf{p}$ denote energy and momentum of the projectile before emission of a photon with momentum $\hbar\boldsymbol{\kappa}$, we have

$$E = [(\mathbf{p} - \mathbf{q} - \hbar\boldsymbol{\kappa})^2 c^2 + m^2 c^4]^{1/2} + \hbar\kappa c \quad . \tag{6.30}$$

Assuming for simplicity that $\mathbf{p}$ and $\boldsymbol{\kappa}$ are parallel, i.e., considering only emission in the beam direction, we obtain

$$\hbar\kappa c\left(1 - \beta + \frac{\mathbf{q}\cdot\mathbf{p}c}{pE}\right) = \frac{\mathbf{q}\cdot\mathbf{p}c^2}{E} - \frac{q^2 c^2}{2E} \quad , \tag{6.31}$$

or, when the notation $\mathbf{q}\cdot\mathbf{p}/p \simeq q\psi$ and the relations $1 - \beta \simeq (2\gamma^2)^{-1}$ and $E = \gamma mc^2$ are introduced,

$$\hbar\kappa c \simeq 2\gamma\left(\frac{pq\psi}{m} - \frac{q^2}{2m}\right)\left(1 + \frac{2\gamma q\psi}{mc}\right)^{-1} \quad . \tag{6.32}$$

This expression may be compared with (6.14,29). Except for the last factor in (6.32), the formulas agree for $q = \hbar g$. The denominator in (6.14), which accounts for refraction, would be included in (6.32) if the momentum of a photon with energy $\hbar\kappa c$ were corrected from $\hbar\boldsymbol{\kappa}$ to $n_r\hbar\boldsymbol{\kappa}$ in (6.30). The last factor in (6.32) is important only

145

when the photon energy is not negligible compared to the projectile energy since it deviates from unity by an amount $\sim\hbar\kappa/\gamma mc$. This correction term is also contained in (6.17).

The photon energy expressed by (6.32) corresponds to emission in the direction of the initial momentum **p** and not parallel to the plane, as in (6.14,17). The last term in (6.17) accounts for this difference and becomes important for angles of incidence comparable to γ^{-1}. In experiments, (6.32) is usually relevant since the angle between the beam and the direction to the photon counter is kept fixed when the crystal is tilted.

Agreement is also found for the radiation intensity between the results obtained in the two pictures of the process, a spontaneous transition between energy bands or radiation induced by perturbation of the plane-wave motion of the projectile. For coherent bremsstrahlung of order $\nu = \Delta n/2$ with frequency $\omega = \kappa c$, one obtains the very simple result for the number of photons emitted in the forward direction [6.28],

$$dN_{ph} \simeq 16\alpha\omega^{-1}(L/c)(mc)^{-2}\gamma^4(\nu g)^2|V_\nu^T|^2 \frac{d\Omega}{4\pi} \quad . \tag{6.33}$$

We may finally mention that results very similar to (6.32,33) are obtained from a classical picture, where the radiation is due to the periodic perturbation of the straight-line trajectory of a particle crossing planes at a small angle, as illustrated in Fig.6.1. The correspondence between different descriptions of coherent bremsstrahlung was discussed in detail in [6.28,30,31].

6.8 Damping and Linewidth

A number of effects may contribute to the width of channeling-radiation lines [6.3,18,32], but for MeV electrons, on which we shall concentrate, the limitation of the coherence length due to single scattering is usually the main source of broadening. The decay of states by radiation is negligible, and the situation is analogous to collisional line broadening for atomic transitions. For positrons, the situation is quite different. A sharp line is observed only for planar channeling, and it is a superposition of many close-lying lines. Since channeled positrons are kept away from atoms, the thermal scattering is weak. The individual lines should therefore be sharper than for electrons, and the width of the composite line be determined mainly by the splitting due to anharmonicity.

In the absence of scattering events, we obtain a photon yield proportional to the square of the expression in (6.12), i.e., for emission in the forward direction,

$$I(\omega) \propto \left(\frac{\sin x}{x}\right)^2 \quad , \qquad x = \Delta\omega(1 - n_r\beta)L/2v \quad , \tag{6.34}$$

146

where $\Delta\omega$ denotes the deviation from the line center given by (6.14). At high projectile energy and without refraction, this leads to a full width at half maximum (FWHM) given by

$$\text{FWHM} \simeq 3.5\pi\gamma^2\hbar c/L \quad , \quad \gamma \gg 1 \quad , \quad n_r = 1 \quad . \tag{6.35}$$

As a result of single scattering, the radiation matrix element decays with depth z as $\exp(-z/2\Lambda)$, and when the mean free path Λ is much shorter than the crystal thickness, $\Lambda < L/2\pi$, the photon spectrum obtained from (6.12) is a Lorentzian with the width

$$\text{FWHM} \simeq 2\gamma^2\hbar c/\Lambda \quad , \quad \gamma \gg 1 \quad , \quad n_r = 1 \quad . \tag{6.36}$$

We first derive the basic formula for the coherence length Λ in terms of the scattering potential δV. Then the explicit expression for this difference between the actual interaction potential and the thermally averaged continuum potential applied in the evaluation of transverse states is introduced, and the thermal and electronic contributions are treated separately. We shall discuss planar channeling of electrons, but a similar description applies for axial channeling.

When the coupling between projectile coordinates $\mathbf{R}$ and lattice coordinates ℓ is included, the states $u_{i,f}(\mathbf{r}) \exp(-iE_{\perp i,f}z/\hbar v)$ in (6.12) must be replaced by solutions of the "Schrödinger equation",

$$i\hbar v \frac{\partial}{\partial z} w(\mathbf{r},z,\ell) = Hw(\mathbf{r},z,\ell)$$

$$= \left[-\frac{\hbar^2}{2\gamma m} \nabla_{\mathbf{r}}^2 + V(\mathbf{r},z,\ell) + H_\ell \right] w(\mathbf{r},z,\ell) \quad . \tag{6.37}$$

We have omitted the degrees of freedom of the radiation field, although the coupling to the charged particles leads not only to radiative decay of the channeling states, but also to excitation of target electrons, which may absorb virtual photons emitted by the projectile. However, both radiation processes and magnetic excitations with sufficient momentum transfer to cause incoherence turn out to be exceedingly rare.

As initial state, we now take the solution w, which matches the incoming wave at the crystal surface, $z = 0$; a complete set of final states w_α can be specified through boundary conditions, for example at the front or back crystal surface. For the amplitude $a_\alpha(\omega)$ for a transition from w to a final state w_α, accompanied by emission of a photon of energy $\hbar\omega$ in the forward direction, we then obtain

$$a_\alpha(\omega) \propto \omega^{-\frac{1}{2}} \int_0^L dz \langle w_\alpha(\mathbf{r},z,\ell) | p_x | w(\mathbf{r},z,\ell) \rangle \exp\left[i \frac{\omega}{v} (1 - n_r\beta)z \right] \quad . \tag{6.38}$$

The functions w and w_α may be expanded in products of eigenfunctions of the projectile Hamiltonian $H_{\perp,T}$, with the continuum potential $V_T(x)$, and eigenfunctions $\varphi_\nu(\ell)$ of the lattice Hamiltonian H_ℓ. We consider only radiation from transitions between bound states, localized in the vicinity of planes, and may, for the purpose of evaluating radiation matrix elements, assume the projectile states to be localized around a single atomic plane. The scattering potential due to the displacement of an atom in the plane is also localized, and hence phase relations between wave-function components around different planes are unimportant for thermal scattering, which gives the dominant contribution to the damping. In order to simplify the notation, we therefore expand in a complete set of localized wave functions $u_n(x)$, $|x| \leq d_p/2$, satisfying, e.g., periodic boundary conditions ($k_x = 0$). The generalization to Bloch waves u_{n,k_x} is straightforward [6.18].

In (6.38), we then expand w and w_α in the form

$$w = \sum_{n, k_y, \nu} \langle n k_y \nu | w \rangle | n k_y \nu \rangle \quad , \tag{6.39a}$$

where

$$|n k_y \nu\rangle = u_n(x) \frac{1}{\sqrt{L_y}} \exp(i k_y y) \varphi_\nu(\ell) \quad . \tag{6.39b}$$

Only terms with the same values of k_y and ν give nonvanishing contributions to the radiation matrix element, and we obtain

$$\langle w_\alpha | p_x | w \rangle = \sum_{n, n'} \langle n | p_x | n' \rangle \sum_{k_y, \nu} \langle w_\alpha | n k_y \nu \rangle \langle n' k_y \nu | w \rangle \quad . \tag{6.40}$$

The z dependence of this quantity, needed in the integral in (6.38), is contained in the states w and w_α. If the Hamiltonian in (6.37) is decomposed into a main part and a perturbation, $H = H_0 + \delta V$, where H_0 corresponds to the continuum potential, we obtain for the change over a small depth interval δz:

$$w(z + \delta z) = \exp\left[- i H_0 \delta z / \hbar v - i \int_z^{z+\delta z} dz' \delta V(\mathbf{r}, z', \ell) / \hbar v \right] w(z)$$

$$= \exp(-i H_0 \delta z / \hbar v) \exp\left[- i \int_z^{z+\delta z} dz' \delta V(\mathbf{r}, z', \ell) / \hbar v \right]$$

$$\times \left\{ 1 + \frac{\delta z}{2(\hbar v)^2} \left[H_0, \int_z^{z+\delta z} dz' \delta V(\mathbf{r}, z', \ell) \right] + \dots \right\} w(z) \quad , \tag{6.41}$$

where [,] denotes a commutator. Here we have introduced the Baker-Hausdorff expansion [6.13] since this enables us to shift the operator $\exp(-i H_0 \delta z / \hbar v)$ to the eigenstates $|n k_y \nu\rangle$ and $|n' k_y \nu\rangle$ in (6.40). The result is an oscillatory factor, $\exp[i(E_{\perp n} - E_{\perp n'}) z / \hbar v]$, and we shall assume that for fixed ω the z integration in (6.38) eliminates all terms except those with, say, $n = 0$ and $n' = 1$. This requires the separation between lines from different transitions to be large compared to the linewidths.

As discussed previously, when corrections to (6.7) are introduced, the oscillatory factors involve not only the photon frequency and the changes in transverse energy corresponding to the x motion, but also depend on the energies connected with motion parallel to the planes and to excitation of the lattice. These correc-

148

tions are usually small, and we may include all states $|k_y\nu>$ in the summation in (6.40). Applying closure, we then obtain for the contribution from the state w_α to the transition from $n'=1$ to $n=0$,

$$<w_\alpha(z+\delta z)|p_x|w(z+\delta z)>_{1\to0} = <0|p_x|1> \exp[-i(E_{\perp 1} - E_{\perp 0})\delta z/\hbar v] \tag{6.42a}$$

$$\times \int d\ell \int dy <e^{-iQ}w_\alpha(z)|u_0><u_1|e^{-iQ}w(z)> \quad,$$

where

$$Q = \frac{1}{\hbar v} \int_z^{z+\delta z} dz' \delta V(\mathbf{r},z',\ell) \quad. \tag{6.42b}$$

The expression within curly brackets in (6.41) has been replaced by unity. This sudden-collision approximation is valid when the correction term is small compared to Q, i.e., when the distance δz is short compared to the length

$$\lambda \sim 2\hbar v/\Delta E_\perp \quad. \tag{6.43}$$

Here $\Delta E_\perp = \Delta E_\perp^x + \Delta E_\perp^y$ is a typical difference in total transverse energy between projectile states connected by the scattering potential. A typical momentum change $\delta p_\perp$ due to a thermal displacement is of the order of $\hbar/\rho_1$, and with $\Delta E_\perp \sim (\delta p_\perp)^2/2m\gamma$, we obtain the estimate

$$\lambda \sim 4\gamma a_0 \alpha^{-1}(\rho_1/a_0)^2 \sim 10\gamma \text{ Å} \quad. \tag{6.44}$$

The sudden-collision approximation is obviously very well fulfilled for scattering by a single atom. Thermal scattering by different atoms may be treated as independent events if the displacements are uncorrelated, as in the Einstein model. But even when correlations are included, for example through the Debye model, one may apply (6.42) to evaluate the combined scattering by several atoms, since correlations usually persist only over distances short compared to λ.

With increasing depth, the amplitude $<w_\alpha|p_x|w>_{1\to0}$ decreases due to scattering processes which lead out of the states u_1 and u_0, or which reduce the overlap of the states w and w_α in the y and ℓ coordinates. A reduction in the damping of the radiation amplitude might be expected from events where an electron at a later stage is scattered back again into one of the original states. Such contributions from 'feeding', however, add with random phases to the part from the depleted original states when the depletion length 2Λ determining the effective limit on the integration in (6.38) is large compared to the length λ introduced in (6.43). As it turns out, Λ is usually larger than λ by at least an order of magnitude. Therefore, the contributions to the radiation from feeding add incoherently and give the same linewidth, determined by the decay of a matrix element. The only exception occurs in thin crystals where the decay length becomes comparable to the crystal thickness [6.18]. Such cases will not be discussed any further here.

In the evaluation of the matrix element in (6.42), we may neglect the parts of the functions w and w_α which, at a depth z, are orthogonal to $u_1(x)$ and $u_0(x)$, respectively, since feeding from ensuing scattering events within the interval $(z, z + \delta z)$ may be disregarded. Similarly, we may neglect the part of $w_\alpha(z)$ which is orthogonal to $w(z)$ in the y,ℓ coordinates. The damping of the matrix element then becomes exponential, with decay length 2Λ, and we may take the functions to be of the form $w = u_1(x)h(y,\ell)$ and $w_\alpha = u_0(x)h(y,\ell)$ at depth z. For the relative change $-\delta z/\Lambda$ of the absolute square of the matrix element over a distance δz, we obtain

$$\delta z/\Lambda = 1 - |[<\exp(iQ)>_0 <\exp(-iQ)>_1]_{y,\ell}|^2 \quad . \tag{6.45}$$

The expectation values are x integrals weighted by u_0^2 and u_1^2, and the subscripts y and ℓ indicate an integration over these variables with weighting function $|h|^2$. The expression (6.45) is expanded to second order in Q, corresponding to an evaluation of the transition amplitudes to first order, and with a slight rearrangement of terms, the result may be written

$$\delta z/\Lambda = (<Q^2>_0 - <Q>_0^2)_{y,\ell} + (<Q^2>_1 - <Q>_1^2)_{y,\ell}$$

$$+ [(<Q>_0 - <Q>_1)^2]_{y,\ell} - [(<Q>_0 - <Q_1>)_{y,\ell}]^2 \quad . \tag{6.46}$$

The two lines in this formula have a simple interpretation. By insertion of the complete set of states $u_n(x)$, e.g., $<Q^2>_0 = \sum_n |<0|Q|n>|^2$, it is seen that the two terms in the first line correspond to the total probability for scattering into states with $n \neq 0$ and $n \neq 1$, respectively (interband scattering).

The second line gives the contribution from intraband scattering, which corresponds to transitions from $h(y,\ell)$ to states orthogonal to h. The effective perturbation is seen to be the difference between the expectation values of Q in the states u_0 and u_1, i.e., intraband scattering contributes to the extent that it is different in the initial and final states u_1 and u_0. The last term, which formally corresponds to 'scattering' to the original state h itself, can always be ignored [6.18].

6.9 Scattering Potential

We now consider the scattering potential δV, i.e., the difference between the interaction potential and the thermally averaged continuum potential $V_T(x)$. Assuming the crystal potential to be a sum of atomic potentials V_a, we may split δV into three parts,

$$\delta V = \delta V_1 + \delta V_2 + \delta V_3$$

$$= \sum_i \left[\left(\frac{-Ze^2}{|R - R_i - U_i|} + \sum_{j=1}^{Z} \frac{e^2}{|R - R_{i,j}|} \right) - V_a(R - R_i - U_i) \right]$$

$$+ \sum_i [V_a(R - R_i - U_i) - V_{a,T}(R - R_i)]$$

$$+ \sum_i V_{a,T}(R - R_i) - V_T(x) \quad . \tag{6.47}$$

Here, the U_i are the coordinates of the atoms connected with the lattice vectors R_i, while j refers to the electrons in the ith atom, with coordinates $R_{i,j}$. Only the first term δV_1 depends on these electronic coordinates and couples to excited electronic states, and, since the expectation value of δV_1 is zero in the electronic ground state, this term does not give scattering without electronic excitation. At zero lattice temperature, there is an analogous distinction between the last two terms since only δV_2 contains the lattice degrees of freedom U_i, and at T = 0, the lattice is in its lowest quantum state, with zero expectation value of δV_2. However, at a finite temperature, the lattice is represented by a statistical distribution over quantum states, and δV_2 also leads to scattering without phonon creation or annihilation. The thermal scattering induced by δV_2 and the 'nonsystematic' reflections caused by δV_3 are then instead distinguished by the momentum transfers, which, for the periodic potential δV_3, are confined to reciprocal-lattice vectors. Owing to the continuous distribution of momentum transfers, electronic and thermal scattering is sometimes characterized as diffuse scattering. In (6.46), we can separate electronic from thermal processes by insertion of a complete set of final lattice states in the first three terms. The electronic contribution then consists of transitions changing the electronic state. Such events involve only the second Coulomb term in the expression for δV_1, and in the evaluation, lattice coordinates other than electronic may be ignored.

6.10 Thermal Scattering

We first consider the remaining scattering processes not affecting electronic states, since normally these events give the dominant contributions to incoherence of channeling radiation. The restriction to excited states without electronic excitation is equivalent to the application of (6.46), with inclusion of atomic (phonon) degrees of freedom for the lattice only.

To obtain the thermal contribution to the incoherence, we therefore insert into (6.46) the expression

$$Q = \frac{1}{\hbar v} \sum_i [\Gamma_a(\mathbf{r} - \mathbf{r}_i - \mathbf{u}_i) - \Gamma_{a,T}(\mathbf{r} - \mathbf{r}_i)] \quad , \tag{6.48}$$

where Γ_a and $\Gamma_{a,T}$ denote the integrals over the longitudinal coordinate of V_a and $V_{a,T}$, respectively, and where the summation is over all atoms within the layer $(z, z + \delta z)$ considered. The vectors $\mathbf{r}_i$ and $\mathbf{u}_i$ refer to the x and y coordinates of $\mathbf{R}_i$ and $\mathbf{U}_i$. The difference $\Gamma_a - \Gamma_{a,T}$ is small outside a region of dimensions $\sim\rho_1$ around the atomic position, and products of differences belonging to different atoms are usually negligible. An exception occurs when the projectile is moving at a small enough angle to an axis in the plane, since the y coordinates of neighboring atoms along a string become nearly equal. Only in these special circumstances is correlation of the motion of neighboring atoms important, and it may then be included into the description [6.18]. Neglecting correlation effects and the last term in (6.46), we obtain for the thermal component of the inverse coherence length,

$$\Lambda_T^{-1} = \frac{Nd_p}{(\hbar v)^2} \int dy \Big\{ <(\Gamma_a^2)_T - \Gamma_{a,T}^2>_0 - (<\Gamma_a - \Gamma_{a,T}>_0^2)_T$$

$$+ <(\Gamma_a^2)_T - \Gamma_{a,T}^2>_1 - (<\Gamma_a - \Gamma_{a,T}>_1^2)_T$$

$$+ [(<\Gamma_a - \Gamma_{a,T}>_0 - <\Gamma_a - \Gamma_{a,T}>_1)^2]_T \Big\} \quad . \tag{6.49}$$

Equation (6.46) contained an integration over atomic (phonon) coordinates weighted by the density distribution in the initial state. Here we have combined this integration with a statistical average over initial states, and the resulting thermal average is indicated by the subscript T.

A formula derived by *Bazylev* and *Goloviznin* [6.33] is equivalent to (6.49) without the last, positive term, i.e., only interband scattering is included in their treatment. The damping is then separable into contributions from the initial and final states. In our initial study of linewidths in the axial case [6.20], we used the same type of estimate of damping from thermal scattering and estimated the error to be $\lesssim 10\%$. As we shall see, however, the nonseparable contribution from intraband scattering is somewhat larger for planes.

The expression for Λ_T^{-1} may alternatively be written as

$$\Lambda_T^{-1} = \frac{Nd_p}{(\hbar v)^2} \int dy \Big[<(\Gamma_a^2)_T - \Gamma_{a,T}^2>_0 + <(\Gamma_a^2)_T - \Gamma_{a,T}^2>_1$$

$$- 2(<\Gamma_a - \Gamma_{a,T}>_0 <\Gamma_a - \Gamma_{a,T}>_1)_T \Big] \quad . \tag{6.50}$$

Here, the first two terms represent the total scattering in the initial and final states, and they correspond to an application to the linewidth problem [6.12,34] of the expressions for damping discussed in treatments of high-energy electron

152

diffraction [6.29,35]. The last term is specific to the radiation process, and it usually reduces the damping.

The expression for the total thermal scattering becomes particularly simple for a plane wave, and one obtains a cross section for incoherent scattering,

$$\sigma_{inc} = \frac{1}{(2\pi\hbar v)^2} \int d^2k [V_a(k)]^2 [1 - \exp(-k^2\rho_1^2)] \quad , \tag{6.51}$$

where $V_a(k)$ is a Fourier component of the atomic potential,

$$V_a(k) = \int d^3R \, V_a(R) \, e^{-ik\cdot R} \quad . \tag{6.52}$$

If thermal scattering is assumed to be localized at the positions of the vibrating atoms, the scattering is modified for a state $u_n(x)$ by the overlap with the Gaussian distribution of atoms, and this leads to a reciprocal mean free path,

$$\Lambda_{T,n}^{-1} = \sigma_{inc} N d_p (2\pi\rho_1^2)^{-\frac{1}{2}} \langle\exp(-x^2/2\rho_1^2)\rangle_n \quad . \tag{6.53}$$

The assumption of complete locality of the interaction results in an overestimation of the scattering for states concentrated around the planes and to an underestimation for states with low density there, i.e., usually for even- and odd-n states, respectively. However, the simple expression (6.53) is often a useful first estimate of coherence lengths [6.20,25].

Using the Doyle-Turner potential, we have obtained explicit expressions for the quantity Λ_T^{-1} with and without correlations between neighboring atoms [6.18]. A few comparisons between experiments and results from such calculations will be given below.

6.11 Nonsystematic Reflections

We discuss only briefly the periodic perturbation δV_3, because it turns out to be important only when the projectile is moving at a small angle φ to an axis in the plane. In this case, the projectile experiences a time-dependent perturbation with frequency $\omega = 2\pi v\varphi/d_s$, where d_s is the spacing of strings in the plane. The perturbation can induce jumps in transverse energy of magnitude $n\hbar\omega$, where n is an integer. The transition probabilities depend strongly on φ since the transition matrix elements are essentially Fourier components of the thermally averaged atomic potential $V_{a,T}(R)$, with a wave vector k_x given by $(\hbar k_x)^2/2m\gamma \simeq 2\pi\hbar c\varphi/d_s$ for n = 1 [6.18]. The Debye-Waller factor leads to an exponential decrease of the matrix element with increasing φ, and the effective cutoff angle corresponding to $k_x \simeq \rho_1^{-1}$ is only of the order of one degree for electrons of a few MeV, and is proportional to γ^{-1} as a function of beam energy. In particular, transitions due to the periodicity along an axis nearly parallel to the beam are therefore always very weak for planar (or axial) channeling of relativistic electrons.

6.12 Electronic Scattering

In general, the contribution to incoherence from scattering by electrons should be most important in light materials since the cross section per atom is proportional to Z^2 and Z for thermal and electronic scattering, respectively. As an indication of the order of magnitude to be expected for the mean free path for scattering by electrons, Λ_{sc}, we may quote the result for a swift electron traversing an electron gas of constant density n, [6.13]:

$$\Lambda_{sc}^{-1} = \frac{e^2}{2\hbar v^2}\, \omega_0 \, \log \frac{4mv^2}{\hbar\omega_0} \quad . \tag{6.54}$$

Here ω_0 denotes the plasma frequency of the gas, $\omega_0^2 = 4\pi n e^2/m$. With plasma frequencies corresponding to typical electron densities in solids, Λ_{sc} becomes of the order of 10^3 lattice spacings for relativistic projectiles, and for low-Z materials, this distance is comparable to the mean free path for thermal scattering. On the other hand, most of the electronic processes are connected with low-momentum (plasmon) excitations, and just as for low-momentum phonon events responsible for intraband transitions, we expect such excitations to contribute to incoherence only to the extent that the scattering differs in the initial and final projectile states. We therefore anticipate a reduction term similar to the last one in (6.50).

As indicated previously, we obtain the electronic contribution to incoherence by inserting intermediate excited electronic states φ_J in the first three terms in (6.46). Atomic coordinates can be ignored, and the states can be chosen to be of the form

$$|n,k_y,J\rangle = u_n(x)\, \frac{1}{\sqrt{L_y}}\, \exp(ik_y y)\varphi_J(R_j) \quad , \quad n = 0,1 \quad , \tag{6.55}$$

where R_j now denotes the target-electron coordinates, and electronic states are denoted by capital letters. For the contribution to the inverse coherence length, we obtain:

$$\Lambda_e^{-1} = \frac{1}{L_y} \int dy \sum_{F \neq I} (\langle 0|\langle I|Q|F\rangle\langle F|Q|I\rangle|0\rangle + \langle 1|\langle I|Q|F\rangle\langle F|Q|I\rangle|1\rangle$$

$$- 2\,\langle I|\langle 0|Q|0\rangle|F\rangle\langle F|\langle 1|Q|1\rangle|I\rangle) \quad . \tag{6.56}$$

The initial and final planar-channeling states are denoted by $|1\rangle$ and $|0\rangle$ as above, and the target electrons are in the state $|I\rangle$ before the excitation. For the operator Q representing the Coulomb interaction between the projectile electron at R

and the target electrons at $\mathbf{R}_j$, we introduce the Fourier representation

$$Q = \frac{e^2}{\hbar v} \sum_j \int dz \left[|\mathbf{r}_j - \mathbf{r}|^2 + (z_j - z)^2 \right]^{-\frac{1}{2}}$$

$$= \frac{e^2}{\hbar v} \sum_j \int dz \, \frac{1}{L_x L_y L_z} \sum_{\mathbf{q}} \frac{4\pi}{q^2} \exp[i\mathbf{q}_\perp \cdot (\mathbf{r}_j - \mathbf{r}) + iq_z(z_j - z)]$$

$$= \frac{e^2}{\hbar v} \sum_j \frac{1}{L_x L_y} \sum_{\mathbf{q}_\perp} \frac{4\pi}{q_\perp^2} \exp[i\mathbf{q}_\perp \cdot (\mathbf{r}_j - \mathbf{r})] \quad . \tag{6.57}$$

The sum over j extends over all electrons within a unit distance in the longitudinal direction.

Insertion of the expression for Q into (6.56) with a subsequent integration over y leads to

$$\Lambda_e^{-1} = \left(\frac{4\pi e^2}{\hbar v} \right)^2 \frac{1}{(L_x L_y)^2} \sum_{q_y} \sum_{q_x} \sum_{q_x'} \, (<0| \exp[i(q_x' - q_x)x]|0>$$

$$+ <1|\exp[i(q_x' - q_x)x]|1> - 2<0|\exp(iq_x'x)|0><1|\exp(-iq_x x)|1>) \tag{6.58}$$

$$\times \frac{1}{q_\perp^2 (q_\perp')^2} \sum_{F \neq I} <I| \sum_i \exp(i\mathbf{q}_\perp \cdot \mathbf{r}_i)|F><F| \sum_j \exp(-i\mathbf{q}_\perp' \cdot \mathbf{r}_j)|I>|_{q_y'=q_y} \quad .$$

The factorization into matrix elements depending only on the projectile motion or on the target degrees of freedom is particularly useful, since the target elements fulfill a sum rule [6.36],

$$\sum_F <I| \sum_i \exp(i\mathbf{q} \cdot \mathbf{R}_i)|F><F| \sum_j \exp(-i\mathbf{q}' \cdot \mathbf{R}_j)|I>(E_F - E_I)$$

$$= \frac{\hbar^2}{2m} \mathbf{q}' \cdot \mathbf{q} <I| \sum_j \exp[i(\mathbf{q} - \mathbf{q}') \cdot \mathbf{R}_j]|I> \quad . \tag{6.59}$$

In a homogeneous electron gas, the Fourier component of the density on the right-hand side is equal to $nL_x L_y \delta_{\mathbf{q},\mathbf{q}'}$. In addition, the excitation energy $E_F - E_I$ may, to a good approximation, be considered a unique function of the momentum transfer $\hbar\mathbf{q}$,

$$E_F - E_I \simeq \hbar\omega_0 + (\hbar q)^2/2m \quad . \tag{6.60}$$

To this approximation, we obtain for the inverse coherence length,

$$\Lambda_e^{-1} = \left(\frac{2e^2}{\hbar v} \right)^2 \frac{\hbar^2}{2m} n \int d^2\mathbf{q}_\perp \left(\frac{1}{q_\perp^2 + (\omega_0/v)^2} \right) \left(\frac{1}{\hbar\omega_0 + (\hbar q)^2/2m} \right)$$

$$\times 2[1 - <0|\exp(iq_x x)|0><1|\exp(-iq_x x)|1>] \quad . \tag{6.61}$$

In the first denominator of the integrand, we have corrected for the minimum longitudinal momentum transfer $\hbar\omega_0/v$, obtained from energy-momentum balance. In the sudden-collision approximation, one neglects this quantity. The term corresponding to unity within the square brackets gives the total scattering out of the initial state, and it correctly leads to (6.54), apart from a factor of two in the large argument of the logarithm. The expression in the brackets becomes proportional to q_x^2 when the x component of the momentum transfer is small compared to $\hbar/a$, where a represents the spatial extent of the planar states $|1\rangle$ and $|0\rangle$. The integral therefore contains an effective lower limit a^{-1}, which is much larger than the limit ω_0/v. This suppression of the low-momentum contributions leads to a coherence length for channeling radiation substantially longer than the mean free path corresponding to the total scattering probability. As a result, scattering by electrons may usually be neglected, compared to the thermal component, in the evaluation of incoherence.

Comparisons between calculated and measured linewidths due to incoherence are shown in Fig.6.13. The theoretical curves are based on thermal scattering only. The short-dashed lines represent interband scattering alone, the long-dashed ones include the total thermal scattering in the initial and final states $n=1$ and $n=0$, and the intermediate solid lines correspond to the expression in (6.49). The small contribution from electronic scattering was estimated to be ~50 eV and does not lead to essential shifts of the theoretical curves. The experimental points are in good agreement with the solid lines and deviate significantly from the dashed ones. This confirms our treatment of intraband scattering.

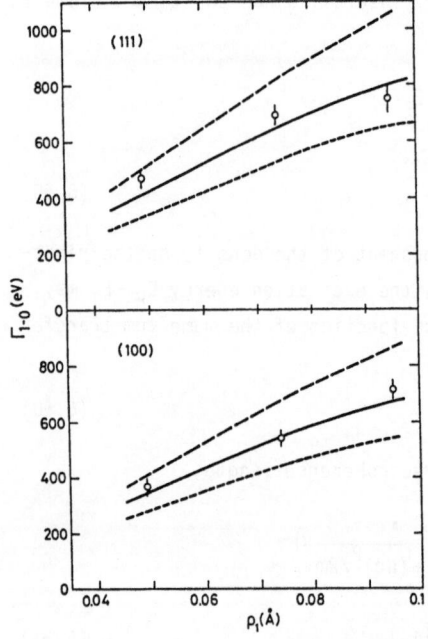

Fig.6.13. Linewidths corresponding to the line energies shown in Fig.6.9 for 4-MeV electrons in Ni are compared with theoretical estimates of the contribution from thermal scattering. The *solid lines* correspond to the expression in (6.49), while the *short-dashed lines* represent interband scattering alone, and the *long-dashed lines* include the total thermal scattering in the initial and final states, $n=1$ and $n=0$ (cf. Fig.6.8). The curves include a small enhancement (~10%) from correlation of thermal vibrations but not the contribution from electronic scattering, which was estimated from (6.61) to be ~50 eV. From [6.18]

It may be mentioned that the simple estimate (6.53) gives results which are fairly close to the upper lines in the figure. As expected (see Fig.6.8), the enhancement from the 'overlap estimate' is too high and too low for the states $n = 0$ and $n = 1$, respectively, and as it turns out, the two errors nearly compensate.

A similar comparison between measured and calculated linewidths for an axial case is shown in the lower part of Fig.6.6. As indicated previously, the relative contribution to incoherence from thermal intraband scattering is lower for axes than for planes. In the case considered, the difference between complete exclusion and indiscriminate inclusion of thermal intraband events amounts to $\lesssim 10\%$, and we show only the results corresponding to the top curves in Fig.6.13. On the other hand, correlation of vibrations is much more important in the axial case, and the calculated values in Fig.6.6 include an enhancement due to correlation of about 25%.

6.13 Multiple Scattering and Line Intensity

Scattering also affects the total intensity of a line. For the number of photons emitted around the forward direction in a transition $1 \rightarrow 0$ within a crystal of thickness L, we obtain from (6.38 and 42) with $\delta z = 0$

$$
dN_{ph,1 \rightarrow 0} \propto \sum_\alpha \int d\omega\, \omega^2 |<0|p_x|1>\omega^{-\frac{1}{2}} \int_0^L dz \exp\left[i\, \frac{z}{v}\, \omega(1 - n_r\beta) \right]
$$

$$
\times \int d\ell \int dy <w_\alpha(z)|u_0><u_1|w(z)>|^2 \quad . \tag{6.62}
$$

The square of the integral over z may be written as a double integral over variables z and z', and for a narrow line the factor ω may be replaced by a constant in the integration over ω, which then gives a delta function in $z - z'$. Generally, this result is obtained for a line shape which is symmetric apart from the factor ω, e.g., a Lorentzian. For fixed z, the functions $w_\alpha(z)$ form a complete set, and applying closure and including the constants given in Sect.6.3, we obtain

$$
dN_{ph,1 \rightarrow 0} = 2(1 + \beta^{-1})(1 + 2\gamma^2\delta)^{-1}\alpha\omega(mc)^{-2}|<0|p_x|1>|^2\, \frac{d\Omega}{4\pi}
$$

$$
\times \frac{1}{c} \int_0^L dz \int d\ell \int dy |<u_1|w(z)>|^2 \quad . \tag{6.63}
$$

Here, the frequency ω is determined by the expression in (6.14), and as before we have assumed that $\delta = 1 - n_r$ is small.

The integrals over ℓ and y give the total population $P_1(z)$ of the transverse state u_1 at the depth z, and the line intensity is proportional to the integral of $P_1(z)$. The various contributions to the projection onto the state u_1 appear with random phases, and in the absolute square of the sum the cross terms are eliminated through the integration over depth. The depth dependence of the populations $P_n(z)$ of the channeling states u_n, including the contributions from feeding, may therefore be obtained from the coupled differential equations

$$\frac{dP_n}{dz} = \sum_{n'} W_{n,n'}(P_{n'} - P_n) \quad , \tag{6.64}$$

where $W_{n,n'}$ denotes the probability per unit length for a transition between bands n and n'. The initial populations at the surface are determined by a Fourier component of the functions $u_n(x)$ as in (6.27). We have in (6.64) ignored the Bloch-vector index on populations and transition probabilities, and, as argued in Sect.6.8, this is a good approximation when we are concerned with localized levels and thermal scattering.

Formulas for the interband transition rates $W_{n,n'}$ may easily be derived through insertion of intermediate states into some of our previous expressions such as the first line in (6.49), which represents the total thermal scattering out of the band $n = 0$, $\sum_{n \neq 0} W_{n,0}$. An example of an explicit evaluation of the scattering rates is shown in Fig.6.14. There is a rapid communication between the deeply bound states $n = 0,1$ and the free states with $n' \geqslant 5$. In the ground state, the scattering to the other deeply bound state $n' = 1$ is much stronger than scattering to individual free states, but around 50% of the total interband scattering is still to states with $n' \geqslant 8$. This is consistent with our estimate $\hbar/\rho_1$ for the typical momentum transfer since each value of n' corresponds to a momentum interval $\hbar\pi/d_p$ in an extended-zone

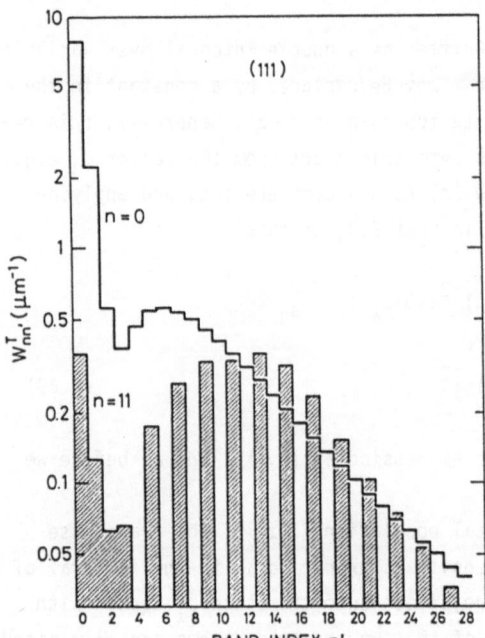

Fig.6.14. Calculated probabilities for thermal scattering between bands n,n' for 4-MeV electrons incident along a (111) plane in Ni at room temperature. The transition rates are shown for electrons populating two different initial states, the ground state $n = 0$ and a continuum state $n = 11$. From [6.18]

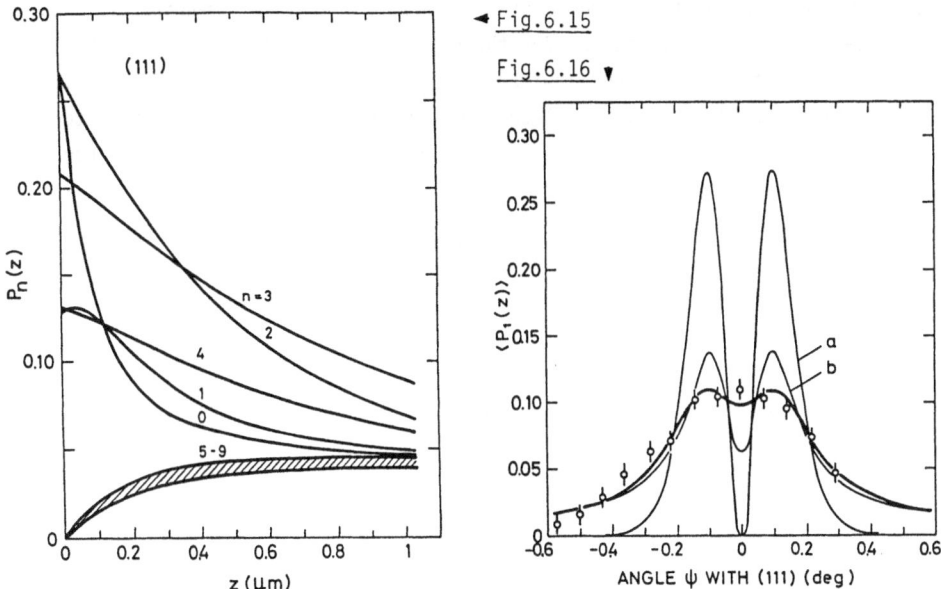

Fig.6.15. Depth dependence of the populations of different bands n for 4-MeV electrons incident at an angle of half the Bragg angle, $\psi = \psi_B/2$, with a (111) plane in Ni at room temperature. Only the thermal scattering, illustrated in Fig.6.14, has been included in the solution of (6.64). From [6.18]

Fig.6.16. Average over depth of the population of the n = 1 level for 4-MeV electrons incident at a small angle ψ to a (111) plane in a 0.5-μm-thick Ni crystal at room temperature. The thin-line curves correspond to the population at the surface (a), determined from (6.27), and the average population (b), obtained from an evaluation of the multiple interband scattering, as illustrated in Fig.6.15. The heavy-line curve includes an average over angles of incidence, corresponding to the mosaic spread in the crystal. The circles are the experimental populations obtained from application of (6.63) to the measured intensity of the 1→0 line. From [6.18]

scheme, and the ratio $\rho_1^{-1}/\pi d_p^{-1}$ is around nine. The large value of this ratio also shows that the variation of the transition rate $W_{n,n'}$ within one band may be neglected even for free states n', as was done in (6.64).

At the crystal surface, only the lowest five bands are significantly populated, but with increasing depth, the higher bands are fed by interband scattering, as illustrated in Fig.6.15. The decay rates for the levels with large initial populations reflect differences in strength of the scattering, which is strongest in the ground state and very weak in the nearly bound states n = 3,4. As could be anticipated from Fig.6.14, the occupation of the free states varies slowly with band index, and for n = 5-9 the populations are indicated by the hatched region. At a depth of 1 μm, statistical equilibrium has been attained within a factor of two for n ≤ 9 and almost half of the total population is in higher levels.

The average population of the n = 1 level may be deduced from the measured 1→0 radiation intensity via (6.63), using the calculated value of the matrix element

$<0|p_x|1>$. Measured populations from a scan through a plane are compared to calculated values in Fig.6.16. Curve (a) for the population at the surface was obtained from (6.27); it reflects the Fourier transform of the n = 1 wave function. Inclusion of the thermal interband scattering leads to curve (b) which is much closer to the measurements. Finally, an average over the distribution of angles of incidence corresponding to the mosaic spread is also carried out; we then obtain the heavy-line curve in Fig.6.16, which is in good agreement with the measurements.

Line intensities which are independent of scattering and beam collimation may be obtained through integration over directions of incidence. According to (6.27), this leads to a uniform population of the bands at the surface, expressed by the equation

$$(K/g) \int d\psi \, P_n(\psi) = 1 \quad . \tag{6.65}$$

It follows from (6.64) that such a uniform population is stable in depth. This point is also illustrated by the curves in Fig.6.16. The integral of the surface population over ψ equals twice the Bragg angle, $2\psi_B \equiv g/K$, and it is not changed by multiple scattering or by the additional average over angle of incidence. In the axial case, a result analogous to (6.65) is obtained, with a two-dimensional integration over angles and a factor K^2 divided by the area of the unit cell of the reciprocal-lattice plane perpendicular to the axis.

When the expression in (6.65) is written as an integral over transverse momenta $\hbar K\psi$, it is seen that a population of one particle per band is obtained for a uniform distribution of angles of incidence, with a density of one particle per transverse momentum interval $\hbar g$. Since each band contains one quantum state per planar spacing, division by d_p reduces the result to a form familiar from statistical mechanics: one particle per phase-space area $h = \hbar g d_p$ corresponds to unit population of quantum states.

The case illustrated in Figs.6.14-16 is characteristic for electrons in the low-MeV region. The change in transverse momentum due to a typical thermal-scattering event is sufficient to promote an electron from a deeply bound channeling state, with a binding energy of the order of a Rydberg, to a continuum state, provided $\hbar/\rho_1$ exceeds $\sim (m\gamma e^2/a_0)^{1/2}$, i.e., $\gamma \lesssim (a_0/\rho_1)^2 \sim 25$. In contrast to this direct communication between channeling states and free states, a diffusion picture applies in the GeV region.

In general for electrons, the enhanced thermal scattering for bound states leads fairly rapidly to an equilibrium between different bound states and between these and the lower states in the continuum. The population can then be estimated from expressions for random multiple scattering since a major fraction of the particles are in free states with normal scattering. In addition, modifications of scattering probabilities due to channeling average out for an equilibrium population.

Scattering in a random medium leads to an angular distribution which may be divided into a central Gaussian part, resulting from many small scattering angles, and a tail caused by single scattering. The fraction of particles in this tail is small for crystals which are not too thin. The two-dimensional Gaussian is characterized by the average-square angle, which increases nearly linearly with depth,

$$\frac{d}{dz} \langle \psi^2 \rangle \simeq \frac{4\pi Z^2 e^4}{(\gamma mc^2)^2} N \log \left(\frac{\psi_{max}}{\psi_{min}}\right)^2 \quad , \quad \gamma \gg 1 \quad . \tag{6.66}$$

The angles ψ_{min} and ψ_{max} are effective limits for the integration of the Rutherford cross section. For scattering in a random medium, the lower limit is determined by screening of the atomic potential, but since we here consider incoherent scattering only, there is, as seen in (6.51), a cutoff determined by a Debye-Waller factor at somewhat larger angles, $\psi_{min} \approx (\rho_1 K)^{-1}$. The upper cutoff excludes the single-scattering events responsible for the tail, $\psi_{max}^2 \approx \langle \psi^2 \rangle$ [6.37]. In one dimension, the Gaussian angular distribution is thus given by

$$P(\psi_x) = \exp(-\psi_x^2/2\langle \psi_x^2 \rangle)/(2\pi \langle \psi_x^2 \rangle)^{\frac{1}{2}} \quad , \tag{6.67}$$

where $\langle \psi_x^2 \rangle = \frac{1}{2} \langle \psi^2 \rangle$.

For a beam incident nearly parallel to a plane, the transverse phase space is structured into bands of channeling states, and according to our discussion above, each band corresponds to an angular interval $\Delta\psi_x = 2\psi_B$. Equilibrium between the bound states and the lower continuum states is established when the angular spread $\langle \psi_x^2 \rangle^{\frac{1}{2}}$ exceeds $n_{max}\psi_B$, where n_{max} is the band index of the highest bound level. The subsequent scattering will be nearly the same as for incidence in a random direction, and for the bound levels, the exponential in (6.67) may be set equal to unity. The population at depth z is then given by

$$P_n \simeq 2\psi_B (2\pi \langle \psi_x^2 \rangle)^{-\frac{1}{2}} \simeq (Z\alpha)^{-1} (d_p^2 zN)^{-\frac{1}{2}} \left[\log \left(\frac{\psi_{max}}{\psi_{min}}\right)^2\right]^{-\frac{1}{2}} \quad . \tag{6.68}$$

For the case illustrated in Fig.6.15, the logarithm is about 2 at a depth of 1 μm, and this leads to $P_n \simeq 5\%$, in good agreement with the detailed calculations, although statistical equilibrium has not quite been established.

In the two-dimensional case, the equilibrium occupation is very small, being of the order of the square of the expression in (6.68), and a description of the initial depopulation by single scattering is more appropriate. For a state with a high population at the surface, the dependence on depth is then nearly exponential, with a decay length equal to the mean free path for interband transitions.

For positrons, the scattering is strongly reduced for states with low transverse energy. Both for electronic scattering and for thermal scattering at large distances from a plane or axis, the momentum transfer is small, and transitions are mainly between neighboring states. The slow depletion of channeling states, "dechanneling", has been described in detail for heavy positive particles [6.1,2], and since the number of bound levels is usually fairly large for positrons, we may expect this classical picture to apply, at least qualitatively. When particles are dechanneled, they quickly scatter to larger transverse energies, and rechanneling can therefore be neglected. The population of channeling states decreases nearly exponentially, with a decay length roughly equal to the depth at which the average increase in transverse energy due to scattering becomes equal to the potential barrier. The latter is independent of γ, and since the average-square angle is proportional to γ^{-2}, as for random scattering described by (6.66), the increase in transverse energy scales with γ^{-1}, $(m\gamma)^2 <\psi_x^2>/2m\gamma \propto \gamma^{-1}$. The characteristic length for dechanneling is, therefore, approximately proportional to γ.

6.14 Applications

Channeling radiation offers a unique possibility for studying the motion of electrons and positrons through crystals. As in atomic spectroscopy, one obtains direct information on individual quantum states, their energies and lifetimes. This is in contrast to other methods such as transmission microscopy [6.29] and observation of wide-angle scattering yields [6.7]. With sufficient measurement accuracy, one may apply channeling radiation to learn about crystal properties such as the detailed charge distribution [6.26] or thermal vibrations and their correlations [6.18,25].

For potential applications of channeling radiation as a photon source, the question of incoherent scattering is crucial since sharp lines with high intensity are desired. For electrons, thermal scattering severely limits the coherence. As can be seen from (6.49-53), there is no explicit dependence of the coherence length on γ for relativistic particles, and hence typical linewidths are proportional to γ^2, (6.36). The line energy for a transition between states with specified indices n,n' scales with a lower power of γ ($\gamma^{3/2}$ for a harmonic potential), and therefore the relative linewidth increases with electron energy. In addition, the number of lines in the spectrum increases, (6.2,4), and as a result, spectra with well-separated discrete lines are produced only for electron energies below about 10 MeV for axial channeling and 100 MeV for planar channeling. These estimates are for targets of low atomic number such as carbon and silicon. The cross section for thermal scattering is roughly proportional to Z^2, and hence the spectra deteriorate rapidly with increasing Z.

For electrons, the useful crystal thicknesses for production of channeling radiation are also limited mainly by thermal scattering. However, in the planar case, a significant population of channeling states is retained over distances much longer than the mean free path for interband scattering, and the population decreases fairly slowly with depth, (6.68).

162

In many respects, the situation is quite different for positrons. A discrete spectrum is obtained only for planar channeling. In this case, the potential is nearly harmonic, and this results in a single line. The broadening is mainly due to anharmonicity, since the incoherent thermal scattering is strongly reduced. The weak scattering also leads to a very slow depopulation of channeling states, with a characteristic length for dechanneling which is proportional to energy and reaches values larger than a millimeter for GeV positrons [6.37]. Channeling radiation from GeV positrons ($\hbar\omega \gtrsim 10$ MeV) is an interesting source of high-energy photons in view of the fact that large crystal thicknesses may be usefully employed in the experiments, and that well-defined radiation frequencies can be obtained. However, the number of bound states is here sufficient to allow a classical description, a detailed discussion of which lies outside the scope of this chapter.

References

6.1   J. Lindhard: K. Dan. Vidensk. Selsk. Mat. Fys. Medd. **34**, No. 14 (1965)
6.2   D.S. Gemmell: Rev. Mod. Phys. **46**, 129 (1974)
6.3   J.U. Andersen, E. Bonderup, R.H. Pantell: Ann. Rev. Nucl. Part Sci. **33**, 453 (1983)
6.4   H. Überall: Phys. Rev. **103**, 1055 (1956)
6.5   M.L. Ter-Mikaelian: *High-Energy Electromagnetic Processes in Condensed Media* (Wiley-Interscience, New York 1972)
6.6   M.A. Kumakhov: Phys. Lett. **57A**, 17 (1976)
6.7   J.U. Andersen, S.K. Andersen, W.M. Augustyniak: K. Dan. Vidensk. Selsk. Mat. Fys. Medd. **39**, No. 10 (1977)
6.8   M.A. Kumakhov: Phys. Status Solidi (b) **84**, 41 (1977)
6.9   M.A. Kumakhov, R. Wedell: Phys. Status Solidi (b) **84**, 581 (1977)
6.10 N.K. Zhevago: Zh. Eksp. Teor. Fiz. **75**, 1389 (1978) [English trans.: Sov. Phys.-JETP **48**, 701 (1978)]
6.11 A.W. Sáenz, H. Überall, A. Nagl: Nucl. Phys. **A372**, 90 (1981)
6.12 D.M. Bird, B.F. Buxton: Proc. Roy. Soc. (London), Ser. A: **379**, 459 (1982)
6.13 P. Lervig, J. Lindhard, V. Nielsen: Nucl. Phys. **A96**, 481 (1967)
6.14 V.A. Bazylev, N.K. Zhevago: Zh. Eksp. Teor. Fiz. **73**, 1697 (1977) [English transl.: Sov. Phys.-JETP **46**, 891 (1977)]
6.15 G. Baym: *Lectures on Quantum Mechanics* (Benjamin, New York 1969)
6.16 J.D. Jackson: *Classical Electrodynamics*, 2nd edn. (Wiley, New York 1975), Chap.11
6.17 W. Heitler: *The Quantum Theory of Radiation*, 3rd ed. (Clarendon, Oxford 1954)
6.18 J.U. Andersen, E. Bonderup, E. Laegsgaard, A.H. Sørensen: Phys. Scr. **28**, 308 (1983)
6.19 A.H. Compton, S.K. Allison: *X Rays in Theory and Experiment* (Van Nostrand, New York 1935)
6.20 J.U. Andersen, E. Bonderup, E. Laegsgaard, B.B. Marsh, A.H. Sørensen: Nucl. Instrum. Methods **194**, 209 (1982)
6.21 R.W. Terhune, R.H. Pantell: Appl. Phys. Lett. **30**, 265 (1977)
6.22 J.U. Andersen, E. Laegsgaard: Phys. Rev. Lett. **44**, 1079 (1980)
6.23 N. Cue, E. Bonderup, B.B. Marsh, H. Bakhru, R.E. Benenson, R. Haight, K. Inglis, G.O. Williams: Phys. Lett. **A80**, 26 (1980)
6.24 P.A. Doyle, P.S. Turner: Acta Crystallogr., Sect. A: **24**, 390 (1968)
6.25 J.U. Andersen, E. Laegsgaard, A.H. Sørensen: Nucl. Instrum. Methods **B2**, 63 (1984)

6.26 J.U. Andersen, S. Datz, E. Laegsgaard, J.P.F. Sellschop, A.H. Sørensen: Phys. Rev. Lett. **49**, 215 (1982)
6.27 E. Laegsgaard, J.U. Andersen: Nucl. Instrum. Methods B**2**, 99 (1984)
6.28 J.U. Andersen, K.R. Eriksen, E. Laegsgaard: Phys. Scr. **24**, 588 (1981)
6.29 P.B. Hirsch, A. Howie, R.B. Nicholson, D.W. Pashley, J.M. Whelan: *Electron Microscopy of Thin Crystals* (Butterworth, London 1965)
6.30 J.U. Andersen: Nucl. Instrum. Methods **170**, 1 (1980)
6.31 A.H. Sørensen: Ph. D. Thesis, University of Aarhus, Denmark (1983)
6.32 R.H. Pantell, M.J. Alguard: J. Appl. Phys. **50**, 798 (1979)
6.33 V.A. Bazylev, V.V. Goloviznin: Radiat. Eff. **60**, 101 (1982); Zh. Eksp. Teor. Fiz. **82**, 1204 (1982) English transl.: Sov. Phys.-JETP **55**, 700 (1982)
6.34 K. Komaki, F. Fujimoto, A. Ootuka: Nucl. Instrum. Methods **194**, 243 (1982)
6.35 P.H. Dederichs: Solid State Phys. **27**, 135 (1972)
6.36 H. Esbensen, J.A. Golovchenko: Nucl. Phys. A**298**, 382 (1978)
6.37 H. Esbensen, O. Fich, J.A. Golovchenko, K.O. Nielsen, E. Uggerhøj, C. Vraast-Thomsen, G. Charpak, S. Majewski, F. Sauli, J.P. Ponpon: Nucl. Phys. B**127**, 281 (1977)

7. Channeling-Radiation Experiments

B. L. Berman and S. Datz

With 19 Figures

In the last five years, many of the properties of channeling radiation (CR) have been delineated experimentally. Channeling radiation is very intense, easily tunable, forward-directed, and for the planar case linearly polarized. Especially since about 1982, owing mainly to rapidly improving experimental techniques, there has been an explosive growth in the amount and quality of data that have been obtained. At present, in fact, the data are sufficiently accurate to call into question the validity of the "standard" potentials used to describe well-known crystals, such as diamond. Also, measurements have been made which demonstrate the sensitivity of CR as a diagnostic probe of impurities and defects in crystalline materials.

7.1 Historical Background

When a relativistic charged particle passes through a single crystal very nearly along a major crystalline plane or axis so that it is channeled in that direction, it undergoes periodic motion in the plane transverse to this direction, and hence it can radiate. Quantum mechanically, this channeling radiation corresponds to a radiative transition between two eigenstates of the transverse crystalline potential; when the transition occurs between two bound states, a sharp spectral line is emitted. In the forward direction in the laboratory frame of reference, the radiation is transformed upwards in energy, because of (a) the relativistic velocity of the charged particle, by a factor of $\gamma = E/mc^2$, where E is the total energy of the particle and m is its rest mass (this can also be thought of as a deepening of the crystalline potential well by a factor of γ) and (b) the Doppler shift, by an additional factor of 2γ. This combined factor of $2\gamma^2$ (equal to 2×10^4 for $\gamma = 100$, corresponding to electrons or positrons of ~50 MeV, for example) brings channeling radiation (CR) up into the interesting keV-to-MeV energy region, and makes it relatively easy to observe using the methods of X- and γ-ray spectroscopy, and relatively easy to tune by varying the incident particle energy. The same relativistic transformation folds the radiation forward in the laboratory into a narrow cone having a characteristic half-angle of $1/\gamma$ (equal to 10 mrad for the above example), and thus makes it very intense within that solid angle. For the case of planar channeling, the radiation is linearly polarized.

In the approximation that the field source of the transverse crystalline potential can be represented by planar sheets or axial strings of charge, the particle-crystal system is equivalent to a one- or two-dimensional hydrogenic atom (for the planar and axial cases, respectively). This establishes selection rules for the radiative transitions, and enables one to predict many of the detailed properties of CR by analogy with these simple quantum-mechanical systems.

CR was first observed at the Lawrence Livermore Laboratory in 1968 [7.1,2], and shortly thereafter at the U.S. Naval Research Laboratory [7.3], as a "low-energy enhancement" in the forward radiation spectrum when beams of either positrons or electrons, in the energy range 16-28 MeV, were channeled along the <110> or <111> axes in silicon. This low-energy enhancement was recognized as channeling radiation in 1975 by researchers at Stanford [7.4], and in 1978 by the Tomsk group [7.5], who showed that the enhancement decreased at still lower energies, i.e., that it had a broad peak structure ($\Gamma/E_p \simeq 1$, where E_p is the energy of the peak in the radiation spectrum and Γ is its width). The first observation of planar CR was made at Livermore (in collaboration with researchers from Stanford and Oak Ridge) in 1978, for positrons [7.6,7], as was the first observation of line spectra, for planar-channeled electrons [7.8]; these spectral lines correspond to single discrete transitions in the transverse planar potential well. These measurements were for 50- and 56-MeV positrons and for 28- and 56-MeV electrons directed along all three major planes in silicon. The first observation of planar CR at high energies was made shortly afterwards at Yerevan [7.9] with 4.7-GeV electrons incident along the (110) plane of diamond, followed by one at SLAC [7.10] with 4-14-GeV positrons.

The experimental measurements of channeling radiation that have been reported to date are summarized in Table 7.1, which classifies the measurements primarily by crystal species. Alternatively, one can divide the measurements into three broad groups by incident beam energy, which also is convenient from the point of view of the physics. At low energies ($E \lesssim 5$ MeV), the axial potential wells contain a manageably small number of bound states, while the planar wells often do not contain the minimum of two bound states necessary to produce a sharp spectral line. Therefore, this energy range is ideal for studying axial channeling radiation, as has been done at Aarhus, Albany, and Illinois. At intermediate energies ($5 \lesssim E \lesssim 150$ MeV), the number of states bound in axial wells is usually so large (at least for $E \gtrsim 15$ MeV) that discrete transitions cannot be resolved because their widths exceed their spacings (although there are a few exceptions), whereas the number of bound states in planar wells is modest for low-Z crystals (at least for $E \lesssim 75$ MeV). Therefore, this energy range is ideal for studying planar channeling radiation, as has been done at Livermore and Saclay. At high energies ($E \gtrsim 150$ MeV), discrete spectral lines can no longer be seen, so that the crystalline potentials are not amenable to study. However, certain ultra-relativistic effects can be studied at these ener-

166

Table 7.1. Channeling-radiation experiments

Crystal	Planar or Axial	Positron or Electron	E	Laboratory and Reference	Comments
Diamond					
	Axial	Electron	4.0 MeV	Aarhus [7.11]	Molecular analogy
	Planar	Electron	17, 31 MeV	Livermore [7.12]	
	Axial	Electron	17, 31 MeV	Livermore [7.12]	
	Planar	Positron	54 MeV	Livermore [7.13]	
	Planar	Electron	54 MeV	Livermore [7.13]	
	Planar	Electron	53 MeV	Saclay [7.14]	
	Planar	Electron	80, 110 MeV	Saclay [7.15]	
	Planar	Electron	600, 750, 900 MeV	Tomsk [7.16]	
	Planar	Electron	900 MeV	Tomsk [7.17]	Angular distribution
	Planar	Electron	900 MeV	Tomsk [7.18]	Polarization
	Axial	Electron	900 MeV	Tomsk [7.19]	Coherent brems-strahlung
	Axial	Electron	1.2 GeV	Kharkov [7.20]	
	Axial	Positron	4 GeV	SLAC [7.10,21]	
	Planar	Electron	4.7 GeV	Yerevan [7.9]	
	Axial	Electron	4.7 GeV	Yerevan [7.9]	
	Planar	Positron	4, 6, 10, 14 GeV	SLAC [7.10,21]	
Silicon					
	Axial	Electron	1.5 to 4.0 MeV	Aarhus [7.22]	Tilt-angle dependence
	Axial	Electron	2.0 to 4.5 MeV	Albany [7.23]	
	Axial	Electron	2.0 to 3.1 MeV	Illinois [7.24]	Coherent brems-strahlung
	Axial	Electron	3.0, 3.5, 4.0 MeV	Aarhus [7.25]	Molecular analogy, tilt-angle dependence
	Axial	Electron	3.5 MeV	Aarhus [7.26]	Temperature dependence
	Planar	Electron	4.0 MeV	Aarhus [7.27]	Coherent brems-strahlung, tilt-angle dependence
	Planar	Electron	17, 31, 54 MeV	Livermore [7.28]	Tilt-angle dependence
	Axial	Electron	17, 54 MeV	Livermore [7.28]	
	Planar	Electron	28, 56 MeV	Livermore [7.8]	
	Axial	Electron	28, 56 MeV	Livermore [7.8]	Tilt-angle dependence
	Planar	Positron	50, 56 MeV	Livermore [7.7]	
	Planar	Positron	54 MeV	Livermore [7.28]	
	Planar	Electron	54 MeV	Livermore [7.29]	$\Delta n = 3$ transitions
	Planar	Electron	54 MeV	Livermore [7.30]	Temperature dependence
	Axial	Positron	54 MeV	Livermore [7.28]	
	Axial	Positron	56 MeV	Livermore [7.7]	
	Planar	Electron	53, 80, 110 MeV	Saclay [7.15]	Angular distribution
	Planar	Electron	350 MeV	Tokyo [7.31]	Dechanneling length
	Axial	Electron	900 MeV	Tomsk [7.17]	Angular distribution
	Planar	Electron	1.2 GeV	Kharkov [7.32]	
	Axial	Electron	1.2 GeV	Kharkov [7.32]	
	Axial	Electron	1.2 GeV	Kharkov [7.20]	Effect of collimation
	Planar	Positron	10 GeV	Serpukhov [7.33]	

Table 7.1 (cont.)

Crystal	Planar or Axial	Positron or Electron	E	Laboratory and Reference	Comments
	Planar	Positron	2, 5, 7, 10, 20, 55 GeV	CERN [7.34,35]	
	Planar	Electron	2, 5, 7, 10 GeV	CERN [7.34,35]	
	Axial	Positron	2, 5 GeV	CERN [7.36]	
	Axial	Electron	2, 5 GeV	CERN [7.36]	
Germanium					
	Planar	Electron	17, 54 MeV	Livermore [7.37]	
	Planar	Positron	54 MeV	Livermore [7.37,38]	Use for nuclear physics
	Planar	Positron	2 to 10 GeV	CERN [7.35]	
	Planar	Electron	2 to 10 GeV	CERN [7.35]	
	Axial	Positron	2, 5 GeV	CERN [7.36]	
	Axial	Electron	2, 5 GeV	CERN [7.36]	
Nickel					
	Axial	Electron	2.2 to 3.0 MeV	Albany [7.39]	
	Planar	Electron	4 MeV	Aarhus [7.40]	Temperature dependence
Gold					
	Axial	Electron	1 to 3 MeV	Illinois [7.24]	No spectral lines
LiF					
	Axial	Electron	17 MeV	Livermore [7.41]	
	Planar	Electron	17, 31, 54 MeV	Livermore [7.41,42]	Tilt-angle dependence
	Planar	Positron	54, 83 MeV	Livermore [7.42,43]	Tilt-angle dependence

gies because the transverse velocity becomes relativistic; also, since the energy of the channeling radiation is now in the 1-50-MeV region, applications to nuclear physics become possible. CR studies at high energies have been carried out at Tomsk, Yerevan, SLAC, Kharkov, Serpukhov, CERN, and Tokyo.

Diamond and silicon crystals have been used for most of the studies of channeling radiation to date, for several reasons: "well-understood" crystal structure, availability of high-quality crystals, and high Debye temperature (since thermal vibrations broaden the CR spectral lines). Some experiments have been done with germanium, nickel, and gold crystals at Livermore, Aarhus, Albany, and Illinois, and recently the first results with a binary crystal (LiF), from Livermore, have been reported. (See Table 7.1 for references and certain other details for all the above.)

There has been progress on the theoretical side as well. Because there still exists a great deal of confusion in the literature, a brief discussion of the exact nature of channeling radiation is in order here. In general, for a given potential well, there are bound states and free (continuum) states. Thus, radiative transitions between these states include bound-to-bound (bb), free-to-bound (fb), and free-to-free (ff) transitions. For the transverse crystalline potentials, CR results from the sum of the bb and fb transitions (the ff transitions constitute the elements of other radiative processes, such as incoherent and coherent bremsstrahlung). Since the density of the continuum states, as well as their widths, increases rapidly with increasing excitation energy, and since the bound states near the top of the potential well become additionally broadened by the periodic nature of the crystal lattice (Bloch-wave broadening), the fb transitions overlap strongly in energy and hence do not give rise to discrete spectral lines, but only to a broad continuum structure (the "bump" underlying the bb peaks). This is not to be confused with the broad peak (referred to above) that results from the superposition of the large number of overlapping bb peaks which are present when there are a large number of *bound* states in the potential well, as for high-γ particles and high-Z crystals (this is usually alluded to as a "smear" rather than a "bump").

Theoretical calculations of CR spectra, particularly of the energies, widths, and transition strengths of the discrete spectral lines, have been carried out in several ways. The best modern technique is the "many-beam" approach, first applied to calculations of channeling radiation by the Aarhus group [7.27], which approximates the crystalline potential and hence the transverse wave function with a Fourier expansion. The number of terms ("beams") used in the expansion determines the number of energy levels and eigenstates computed; the calculation of these quantities converges rapidly (as a function of the number of beams) for the one-dimensional planar case (but not so rapidly for the two-dimensional axial case). This approach also takes into account the periodicity of the potential, so that the energy levels near the tops of the wells broaden into bands.

No matter how the computation is made, a transverse crystalline potential must be chosen to insert into the Schrödinger equation. (Except for very high incident-particle energies, the transverse motion is non-relativistic.) What has come to be the "standard" procedure is the use of the Doyle-Turner approximation [7.44] to tabulated electron-scattering factors [7.45] whose coefficients were obtained by fitting to relativistic Hartree-Fock calculations. Thermal vibration of the atoms is accounted for by multiplying the Fourier coefficients by the appropriate Debye-Waller factors.

Further discussion of calculational techniques is beyond the scope of this chapter; extensive discussions of the theoretical underpinnings of channeling radiation, both semiclassical and quantum-mechanical, are given in other chapters of this book. This chapter is concerned primarily with the experimental techniques that have been developed and the experimental results that have been obtained in recent years.

7.2 Experimental Apparatus and Techniques

Most of the CR experiments that have been carried out to date have made use of principal components that are very similar: an accelerator which serves as the source
of relativistic charged particles (to date, limited to positrons or electrons), a
beam-transport, energy-analysis, and collimation system which directs a monoenerge-
tic, low-divergence charged-particle beam onto a thin crystal, mounted in a goniometer capable of small angular steps around two or three orthogonal axes, and a
photon detector positioned at or near 0° with respect to the direction of the incident beam. The photon beam seen by the detector, typically a solid-state spectrometer, is usually (but not always) collimated to limit the field of view of the detector. This is done partly because of background considerations and partly to restrict the solid angle it subtends at the crystal in order to limit the Doppler
broadening resulting from multiple scattering of the beam particles in the crystal.
Because of the essential similarities between the experimental apparatus and pro-
cedures used at several laboratories, only one such arrangement, namely, the one
used at Livermore, will be described in detail here. Earlier accounts of this arrangement have been given in [7.6,46,47].

The Lawrence Livermore National Laboratory Electron-Positron Linear Accelerator
is a high-current, five-section, s-band linac, capable of operating between 5 and
170 MeV. When fully loaded (at ~70 MeV), its average electron beam current can
reach 700 µA. Its maximum (short-) pulse repetition frequency for normal operation
is 1440 s^{-1}, which is used for CR (or other) experiments for which the counting
rates are limited by pileup considerations. Positrons are produced by pair production in a thick, water-cooled, W-Re converter positioned several meters downstream
from the accelerator, upon which a 120-MeV, 180-µA (average) electron beam is
directed and focused (by steering coils and a quadrupole triplet). The positron or
electron beam is energy-analyzed with a bending magnet and slit to $\Delta E/E \simeq 0.1$-0.2%
for electrons or to 0.2-0.4% for positrons. Its angular divergence is then limited
by directing it through a thick copper collimator of diameter 2.4 mm for electrons
or 4.9 mm for positrons. The resulting beam current is limited further with the
linac gun to a level which results in a counting rate (in the photon detector) of
≲0.3 counts per beam burst (≲400 s^{-1}); the final beam current ranges from a few pA
to a few tens of pA, depending upon the species of crystal under study, its diameter, thickness, and orientation, and the beam energy and polarity. The beam-
transport system (other than steering coils) up to the point where the beam enters
the experimental cave consists of four dipoles, four quadrupole singlets, and three
quadrupole doublets (see [7.48]) for further details).

Figure 7.1 shows a schematic diagram of the experimental arrangement used for
radiation measurements. After the energy-analyzed and collimated beam of positrons
or electrons is transported through a heavy shielding wall into the experimental

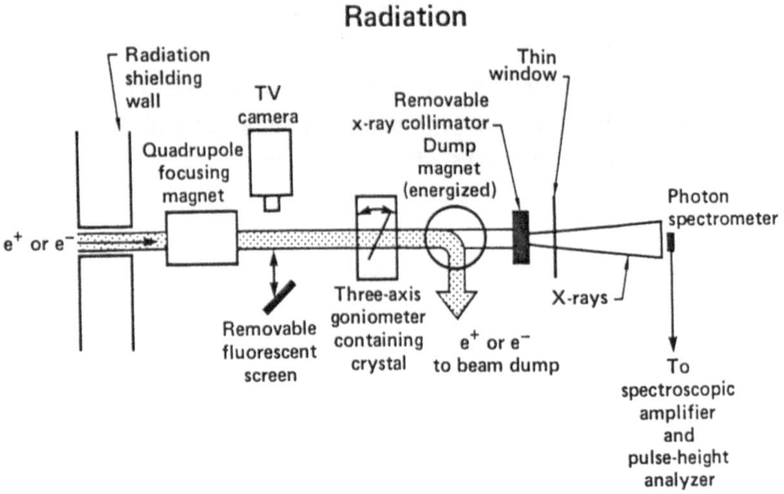

Radiation

Fig.7.1. Schematic diagram (not to scale) of the experimental arrangement at the Livermore linac for the measurement of channeling-radiation spectra from positrons or electrons (see text)

cave, it is defocused by an asymmetrically split quadrupole triplet to give a low-divergence (nearly parallel) beam incident upon the crystal in its goniometer. After it has passed through the crystal, the charged-particle beam is swept by a magnet into a 5-m-deep hole in the floor, through a large paddle-shaped plastic scintillator which serves as a beam current monitor. (The paddle was calibrated against a Faraday cup.) A thick, 4.9-mm-diameter tantalum collimator positioned approximately one third of the way from the crystal to the photon detector limits the angular divergence of the forward photon beam and also prevents the photon spectrometer [a large Ge(Li) or intrinsic-germanium detector] from viewing the crystal holder and other potential sources of background. Another, larger, brass collimator (19 mm in diameter) is positioned just upstream of the photon detector, and additional lead shielding surrounds the detector. With this arrangement, background data taken with no crystal in place are negligible.

The only major difference in the accelerators used at various laboratories from the point of view of CR experiments, other than their obvious differences in energy, is their duty factor. Pulsed machines, such as conventional linacs, have short duty factors (typically 10^{-4} to 10^{-3}), while Van de Graaff accelerators, dynamitrons, and microtrons, or storage rings operated in the pulse-stretcher mode, have long duty factors (sometimes approaching unity). The obvious advantage of long-duty-factor machines is in the rate of data collection, which is not so severely limited by pileup; thus, as has been done at Aarhus, data sufficient for the production of three-dimensional contour maps can be obtained routinely, making it possible for one to obtain a more complete representation of the phenomenon under study. Another advantage of a long duty factor is the ability to perform coincidence experiments, e.g., tagging the radiated photon with the electron that radiated it, as has been done at Tokyo and CERN. On the other hand, the advantage of pulsed machines lies in the time structure of the beam; intense bursts of positrons or electrons can be made as short as a few picoseconds, and trains of these pulses as short as 2 ns long are available routinely. A short pulse makes it possible to use channeling radiation to study transient phenomena, such as relaxation processes and the like. Although no such measurements have been reported to date, the very large cross sections for producing channeling radiation make such measurements possible (recall that when studying the channeling radiation itself, a few picoamperes of beam current are used when many microamperes are available).

The most critical factor in performing CR experiments is the divergence of the incident beam. Since the characteristic angle for the process is $1/\gamma$, an angular resolution at least an order of magnitude smaller is required in order to obtain data of sufficient precision to compare with the results of theoretical calculations. For $\gamma = 100$, for example, a beam divergence larger than 1 mrad is inadequate; for $\gamma = 10^4$ (~5 GeV), a divergence $\lesssim 10^{-5}$ rad is required. It has been largely through improvements in the beam divergence (now $\lesssim 0.1$ mrad for $\gamma = 100$) that recent progress in the field has been made. Of course, it should be clear that careful alignment (using a telescope or laser) of the experimental components along the beam line is absolutely essential for channeling-radiation experiments.

Although at high energies it is possible to track the trajectory of the radiating particle with multiwire chambers (see [7.34-36], for example), this technique depends upon low multiple scattering as well as long duty factor. Therefore, at lower energies, the beam itself must be collimated and tuned to low divergence. The experimental arrangement for accomplishing this at Livermore is shown in Fig. 7.2; this arrangement is also used for measurements of the transmission of positrons or electrons through crystals. A CsI scintillator, ruled with grid lines, is placed at the exit window of the vacuum pipe and viewed (through a mirror) with a television camera equipped with an image intensifier for high gain. The television signal is processed by a color quantizer, which assigns a different color to each of ten intervals of intensity. The upper and lower thresholds for each color are independently adjustable. When viewed by a high-quality color-television monitor, this gives a characteristic, multicolored bull's-eye pattern that greatly facilitates beam tuning. The beam is tuned through a removable collimator 9.6 mm in diameter positioned just upstream of the goniometer. With the dump magnet off, the beam pipe between this collimator and the CsI scintillator degaussed, and no crystal in the goniometer, the beam is tuned for minimum divergence. This is done by requiring a beam spot on the CsI scintillator that is as nearly as possible the same size as the collimator diameter when nearly 100% of the beam passes through the collimator, as measured with a plastic scintillator paddle positioned just downstream of the CsI

172

Transmission

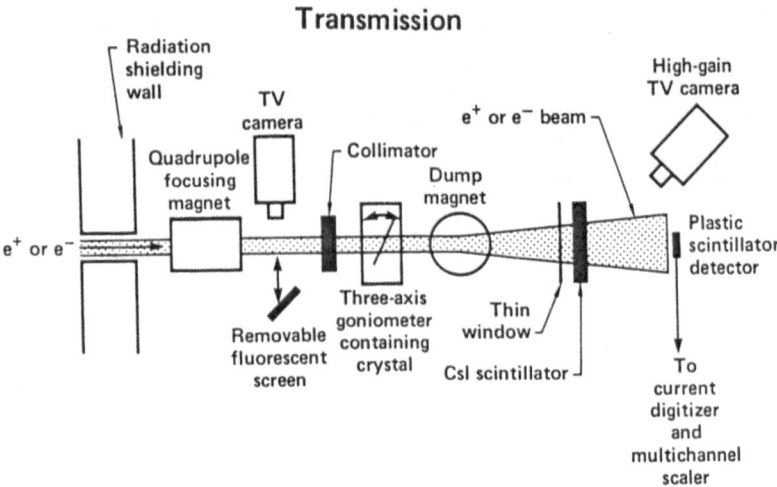

<u>Fig.7.2.</u> Schematic diagram (not to scale) of the experimental arrangement at the Livermore linac for the measurement of the transmission of positrons or electrons through a crystal (see text)

scintillator. The actual beam size and shape are measured subsequently with a small plastic scintillator button positioned at the same location. With this scintillator button, the beam is scanned both horizontally and vertically (in the transverse plane), with the CsI scintillator removed. By scanning the beam both with and without the collimator in place, the beam divergence (or convergence) is measured directly. For recent experimental runs, the beam divergence, both for positrons and for electrons, has been measured to be at the limit of sensitivity of this apparatus ($\lesssim$0.1 mrad), and hence is no longer a factor in considerations of angular or energy resolution.

With the scintillator button centered on the beam, the tuning collimator and the CsI scintillator removed, a crystal mounted in the goniometer, and the X-ray collimator (see Fig.7.1) inserted along the beam line, the arrangement of Fig.7.2 is used to make positron- or electron-transmission measurements. The crystal mapping is achieved most quickly and easily with positron-transmission scans, an example of which is shown in Fig.7.3a. One sees transmission peaks corresponding to planar channeling of the positrons (since the channeling directions are characterized by reduced scattering), from which a map of the crystal, like the one shown in Fig. 7.4, is constructed. One also can map the crystal, if it is of sufficient size and quality, with electrons, but the transmission dips for electrons (directions of increased scattering) are much smaller in their relative channel-to-random signal ratios than are the corresponding peaks for positrons.

Once the crystal is mapped by planar scans, the locations of the axes can be determined from the intersections of the planes, as can be seen in Fig.7.4. When transmission scans are made through an axis, the channel-to-random signal ratios

173

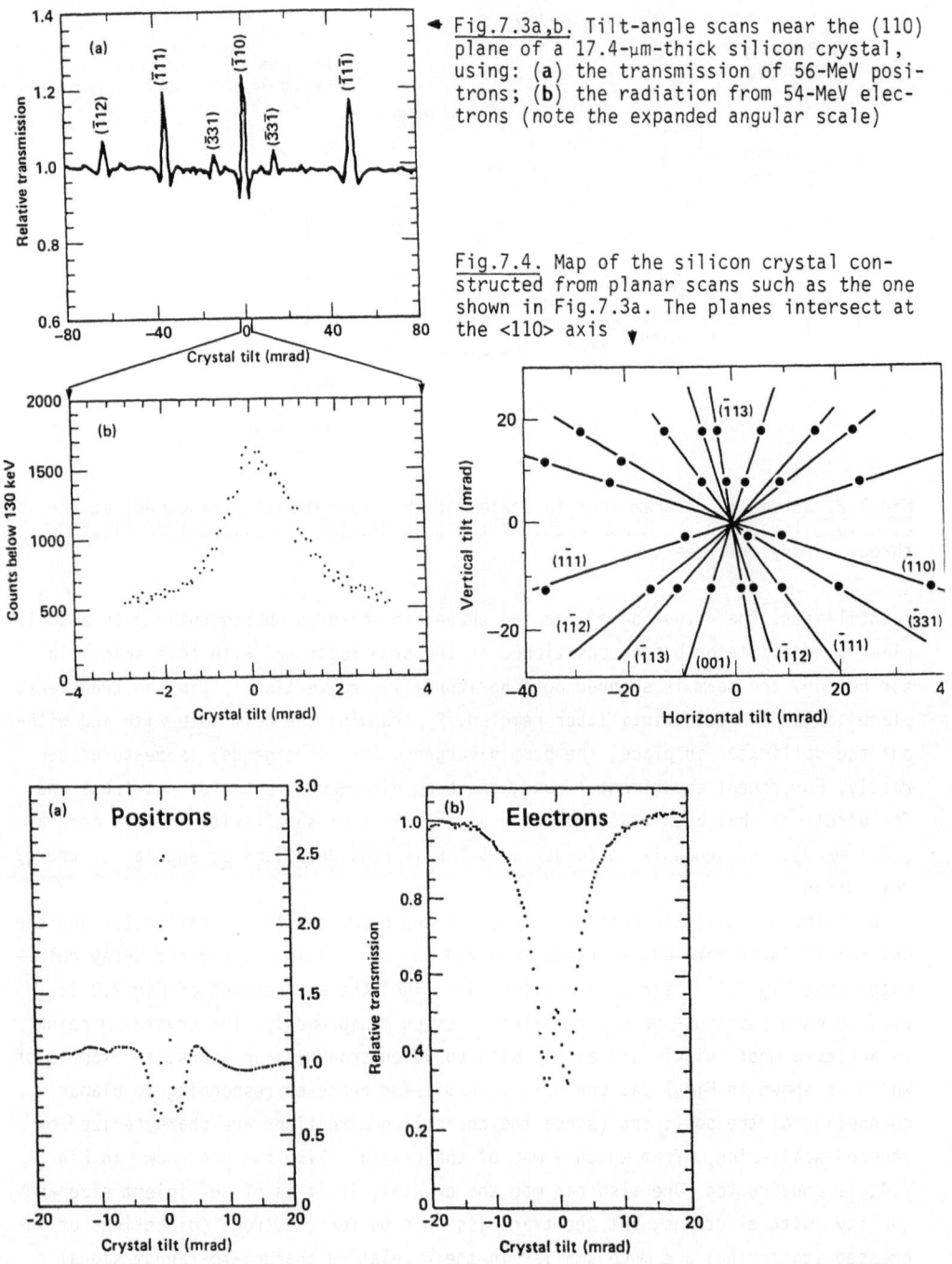

Fig.7.3a,b. Tilt-angle scans near the (110) plane of a 17.4-μm-thick silicon crystal, using: (a) the transmission of 56-MeV positrons; (b) the radiation from 54-MeV electrons (note the expanded angular scale)

Fig.7.4. Map of the silicon crystal constructed from planar scans such as the one shown in Fig.7.3a. The planes intersect at the <110> axis ▼

Fig.7.5a,b. Transmission scans through the <110> axis of the silicon crystal whose map is shown in Fig.7.4: (a) for 56-MeV positrons; (b) for 56-MeV electrons

are much larger than for a plane, as shown in Fig.7.5. Figure 7.5a shows a positron scan, with its characteristic prominent compensation shoulders (because of the conservation of the number of charged particles) just astride the large channeling peak. Figure 7.5b shows the characteristic "flying-W" pattern for an electron scan; the central peak results from the capture of incident electrons into bound (channeling) states of the deep axial string potential. (The compensation shoulders for electrons are much broader.) With a very thin crystal (and a highly collimated beam), a dip in the middle of this central peak has been seen [7.15,49], resulting from electron capture into s states, which have a large overlap with the atoms in the string; an electron in an s state is therefore much more likely to suffer a large-angle scattering event (i.e., it has a much shorter coherence length) than one in a bound state with nonzero angular momentum.

Once a crystal has been mapped, photon spectra can be obtained (with the experimental arrangement shown in Fig.7.1). Prior to this, however, it is important to scan the crystal orientation using the photon detector itself, in order to verify that the direction of the crystal plane or axis under study is truly along the beam line, since the positron (or electron) beam might have been deflected slightly by residual magnetic fields during the mapping scans. Figure 7.3b shows the results of such a photon scan for the (110) plane in silicon, where the detected photons between 20 and 130 keV (for incident 54-MeV electrons) are plotted against the crystal tilt angle on a greatly expanded scale. [The width of this peak exceeds the critical angle here because fb transitions increase the low-energy photon yield in an angular range wider than that over which bound-state channeling occurs; moreover, coherent bremsstrahlung (which corresponds to ff transitions) becomes important in the angular region just outside the critical angle.] In favorable cases, the crystal can be mapped entirely by means of photon scans, with no need for transmission scans; but because the data-collection rate for photon spectra is limited by pileup, such a procedure is tedious and time-consuming (except with long-duty-factor accelerators). Finally, it should be noted that measurements with fine angular resolution require a goniometer capable of small angular steps [the data of Fig.7.3b were obtained in 0.07-mrad steps]; at high energies, where all the angles and solid angles become exceedingly small, this requirement is very difficult to meet, which in turn *requires* precise tracking of the particle trajectories and hence both a long-duty-factor facility and very long flight paths.

Where fine photon-energy resolution (of the order of 1%) is required, the photon detectors of choice are germanium or silicon solid-state detectors, as have been used for all the experiments with incident beams below 150 MeV. For high-energy measurements, where no single-line spectra are visible, the resolution capability of large NaI detectors (~5 to 10%) is sufficient (see, for example, [7.34]). The photon detectors are generally preceded by a collimator, for the reasons given above; examples of spectra obtained both with and without such a collimator are given in [7.20].

Angular mapping of channeling radiation can be done in several ways, each giving different, but complementary information. When the photon detector is positioned along the incident beam line, it is said to be "at 0°". When, in addition, the incident beam is directed along a channeling direction in the crystal, the radiation spectrum observed is said to be "normal". For a very low-divergence beam and a very thin crystal, the population distribution will be dominated by low-n eigenstates, where n is the principal quantum number of a state bound in the transverse crystalline potential well; low-n eigenstates correspond to particle trajectories in the crystal which are flat with respect to the channeling direction. (High-n eigenstates correspond to oblique trajectories.) Beginning from the normal case, the angle of the detector, that of the crystal, or that of the beam can be varied. Variation of the detector angle (only) results in the mapping of the angular dependence of normal channeling radiation; this is called the "angular distribution" and has been measured by the authors of [7.15,17]. Variation of the beam angle (only) results in the mapping of the n-dependence of channeling radiation at 0°; no such measurements have yet been reported. Variation of the crystal angle results in the mapping of a convolution of the angular distribution and n dependence; this is called the "tilt-angle dependence" and has been measured for several of the cases studied at Aarhus, Livermore, and Saclay. Of course, the remarks above regarding angular resolution and goniometer step size apply equally to beam-direction and detector-angle variation as well.

A measurement of the degree of linear polarization of planar CR has been reported in [7.18]. For this experiment, a nuclear technique was used; the yield of photoneutrons from a sample of D_2O positioned at 0° was measured by two neutron detectors, both positioned in the plane of the D_2O sample transverse to the photon (channeling·radiation) beam. The CR "smear" for 900-MeV electrons along the (110) plane of diamond peaks at ~4 MeV, which nearly coincides with the peak of the photoneutron cross section for deuterium. This cross section is dominated by dipole photon absorption and direct neutron (and proton) emission, so that the photoneutron yield peaks at 90°. Therefore, deuterium acts as a nearly perfect polarization analyzer at 4 MeV, and the asymmetry of the photoneutron yields for the crystal plane oriented first in the plane of one neutron detector and then in the plane of the other gives a direct measure of the polarization of the channeling radiation. [It was reported in [7.18] as equal to 0.80 ± 0.15 for the (110) plane and 0.65 ± 0.15 for the (100) plane, in keeping with the imperfect alignment and divergence conditions of the measurement.]

176

7.3 Experimental Results

7.3.1 Perspective

Channeling radiation is generally described as arising from radiative transitions between bound states in a continuum potential. The concept of the continuum potential was first described by *Lindhard* [7.50] in connection with the channeling (directed motion) of heavy, positively charged particles penetrating crystals along a low-index axial or planar direction.

If a positively charged particle is injected into a crystal at a small angle with respect to an atomic row (a crystal axis), it will undergo a set of small-angle correlated collisions which steer the particle away from that row, i.e., each atom in the row acts to shadow (partially) the atom behind it. Lindhard showed, for positively charged projectiles of charge Z_1 injected with energy E into crystals made up of atoms of charge Z_2 with spacing d along the rows, that classically this steering effect takes place for angles less than the critical angle,

$$\psi_c = (2Z_1 Z_2 \, e^2/E \, d)^{\frac{1}{2}} \quad , \tag{7.1}$$

and, moreover, that the potential experienced by such particles is described by that of a continuous charge distribution made up from the atomic potentials in the row. The two-dimensional row potential $\bar{V}_R(\rho)$ is cylindrically symmetric and is a function only of ρ, the distance from the row. For any atomic screened Coulomb potential with screenign constant a,

$$\bar{V}_R(\rho) = (2Z_1 Z_2 \, e^2/d) K_R(\rho/a) \quad , \tag{7.2}$$

where the function K_R is defined in [7.50]. If one rotates a crystal about an axis into some arbitrary angle, the axial (row) symmetry disappears, but planar symmetry is retained. Particles injected within an angle to the plane

$$\psi_p = (2\pi n_p Z_1 Z_2 \, e^2 \, a/E)^{\frac{1}{2}} \quad , \tag{7.3}$$

where n_p is the areal atomic density in the plane, experience a continuum one-dimensional planar potential

$$\bar{V}_p = 2\pi n_p Z_1 Z_2 \, e^2 \, K_R(\rho/a) \tag{7.4}$$

where ρ here is the distance from the plane.

For massive particles, e.g., $m \gtrsim m_p$ (the proton mass), the number of states bound to such potentials is very large and their motion can be treated classically; but for lighter particles this is not the case.

At the time of writing, even with only five years of development in the field of CR, the body of accumulated data is already much too large (see Table 7.1) to cover the subject in detail. Instead, in this section we outline the various kinds of CR which have been observed and give some examples of each type.

177

7.3.2 Electron Channeling Radiation

When electrons are incident upon a crystal at a small angle with respect to an axis or a plane they may be captured into localized bound states which, for each axial direction (say), are eigenstates of the Hamiltonian

$$[p_\perp^2/2M + \bar{V}_R(\boldsymbol{\rho})]\psi_i(\boldsymbol{\rho}) = E_{\perp,i}\psi_i(\boldsymbol{\rho}) \quad , \qquad (7.5)$$

where $p_\perp$ and $\boldsymbol{\rho}$ are the projections of the momentum and position vectors of the particle on a plane perpendicular to the axis, and $M = \gamma m$ with $\gamma = (1 - \beta^2)^{-\frac{1}{2}}$ and $\beta = v/c$. Because of the relativistic increase in M the number of bound states increases with γ. (For planar channeling, because $\bar{V}_p$ is one dimensional, the number of bound states increases as $\gamma^{\frac{1}{2}}$.) Transitions can occur between bound states, giving rise to radiation which, for $\beta \simeq 1$, has an energy

$$\hbar\omega \simeq 2\gamma^2(E_{\perp,j} - E_{\perp,i}) \qquad (7.6)$$

when viewed in the forward direction in the laboratory frame.

a) Electron Planar-Channeling Radiation

For the case of planar-channeled electrons we have a one-dimensional potential and only one quantum number, n; the dipole selection rules dictate that Δn be odd. An example of the photon spectrum observed in the forward direction for 54-MeV electrons incident along the (110) planar direction in silicon (17-μm thick) is shown in Fig.7.6a. The bound-state $\Delta n = 1$ transitions are evident up to $n = 7 \rightarrow n = 6$, and at higher energies $\Delta n = 3$ transitions also are evident. Now, one can compare the observed spectrum with calculations based upon (for example) a Hartree-Fock description of the silicon atom. This can be done either directly by solving the

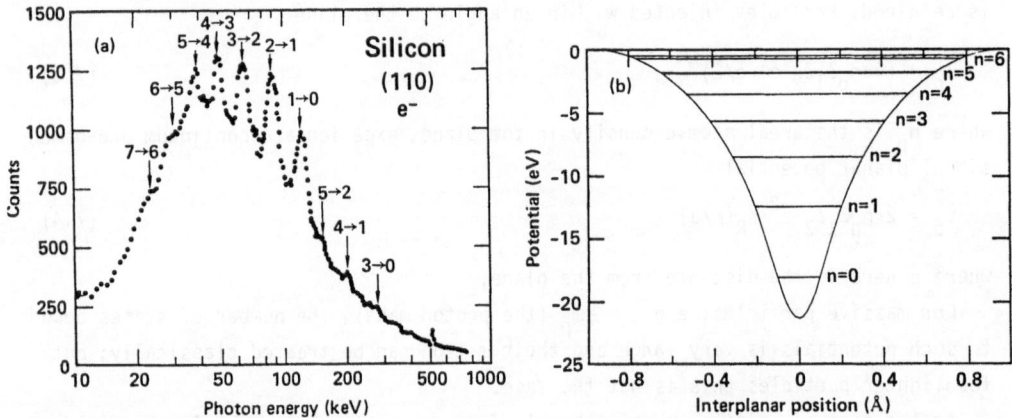

Fig.7.6a,b. For 54.5-MeV electrons channeled along the (110) planar direction in silicon: (a) measured photon spectrum for a 17-μm thick crystal and (b) calculated planar potential and energy levels

Table 7.2. Experimental and calculated spectral-line energies for 54.5-MeV electrons channeled along the (110) plane of silicon

Transition $i \rightarrow j$	Photon energy [keV]	
	Experiment	Theory
$1 \rightarrow 0$	122.3 ± 1	125.3
$2 \rightarrow 1$	88.4 ± 1	89.0
$3 \rightarrow 2$	64.2 ± 1	64.5
$4 \rightarrow 3$	49.1 ± 1	49.0
$5 \rightarrow 4$	38.1 ± 1	38.4

one-dimensional Schrödinger equation for the transverse motion, or via the many-beam formulation of the Schrödinger equation for this motion [7.27]. The planar potential and eigenstates derived from a many-beam calculation for the (110) plane in silicon are shown in Fig.7.6b; a comparison of the calculated and measured line energies is given in Table 7.2. The agreement for transitions with states having $n > 1$ is indeed excellent; the slight disagreement for transitions to the $n = 0$ state perhaps can be attributed to the sensitivity of the potential to the magnitude of the thermal vibration amplitude used in the calculation.

The thermal vibration amplitude also affects the effective coherence length and hence the linewidth (see Chap.6); hence, diamond, which has an extremely high Debye temperature, would be expected to display a sharper spectrum. This anticipation is amply borne out by the results, shown in Fig.7.7, reported by *Gouanère* et al. [7.14, 15] using a 20-μm-thick diamond. The changes in the spectra with increasing energy demonstrate (1) the scaling of the transition energies with incident electron energy (the line positions here are in good agreement with many-beam calculations), (2) the increase in the number of bound states with $\gamma^{\frac{1}{2}}$, and (3) the linewidth increasing with γ^2, observed in a detector with a fixed-acceptance aperture (4.3×10^{-6} sr).

For either the planar or the axial case the binding potential is extended in space and can be aligned with the incoming electron beam direction. The initial population of states depends upon the transverse momentum and hence upon impact parameter and incidence angle with respect to the axis or plane. Thus, by tilting the crystal with respect to the beam direction, one can populate specific states selectively and essentially map out the square of the wave function for that state. This has been demonstrated elegantly by *Andersen* et al. [7.27], and an example is shown in Fig.7.8 for the (110) plane in silicon. This two-dimensional plot shows a series of photon-energy spectra corresponding to the variation of the incidence angle of a 4-MeV electron beam in steps of $0.01°$ across the (110) plane of a 0.47-μm-thick silicon crystal. At this energy, only two states ($n = 0$ and $n = 1$) are bound. The variation in intensity of the line at ~1.6 keV represents the angular dependence of the population of the $n = 1$ state; this variation was found to agree well with calculations [7.27]. The small bumps in the energy spectra which increase in energy

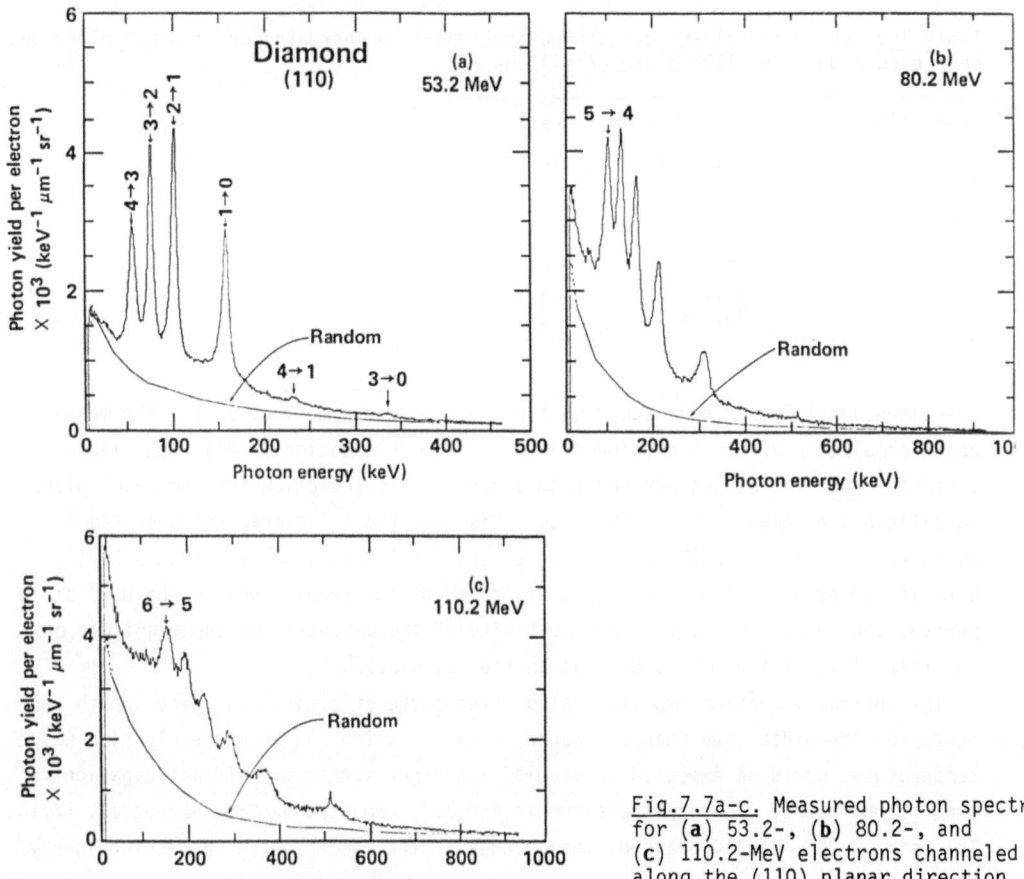

Fig.7.7a-c. Measured photon spectr for (**a**) 53.2-, (**b**) 80.2-, and (**c**) 110.2-MeV electrons channeled along the (110) planar direction in diamond

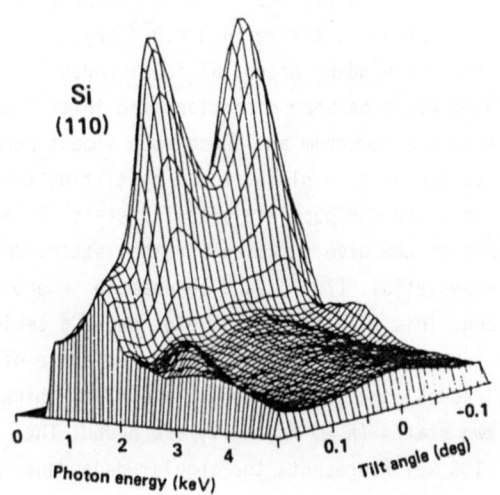

Fig.7.8. Measured photon spectra for 4-MeV electrons channeled along the (111) planar direction in silicon as a function of the tilt angle with respect to the plane

with tilt angle and extend out to ~3.5 keV result from coherent bremsstrahlung. (The connection between this radiation and channeling radiation is discussed in [7.27] and in Chaps.2,3.) With regard to experiments of this type, two things should be noted. First, the crystal specimen that is used should be thin enough so that multiple scattering within it does not obscure the effect of the initial entrance angle. Second, in order to carry out studies of this kind using photon-counting techniques in a reasonable time span, one requires a long-duty-factor electron beam from, e.g., a Van de Graaff accelerator or a microtron. (Using a pulsed accelerator, such as a linac, at ~1000 pulses per second requires that count rates be held down to $\lesssim 300$ s^{-1} to avoid excessive pileup.)

b) Electron Axial-Channeling Radiation

For the case of axially channeled electrons (e.g., as described by the continuum model) we have a cylindrically symmetric (two-dimensional) potential. Classically, the electron spirals around the row, and the projection of its path upon a plane perpendicular to the row forms a rosette with an ellipticity determined by the angular momentum. Quantum mechanically this corresponds to capture into bound states of the row potential (7.2). The eigenstates then are described in terms of a principal quantum number n and an angular-momentum quantum number ℓ.

Since the linear atomic density in a low-index axial direction is greater than the areal density in a low-index planar direction, the potential is stronger and, in general, the number of axial bound states is higher than the number of planar bound states for a given γ. Thus, e.g., for 54-MeV electrons in silicon, where planar line structure is clearly resolvable (Fig.7.6), axial channeling radiation is seen as an unresolvable continuum [7.8], where the linewidths are greater than the line spacings. In order to generate a spectrum containing resolved spectral lines, it is necessary to use lower incident-electron energy. For example, channeling of 16.9-MeV electrons along the <100> direction (the weakest of the principal axes) in diamond yields the spectrum shown in Fig.7.9a. The energy levels generated by a many-beam calculation are shown in Fig.7.9b; comparison of the theoretical and experimental line energies is given in Table 7.3. Here again, the agreement is excellent.

Because of the high density of states in axial channeling with high-energy particles, most of the detailed studies have been carried out in the low-energy region (1-4 MeV). *Andersen* et al. [7.22] have shown that the line energies in the longitudinal system of the electron ($h\nu_{lab}/2\gamma$) observed in the <100>, <110>, and <111> directions of silicon scale linearly with γ/d, where d is the interatomic spacing along a row, as can be seen in Fig.7.10. An example of a two-dimensional plot of photon spectra vs. tilt angle is shown in Fig.7.11 for 4-MeV electrons incident along the <111> axis of silicon (0.5 μm thick). The peaks, in order of increasing energy, are, as in the diamond case discussed above, due to 2s→2p, 3d→2p, 2p→1s,

181

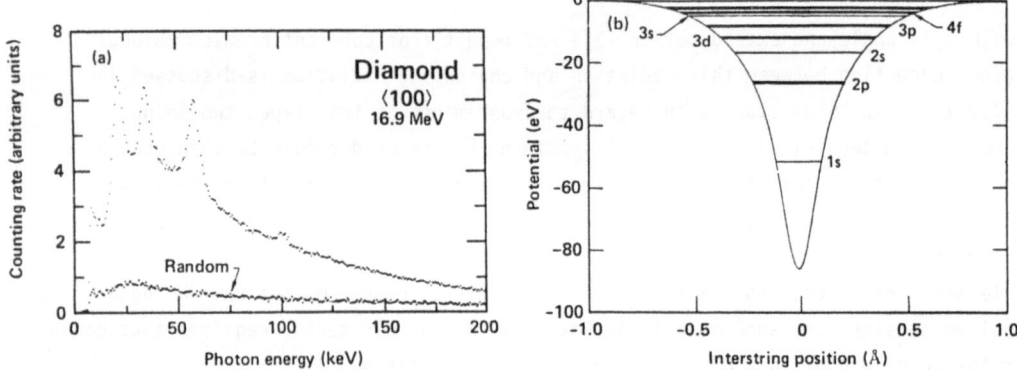

Fig.7.9a,b. For 16.9-MeV electrons channeled along the <100> axial direction in diamond: (a) measured photon spectrum and (b) calculated axial potential and energy levels

Table 7.3. Experimental and calculated spectral-line energies for 16.9-MeV electrons channeled along the <100> axis of diamond

Transition i → j	Photon energy [keV] Experiment	Theory
3p → 1s	101.5 ± 1.0	100.0
2p → 1s	58.3 ± 0.5	57.9
3d → 2p	35.0 ± 0.5	33.9
Several	21.9 ± 1.0	20.5[a]

[a]Strength-weighted mean of 12 closely spaced transitions

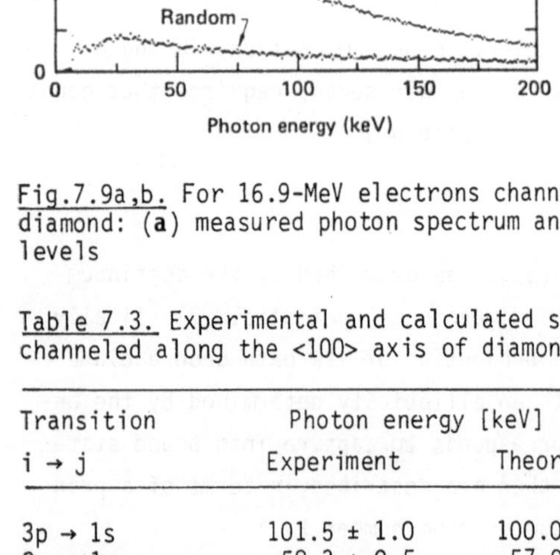

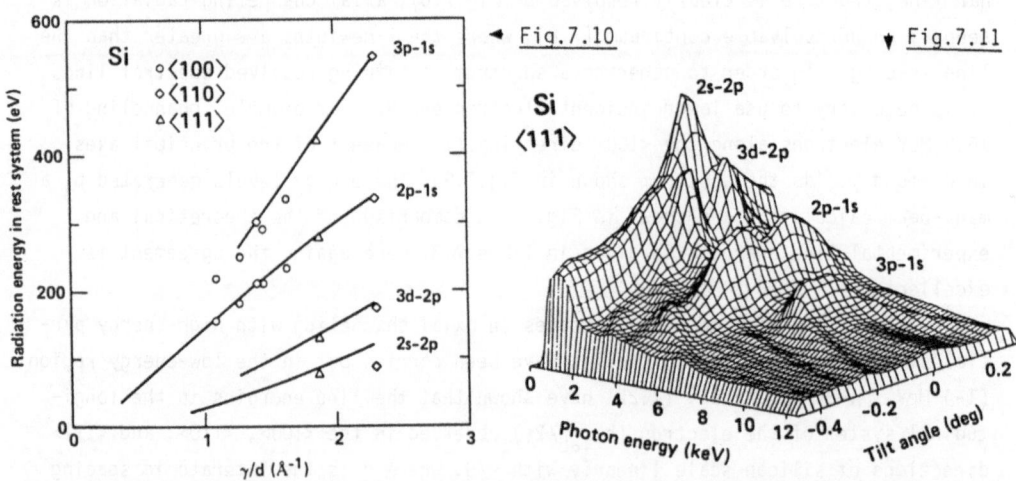

◄ Fig.7.10   ▼ Fig.7.11

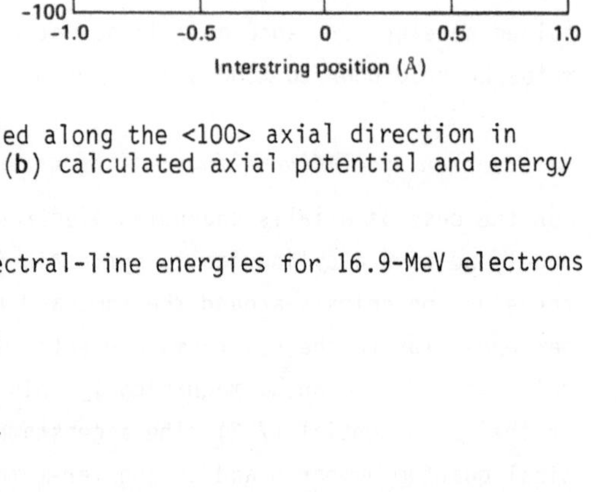

Fig.7.10. Variation with γ/d of the axial-channeling-radiation transition energies in the longitudinal rest frame of the incident electron for the three principal axes in silicon, compared with theoretical predictions using the single-row approximation and the Doyle-Turner potential

Fig.7.11. Measured photon spectrum for 4-MeV electrons channeled along the <110> axial direction in silicon as a function of the tilt angle with respect to the axis

and 3p →1s transitions. Note that the only line which peaks at zero angle with
respect to the axis is that arising from the 2s →2p transition, because it is the
only one which depends upon the initial population of an s state. An analysis of
the angular dependence of the 2p →1s intensity shows excellent agreement between
theory and experiment.

c) Two-Dimensional "Molecular" Bound States for Channeled Electrons

As discussed above, axially channeled electrons are captured into bound states of
the row potential. When two rows lie in close proximity as in the case of the <110>
direction in diamond, the potentials from these rows overlap, forming a saddle point
between the rows. The potential for 4-MeV electrons channeled along the <110> axis
in diamond is shown in Fig.7.12b. Solving for the eigenstates of this potential,
one finds that the 1s states are well localized around a single row. However, the
2p states lie above the saddle point and split into four molecular-type levels.
The resultant spectrum (from [7.11]), shown in Fig.7.12a, gives evidence for this
effect. Its qualitative features are obtained in a simple treatment, which is analogous to the linear-combination-of-atomic-orbitals (LCAO) method in chemistry.
When the transverse Hamiltonian is diagonalized in the subspace spanned by the
four single-row 2p states for a pair of rows, four eigenstates are obtained. These
eigenstates can be classified according to their symmetry under reflection in the
mid-plane between the rows, gerade and ungerade, and under reflection in a normal
line connecting the two rows, σ and π. The line energies for transitions between

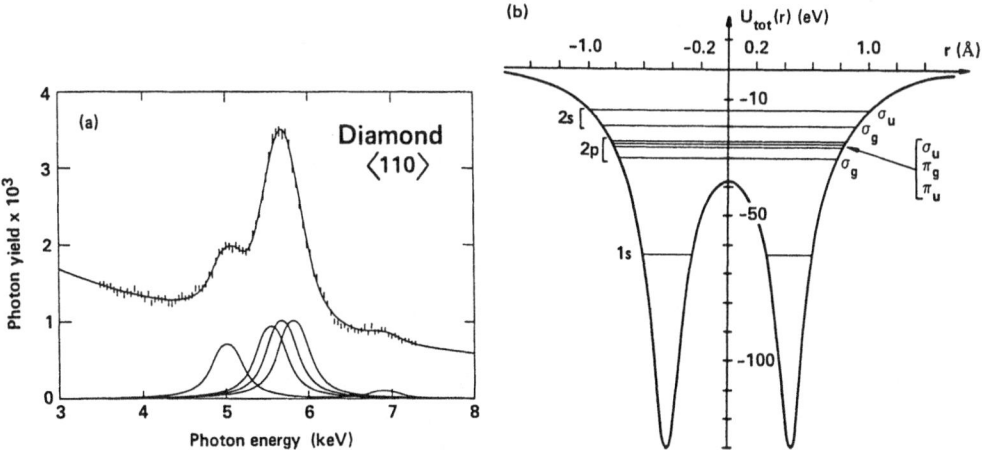

Fig.7.12a,b. For 4-MeV electrons channeled along the <110> axial direction in diamond: (a) measured photon spectrum and (b) calculated potential and energy levels.
The fitting of the 2p →1s lines in (a) involves assumptions of equal populations
of the 2p levels, which is expected because of incoherent scattering, and of equal
widths, which is anticipated because of the dominance of the 1s state in the determination of the coherence lengths

Table 7.4. Photon energies (in eV) for the 2p →1s and 2s →1s transitions for 4-MeV electrons channeled along the <110> axis in diamond

Transition	Single-string	First-order LCAO	With 2s - 2p mixing	Many-beam	Experiment
$\sigma_g 2s \rightarrow \sigma_u 1s$			7253	7103	6933
2p →1s	5801				
$\sigma_u \rightarrow \sigma_g$		6025	6015	6019	5897
$\pi_g \rightarrow \sigma_u$		5888		5887	5751
$\pi_u \rightarrow \sigma_g$		5771		5742	5624
$\sigma_g \rightarrow \sigma_u$		5279	5115	5090	5084

molecular energy levels obtained from the LCAO treatment are given in Table 7.4, together with the single-row values obtained from a solution of the two-dimensional Schrödinger equation. Mixing with the near-lying 2s levels, which turns out to be quite strong, particularly for the $\sigma_g 2p$ level (which is lowered considerably when mixed with the $\sigma_g 2s$ level), is introduced. The splitting between the lines is seen to increase by ~150 eV so that it is now ~100 eV larger than the observed splitting. Also, a dipole 2s →1s transition becomes allowed, and a corresponding weak line is visible in the spectrum (see Fig.7.12a). However, while the LCAO-type model is very instructive for qualitative purposes, its accuracy is limited and difficult to assess.

The values obtained for CR energies from the many-beam formulation are listed in Table 7.4. For the 2p →1s transitions, there is reasonable agreement with the result of the LCAO-type calculations. The upper three lines and the 2s →1s line are higher than the measured values by ~120 eV. However, the most significant deviation is for the separation of the $\sigma_g \rightarrow \sigma_u$ line, which is calculated to be larger by ~130 eV than the experimental value. This has been attributed (in [7.11]) to the accumulation of charge in the tetrahedral bonds in diamond.

Each atom in a <110> row is bonded to two atoms in a neighboring row and to one atom in each of the next-nearest neighboring row. Hence, the accumulation of charge in the bonds increases the electron density between close-lying pairs of <110> rows and increases the potential energy of electrons channeled in this region. Since the $\sigma_g 2p$ state has a high density here, this level will increase in energy relative to that of other 2p levels, i.e., the splitting will be reduced, as required by experiment. Using the many-beam method with the Gaussian parameters obtained from X-ray analysis, the energy levels were recalculated as a function of the electron-density enhancement; the closest match with the observed splitting is obtained with an electron density of 1.7 $\AA^{-3}$ in the center of the bonds. Although this determination is probably less accurate than that obtained from the X-ray data, it does demonstrate that information on charge distributions and potentials in crystals can be obtained from channeling radiation.

7.3.3 Positron Channeling Radiation

a) Positron Axial-Channeling Radiation

Unlike electrons, positively charged particles are repelled from the positively
charged cores (nuclei) which are the source of the axial and planar potentials. In
axial channeling we can distinguish two kinds of motion. In the first kind, within
the critical angle ψ_c, the particle experiences a repulsive row potential but has
enough transverse energy to pass over the small potential col between two adjacent
rows. The particle, although channeled, may then wander about between strings in an
unstable trajectory, and its radiation would appear as a broad spectral band. Particles with even larger transverse energy may be scattered into one of the inter-
secting planes that make up the axis [7.29] or into a random trajectory. In the
second kind of motion, the positively charged particle is contained between a set
of rows that make up the axis (e.g., the four equally spaced rows in the <100> direction of a face-centered cubic crystal). The particle is then said to be "hyperchanneled" [7.30], is bound in a stable trajectory, and has a sufficiently large
coherence length to radiate in a narrow band. This type of motion requires that the
incident beam be extremely parallel, since only a small amount of transverse energy
is required to pass over the barrier to the next set of rows. Recently, *Bak* et al.
[7.36], using 5-GeV/c positrons in silicon and germanium with an angular resolution
of 20 μrad (compared to axial-channeling critical angles ψ_c ranging from 130 to
320 μrad), have observed features in the photon spectrum identifiable with stable
hyperchanneled trajectories. At the time of writing, however, the analysis of these
data has not yet been completed.

b) Positron Planar-Channeling Radiation

While planar channeled electrons are bound to a single atomic plane, positrons
can be captured in the potential trough (for positively charged particles) between
two adjacent planes. Here we define the interplanar potential $\bar{V}$ in terms of the displacement x from the midpoint between the two planes which lies at a distance ℓ from
each plane:

$$\bar{V}_2(x) = \bar{V}_p(\ell + x) + \bar{V}_p(\ell - x) \quad . \tag{7.7}$$

Positively charged particles undergo an oscillatory motion in this potential,
which, from studies of trajectories of planar-channeled positive ions [7.52],
has been shown to be describable in terms of atomic Hartree-Fock potentials.

In single-element crystals the interplanar potentials are approximately harmonic
and the positron CR spectrum consists of a set of closely spaced lines because of
the nearly equal spacing of the energy levels; i.e., if the potential were purely
harmonic, the transition frequency would be

$$\omega_0 = (K/m)^{\frac{1}{2}} \quad , \tag{7.8}$$

where the force constant $K = \bar{V}_2(0)$, and m is the rest mass of the positron. Since $M = \gamma m$, the frequency in the laboratory frame becomes $\omega = \omega_0 \gamma^{-\frac{1}{2}}$, and the energy of the photon emitted in the forward direction becomes

$$\omega_{max} \simeq 2\gamma^2 \omega = 2\gamma^{3/2} \omega_0 \quad . \tag{7.9}$$

This type of radiation was predicted by *Kumakhov* [7.53], and was first observed experimentally by *Alguard* et al. [7.7], for positron channeling in silicon.

In reality, of course, the potential $\bar{V}_2(x)$ is not exactly harmonic; it rises more steeply as it approaches the plane and then rounds off because of the influence of the adjacent interplanar potential. A plot of the potential for 54-MeV positrons channeled along the (110) plane in silicon is shown in Fig.7.13b, together with the energy levels obtained from a many-beam calculation. In Fig.7.13a we show a measured spectrum from a 17-μm silicon crystal [7.28] and compare it with the spectrum predicted by the theory. The predicted spectrum shown here includes the

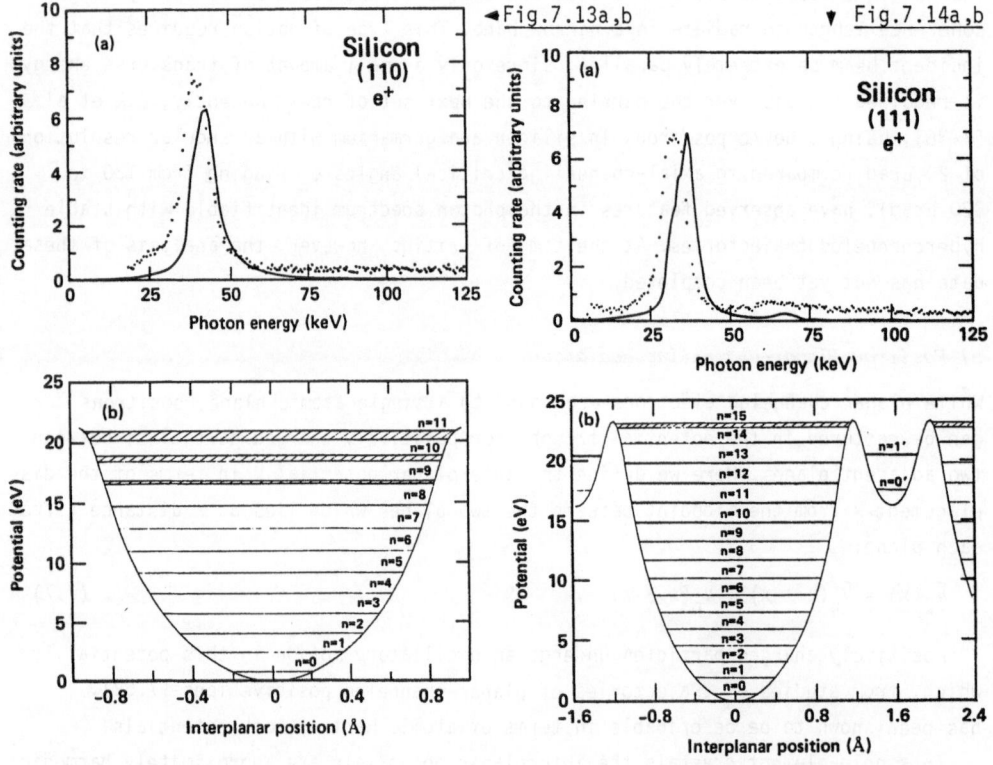

Fig.7.13a,b. For 54.4-MeV positrons channeled along the (110) planar direction in silicon: (a) measured and calculated photon spectra and (b) calculated interplanar potential and energy levels

Fig.7.14a,b. For 54.4-MeV positrons channeled along the (111) planar direction in silicon: (a) measured and calculated photon spectra and (b) calculated interplanar potential and energy levels

line positions, their intensities, and all of the important line-broadening effects. In Fig.7.14 we show the same calculations and comparison for the (111) plane. The difference to be noted here is a small secondary peak at higher energies, which results from a transition in the narrow (111) plane (1' →0'). This is predicted to appear at ~68 keV and appears in the data at ~62 keV. For 54-MeV positrons in diamond there is only a single bound state in the narrow (111) plane, shown in Fig. 7.15b, and this secondary peak does not appear in the measured spectrum of Fig.7.15a. Here the predicted spectrum is shown as a set of lines associated with the allowed transitions, without the broadening effects included.

One noteworthy feature which appears in three of these figures is the theoretical prediction of values for the peak energies which are higher than the observed values. The same potential which binds electrons to planes confines positrons to bound states between adjacent planes; hence the radiation spectra from both electron and positron channeling should be calculable from the same potential. Yet, the same potential which accurately predicts the energies of the peaks in the electron channeling-radiation spectra fails to predict those for positron channeling radiation. The origin of this discrepancy has not yet been found.

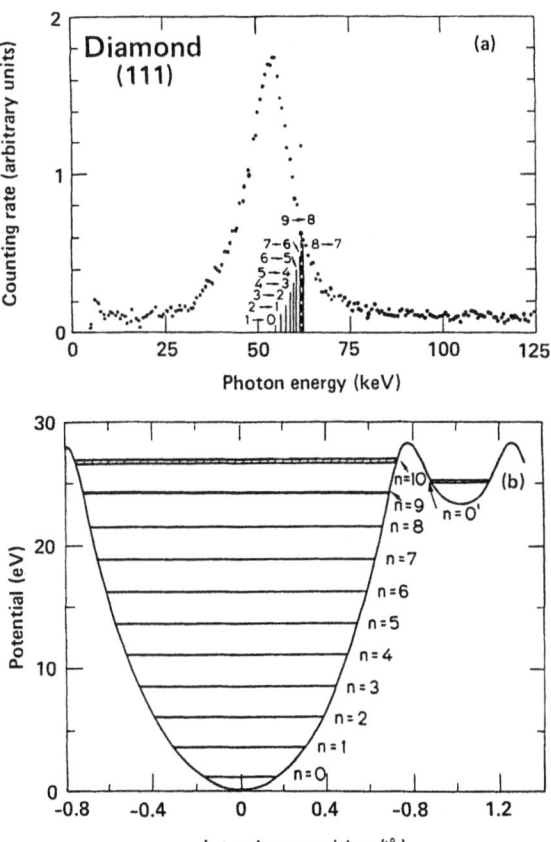

Fig.7.15a,b. For 54.4-MeV positrons channeled along the (111) planar direction in diamond: (a) measured and calculated photon spectra and (b) calculated interplanar potential and energy levels

c) High-Energy (≳1 GeV/c) Planar-Channeling Radiation

For electron channeling in the high-energy region the number of planar bound states is much too high to permit observation of individual transitions, and one expects only a continuum (a "smear"). This expectation has been borne out in the experiments, as exemplified in Fig.7.16b, which shows the photon spectrum obtained with 7-GeV/c electrons channeled along the (110) planar direction in a 135-μm-thick silicon crystal [7.35].

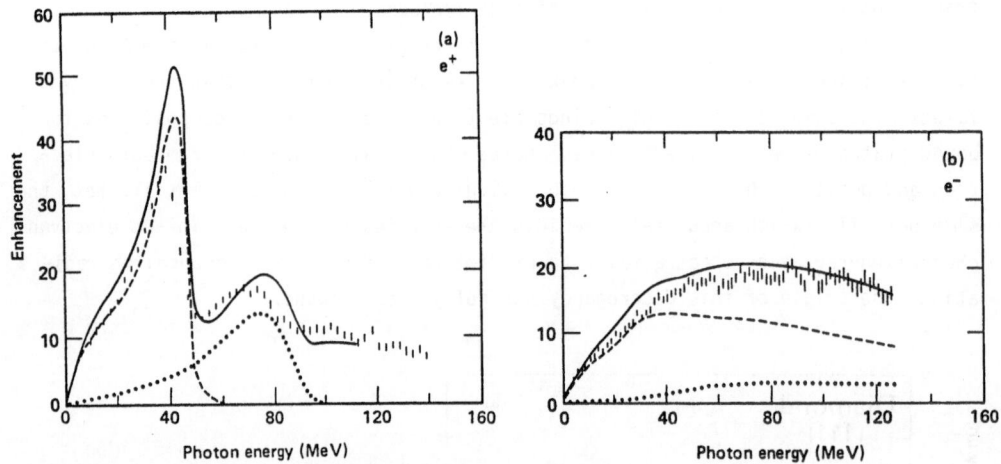

Fig.7.16a,b. Measured and calculated photon spectra for (**a**) 7-GeV/c positrons chan-neled along the (110) planar direction in silicon and (**b**) 7-GeV/c electrons chan-neled along the same direction

More interesting is the spectrum obtained with 7-GeV/c positrons shown in Fig. 7.16a. The expected $\gamma^{3/2}$ scaling of the photon energies has been borne out well up to energies of 83 MeV [7.28]. This simple scaling is expected as long as the transverse energy is nonrelativistic. More generally, the formula for the maximum photon energy in the Kth harmonic is

$$\omega_{max} = 4\pi K \gamma^{3/2}/[A'(E_\perp) + \gamma A(E_\perp) m_e c^2] \quad , \tag{7.10}$$

where the derivative of the action $A(E_\perp)$ is $A'(E_\perp) = P(E_\perp,\gamma)/\sqrt{\gamma}$, $A(0) = 0$, and $P(E_\perp,\gamma)$ is the period of the oscillation.

The denominator of (7.10) tends to diminish the γ dependence of the photon energy and its effect is greater on the higher-amplitude (higher-$E_\perp$) trajectories. Its ef-fect, as long as the lower-amplitude trajectories are not too much affected, is to compress the high-energy edge of the spectrum. This can be seen in the data of Fig. 7.16a for 7-GeV/c positrons. At higher momenta (10-55 GeV/c) the broadening effect on each transition smears out the spectrum. The theoretical curve (solid line) shown in Fig.7.16a is derived from numerical calculations using a thermally averaged Hartree-Fock potential. It is the sum of the contributions from the first- and second-harmonic contributions. The agreement in shape is quite good but, as was observed at lower energies, the theory predicts significantly (5-10%) higher energies than those observed.

7.3.4 Binary Crystals

Additional information on CR can be obtained using compound crystal targets. Here we use LiF as an example [7.41-43]. Lithium fluoride has a simple cubic structure. The (100) and the (110) planes are made up of equal numbers of Li^+ and F^- ions, so that the planar potential smooths over individual atom differences and one obtains an averaged potential characteristic of $Z = 6$ atoms with the areal density characteristic of the LiF planes. Using this assumption one obtains reasonable agreement for both electron and positron channeling radiation for 54.4-MeV particles in LiF.

Of particular interest is the (111) plane, because in this direction the crystal consists of alternating planes of Li^+ and F^- ions. The potential and eigenstates for positron channeling in this plane are shown in Fig.3.17b. In this figure, a Li^+ plane (at the zero position) lies halfway between two F^- planes, which give rise to a double (nearly harmonic) well within a larger (nearly harmonic) well. A comparison of the calculated line energies with the measured spectrum [7.43] is shown in Fig.7.17a. Qualitative agreement is evident, but it is clear that greater refinement in the theory is necessary.

Somewhat better accord with theory is obtained with electron channeling [7.41, 42], as can be seen in Fig.7.18. In Fig.7.18b the zero is taken at the center of an F^- plane and the secondary well is centered on the Li^+ plane. At 54.4 MeV the $1' \rightarrow 0'$ transition is not distinguishable from the $3 \rightarrow 2$ transition. However, they can be separated by changing the electron energy, and the $1' \rightarrow 0'$ transition has been observed as a distinct line using 30.5-MeV electrons [7.41].

7.3.5 Crystal Defects

As can be seen from the above, the study of channeling radiation is at a stage of development where small differences between theory and experiment in well-ordered crystals become interesting and where the effects of crystal imperfections on channeling-radiation spectra might be used as a diagnostic tool to characterize crystal defects.

A start in this direction has been made, with a study of the effect of platelet defects in diamond on channeling radiation from 54-MeV positrons and electrons [7.13,54]. In natural diamond, one type which occurs often (classified Type Ia) contains nitrogen impurities in the form of aggregates which are responsible for an infrared absorption peak at 7.8 μm. Some Type-Ia diamonds have an absorption peak at 7.3 μm resulting from the presence of platelets. These are thought to be

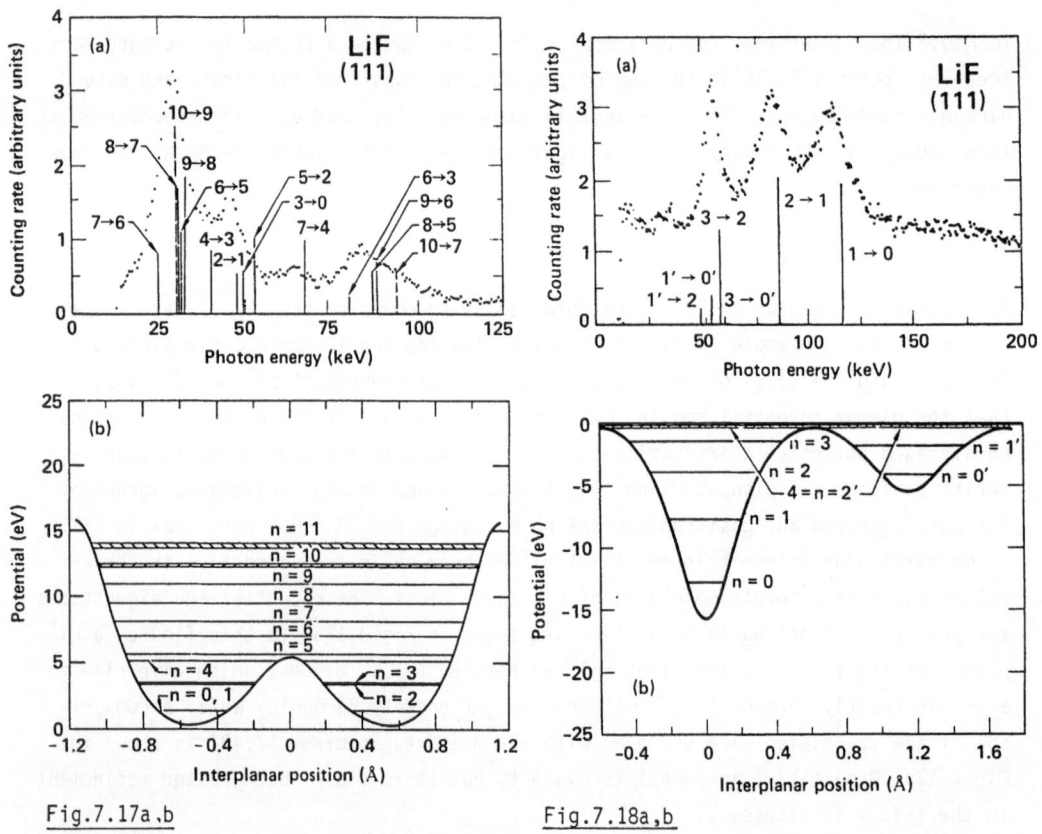

Fig.7.17a,b

Fig.7.18a,b

Fig.7.17a,b. For 54-MeV positrons channeled along the (111) planar direction in LiF: (a) measured and calculated photon spectra and (b) calculated interplanar potential and energy levels

Fig.7.18a,b. For 54-MeV electrons channeled along the (111) planar direction in LiF: (a) measured and calculated photon spectra and (b) calculated interplanar potential and energy levels

single or double layers of atomic nitrogen precipitated along (100) planes. When viewed along this (100) direction, the defect has the appearance of a dislocation; but when viewed along the (110) or the (111) planar direction, the defect appears as a stacking fault. Hence one might expect anisotropic differences in the effect of these defects on channeling radiation from diamond. Strong effects have indeed been found, which depend not only upon the crystallographic direction, but also upon whether electrons or positrons are used as the probe. An example is given in Fig.7.19, which shows spectra from positrons and electrons channeled along the (110) direction of platelet-free (Type-IIa) and platelet-containing (Type-Ia) diamonds. Comparing the positron spectra, shown in Figs.7.19a,c, one finds a reduction in intensity by a factor of about five; the remaining peak is at the same position but it is reduced in width by almost a factor of two! Comparing the electron spectra,

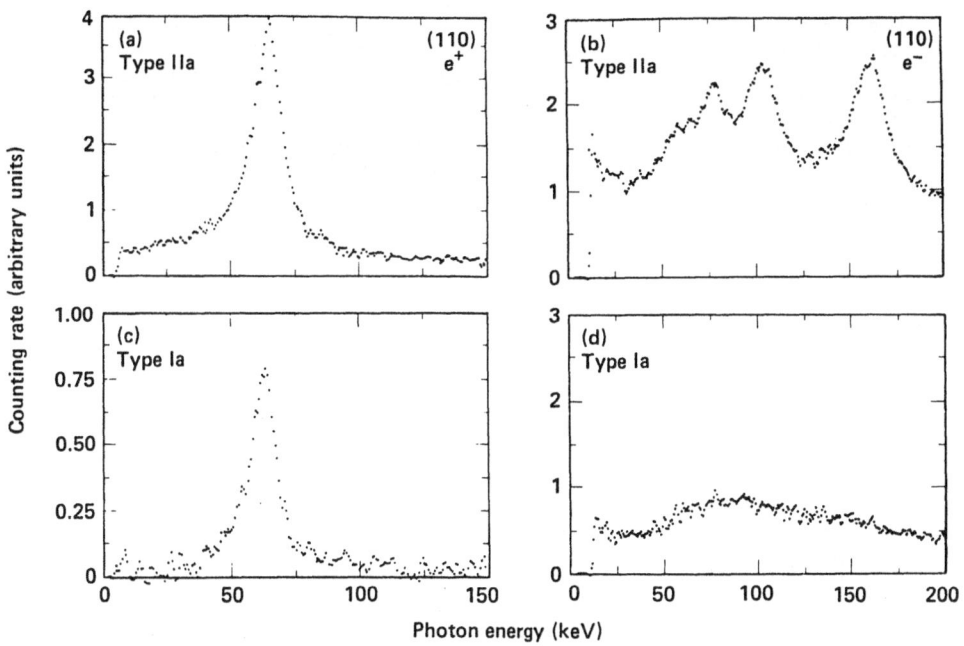

Fig.7.19a-d. Comparison of measured 54.5-MeV positron and electron CR spectra for the (110) planar direction in diamond for Type-IIa (platelet-free) and Type-Ia (platelet-containing) crystals: (a) for positrons, Type IIa, (b) for electrons, Type IIa, (c) for positrons, Type Ia, and (d) for electrons, Type Ia

shown in Figs.7.19b,d, one finds that the defects cause a complete obliteration of the channeling-radiation structure. This stands in sharp contrast to the (111) spectrum, which is almost unaffected by the presence of platelets. Possible causes of these effects are discussed in [7.54]. Regardless of whether or not these phenomena are currently understood, it does appear that channeling radiation might have an important use in delineating types and concentrations of impurities or defects in crystals.

Acknowledgments. We thank our colleagues, especially those at Stanford, Lyon, and Aarhus, for many valuable comments and discussions. This work was performed under the auspices of the U.S. Department of Energy under contract numbers W-7405-ENG-48 (LLNL) and W-7405-ENG-26 (ORNL).

References

7.1   R.L. Walker, B.L. Berman, R.C. Der, T.M. Kavanagh, J.M. Khan: Phys. Rev. Lett. **25**, 5 (1970). This experiment was essentially duplicated, with 1-GeV positrons and electrons, by
       V.L. Morikhovskii, G.D. Kovalenko, I.A. Grishaev, A.N. Fisun, V.I. Kasilov, B.I. Shramenko, A.N. Krinitsyn: Pis'ma Zh. Eksp. Teor. Fiz. **16**, 162 (1972)

[English transl.: JETP Lett. **16**, 112 (1972)]. They, too, observed channeling radiation, but this fact has not been recognized until now.

7.2 R.L. Walker, B.L. Berman, S.D. Bloom: Phys. Rev. **A11**, 736 (1975). Low-energy enhancement like that seen in this experiment has also been seen with 800-MeV electrons by
S.A. Vorobiev, B.N. Kalinin, V.V. Kaplin, A.P. Potylitsin: Pis'ma Zh. Tekh. Fiz. **4**, 1340 (1978) [English transl.: Sov. Phys.-Tech. Phys. Lett. **4**, 539 (1978)], and with 1.2-GeV positrons by
B.I. Shramenko, V.I. Vit'ko, I.A. Grishaev: Pis'ma Zh. Tekh. Fiz. **4**, 1423 (1978) [English transl.: Sov. Phys.-Tech. Phys. Lett. **4**, 576 (1978)]

7.3 T.F. Godlove, M.E. Toms: Private communication (1969); U.S. Naval Research Laboratory, Nuclear Physics Division, Annual Report (1969) p.96

7.4 R.W. Terhune, R.H. Pantell: Appl. Phys. Lett. **30**, 265 (1977);
R.H. Pantell: Private communication (1975)

7.5 B.N. Kalinin, V.V. Kaplin, A.P. Potylitsin, S.A. Vorobiev: Phys. Lett. **70A**, 447 (1979); see also
S.A. Vorobiev, B.N. Kalinin, V.V. Kaplin, A.P. Potylitsin: Izv. Vyssh. Uchebn. Zaved., Fiz. No.11, 117 (1978) [English transl.: Sov. Phys. J. **21**, 1483 (1979)]

7.6 M.J. Alguard, R.L. Swent, R.H. Pantell, B.L. Berman, S.D. Bloom, S. Datz: IEEE Trans. Nucl. Sci. NS-**26**, 3865 (1979)

7.7 M.J. Alguard, R.L. Swent, R.H. Pantell, B.L. Berman, S.D. Bloom, S. Datz: Phys. Rev. Lett. **42**, 1148 (1979)

7.8 R.L. Swent, R.H. Pantell, M.J. Alguard, B.L. Berman, S.D. Bloom, S. Datz: Phys. Rev. Lett. **43**, 1723 (1979)

7.9 A.O. Agan'yants, Yu.A. Vartanov, G.A. Vartapetyan, M.A. Kumakhov, Kh. Trikalinos, V.Ya. Yaralov: Pis'ma Zh. Eksp. Teor. Fiz. **29**, 554 (1979) [English transl.: JETP Lett. **29**, 505 (1979)]

7.10 I.I. Miroshnichenko, J.J. Murray, R.O. Avakyan, T.Kh. Figut: Pis'ma Zh. Eksp. Teor. Fiz. **29**, 786 (1979) [English transl.: JETP Lett. **29**, 722 (1979)], in which the authors make the erroneous claim of the first observation of the radiation of channeled relativistic positrons. Unfortunately, this error has been propagated in several subsequent review papers

7.11 J.U. Andersen, S. Datz, E. Laegsgaard, J.P.F. Sellschop, A.H. Sørensen: Phys. Rev. Lett. **49**, 215 (1982)

7.12 R.K. Klein, J.O. Kephart, R.H. Pantell, H. Park, R.L. Swent, S. Datz, R.W. Fearick, B.L. Berman: Phys. Rev. B**31**, 68 (1985)

7.13 S. Datz, R.W. Fearick, H. Park, R.H. Pantell, R.L. Swent, J.O. Kephart, R.K. Klein, B.L. Berman: Phys. Lett. **96**A, 314 (1983); *Proc. Int. Conf. Atomic Collisions in Solids*, Bad Iburg, 1983, p.8 and Nucl. Instrum. Methods. **230**, 74 (1984)

7.14 M. Gouanère, D. Sillou, M. Spighel, N. Cue, M.J. Gaillard, R.G. Kirsch, J.-C. Poizat, J. Remillieux, B.L. Berman, P. Catillon, L. Roussel, G.M. Temmer: Nucl. Instrum. Methods **194**, 225 (1982)

7.15 M. Gouanère, D. Sillou, M. Spighel, N. Cue, M.J. Gaillard, R.G. Kirsch, J.-C. Poizat, J. Remillieux, B.L. Berman, P. Catillon, L. Roussel, G.M. Temmer: *Proc. Int. Conf. Atomic Collisions in Solids*, Bad Iburg, 1983, p.8, and private communication

7.16 Yu.N. Adishchev, A.N. Didenko, V.V. Kaplin, A.P. Potylitsin, S.A. Vorobiev: Phys. Lett. **83**A, 337 (1981)

7.17 Yu.N. Adishchev, P.S. Anan'in, S.A. Vorobiev, V.N. Zabaev, B.N. Kalinin, V.V. Kaplin, A.P. Potylitsin, V.K. Tomchakov, E.I. Rozum: Pis'ma Zh. Eksp. Teor. Fiz. **30**, 430 (1979) [English transl.: JETP Lett. **30**, 402 (1979)]

7.18 Yu.N. Adishchev, I.E. Vnukov, S.A. Vorobiev, V.M. Golovkov, V.N. Zabaev, V.I. Lunev, A.A. Kurkov, B.N. Kalinin, A.P. Potylitsin: Pis'ma Zh. Eksp. Teor. Fiz. **33**, 478 (1981) [English transl.: JETP Lett. **33**, 462 (1981)]

7.19 S.A. Vorobiev, V.N. Zabaev, B.N. Kalinin, V.V. Kaplin, A.P. Potylitsin: Pis'ma Zh. Eksp. Teor. Fiz. **29**, 414 (1979) [English transl.: JETP Lett. **29**, 376 (1979)]; see also
S.A. Vorobiev, A.N. Didenko, V.N. Zabaev, B.N. Kalinin, V.V. Kaplin, A.A. Kurkov, A.P. Potylitsin, V.K. Tomchakov: Pis'ma Zh. Eksp. Teor. Fiz. **32**, 261 (1980) [English transl.: JETP Lett. **32**, 241 (1980)], and [7.5]

7.20 V.B. Ganenko, I.I. Miroshnichenko, E.V. Pegushin, V.M. Sanin, S.V. Shalatskij: Radiat. Eff. **62**, 167 (1982); see also
V.B. Ganenko, L.E. Gendenshtein, I.I. Miroshnichenko, V.L. Morokhovskii, E.V. Pegushin, V.M. Sanin, S.V. Shalatskii: Pis'ma Zh. Eksp. Teor. Fiz. **32**, 397 (1980) [English transl.: JETP Lett. **32**, 373 (1980)]

7.21 N.F. Shul'ga, L.E. Gendenshtein, I.I. Miroshnichenko, E.V. Pegushin, S.P. Fomin, R.O. Avakyan: Zh. Eksp. Teor. Fiz. **82**, 50 (1982) [English transl.: Sov. Phys.-JETP **55**, 30 (1982)], which reiterates the erroneous claim made in 7.10.

7.22 J.U. Andersen, E. Laegsgaard: Phys. Rev. Lett. **44**, 1079 (1980);
J.U. Andersen, E. Bonderup, E. Laegsgaard, B.B. Marsh, A.H. Sørensen: Nucl. Instrum. Methods **194**, 209 (1982)

7.23 N. Cue, E. Bonderup, B.B. Marsh, H. Bakhru, R.E. Benenson, R. Haight, K. Inglis, G.O. Williams: Phys. Lett. **80A**, 26 (1980)

7.24 J.E. Watson, J.S. Koehler: Phys. Rev. A24, 861 (1981); Phys. Rev. B25, 3079 (1982)

7.25 E. Laegsgaard, J.U. Andersen: *Proc. Int. Conf. Atomic Collisions in Solids*, Bad Iburg, 1983, p.143 and Nucl. Instrum. Methods **230**, 99 (1984)

7.26 J.U. Andersen, E. Laegsgaard, A.H. Sørensen: *Proc. Int. Conf. Atomic Collisions in Solids*, Bad Iburg, 1983, p.135 and Nucl. Instrum. Methods **230**, 63 (1984)

7.27 J.U. Andersen, K.R. Eriksen, E. Laegsgaard: Phys. Scr. **24**, 588 (1981)

7.28 M.J. Alguard, B.L. Berman, S. Datz, J.O. Kephart, R.K. Klein, R.H. Pantell, H. Park, R.L. Swent: To be submitted to Phys. Rev. B

7.29 B.L. Berman, S.D. Bloom, S. Datz, M.J. Alguard, R.L. Swent, R.H. Pantell: Phys. Lett. **82A**, 459 (1981)

7.30 R.L. Swent, R.H. Pantell, S. Datz, R. Alvarez: Nucl. Instrum. Methods **194**, 235 (1982)

7.31 K. Komaki, A. Ootuka, F. Fujimoto, N. Horikawa, T. Nakanishi, C.Y. Gao, T. Iwata, S. Fukui, M. Mutou, H. Okuno: Univ. of Tokyo preprint (1983), *Proc. Int. Conf. Atomic Collisions in Solids*, Bad Iburg, 1983, p.7, and Nucl. Instrum. Methods 230, 71 (1984)

7.32 V.I. Vit'ko, I.A. Grishaev, G.D. Kovalenko, B.I. Shramenko: Pis'ma Zh. Tekh. Fiz. **5**, 1291 (1979) [English transl.: Sov. Phys.-Tech. Phys. Lett. **5**, 541 (1979)]

7.33 N.A. Filatova, V.M. Golovatyuk, A.N. Iskakov, I.M. Ivanchenko, R.B. Kadyrov, N.N. Karpenko, T.S. Nigmanov, V.V. Palchik, V.D. Riabstov, M.D. Shafranov, E.N. Tsyganov, I.A. Tyapkin, D.V. Uralski, A. Forycki, Z. Guzik, J. Wojtkowska, R.A. Carrigan, Jr., T.E. Toohig, C. Carmack, W.M. Gibson, I.-J. Kim, C.-R. Sun, M.D. Bavizhev, N.K. Bulgakov, N.I. Zimin, I.A. Grishaev, G.D. Kovalenko, B.I. Shramenko, E.I. Denisov, V.I. Glebov, V.V. Avdeichikov: Phys. Rev. Lett. **48**, 488 (1982); Nucl. Instrum. Methods **194**, 239 (1982)

7.34 M. Atkinson, J.F. Bak, P.J. Bussey, P. Christensen, J.A. Ellison, R.J. Ellison, K.R. Eriksen, D. Giddings, R.E. Hughes-Jones, B.B. Marsh, D. Mercer, F.E. Meyer, S.P. Møller, D. Newton, P. Pavlopoulos, P.H. Sharp, R. Stensgaard, M. Suffert, E. Uggerhøj: Phys. Lett. **110B**, 162 (1982)

7.35 J.F. Bak, J.A. Ellison, B.B. Marsh, F.E. Meyer, S.P. Møller, J.B.B. Petersen, M. Suffert, A.H. Sørensen, E. Uggerhøj, K. Østergaard: *Proc. Int. Conf. Atomic Collisions in Solids*, Bad Iburg, 1983, p.48

7.36 J.F. Bak, J.A. Ellison, B.B. Marsh, F.E. Meyer, S.P. Møller, J.B.B. Petersen, M. Suffert, A.H. Sørensen, E. Uggerhøj, K. Østergaard: *Proc. Int. Conf. Atomic Collisions in Solids*, Bad Iburg, 1983, p.144

7.37 H. Park, R.L. Swent, J.O. Kephart, R.H. Pantell, B.L. Berman, S. Datz, R.W. Fearick: Phys. Lett. **96A**, 45 (1983)

7.38 B.L. Berman: CEN Saclay Report No. DPhN/HE/81/2 (1981), p.5 (unpublished)

7.39 N. Cue, B.B. Marsh, R.E. Benensen, Jin Han-sheng, Wang Guang-hou, Long Xiang-guan, Wang Ke-ming: *Proc. Int. Conf. Atomic Collisions in Solids*, Bad Iburg, 1983, p.144, and Nucl. Instrum. Methods **230**, 104 (1984)

7.40 J.U. Andersen, E. Bonderup, E. Laegsgaard, A.H. Sørensen: Phys. Scr. (to be published)

7.41 R.L. Swent, R.H. Pantell, H. Park, J.O. Kephart, R.K. Klein, S. Datz, R.W. Fearick, B.L. Berman: Phys. Rev. B29, 52 (1984)

7.42 B.L. Berman, S. Datz, R.W. Fearick, R.L. Swent, R.H. Pantell, H. Park, J.O. Kephart, R.K. Klein: *Proc. Int. Conf. Atomic Collisions in Solids*, Bad Iburg, 1983, p.49, University of California Lawrence Livermore National Laboratory Report No. UCRL-89454 (1984) (unpublished), and Nucl. Instrum. Methods **230**, 90 (1984)

7.43 B.L. Berman, S. Datz, R.W. Fearick, J.O. Kephart, R.H. Pantell, H. Park, R.L. Swent: Phys. Rev. Lett. **49**, 474 (1982)

7.44 P.A. Doyle, P.S. Turner: Acta Crystallogr. A**24**, 390 (1968)

7.45 *International Tables for X-Ray Crystallography*, ed. by N.F.M. Henry, K. Lonsdale (Kynoch Press, Birmingham 1959)

7.46 B.L. Berman, S.D. Bloom: Energy Tech. Rev. 81-1, 1 (1981)

7.47 R.L. Swent: Ph. D. Thesis, Stanford University (1982)

7.48 B.L. Berman, S.C. Fultz: University of California Lawrence Livermore Laboratory Report No. UCRL-75383 (1974) (unpublished); Rev. Mod. Phys. **47**, 713 (1975)

7.49 G.M. Temmer: Chin. J. Nucl. Phys. **2**, 353 (1980) [Chin. Phys. **1**, 1024 (1981)]

7.50 J. Lindhard: Phys. Lett. **12**, 126 (1964); K. Dan. Vidensk. Selsk. Mat. Fys. Medd. **34**, No.14 (1965)

7.51 M.J. Alguard, R.L. Swent, R.H. Pantell, S. Datz, J.H. Barrett, B.L. Berman, S.D. Bloom: Nucl. Instrum. Methods **170**, 7 (1980)

7.52 S. Datz, C.D. Moak: "Heavy Ion Channeling" in *Heavy Ion Physics*, ed. by D.A. Bromley (Plenum Press, in press)

7.53 M.A. Kumakhov: Phys. Lett. **57**A, 17 (1976); Phys. Status Solidi (b) **84**, 581 (1977)

7.54 H. Park, R.H. Pantell, R.L. Swent, J.O. Kephart, B.L. Berman, S. Datz, R.W. Fearick: J. Appl. Phys. **55**, 358 (1984)

Additional References

Beezhold, W., Sanford, T.W.L., Park, H., Kephart, J.O., Klein, R.K., Pantell, R.H., Berman, B.L., Datz, S.: Channeling radiation from tungsten. Bull. Am. Phys. Soc. **30**, 374 (1985)

Berman, B.L.: Research with channeling radiation. Energy Tech. Rev. **85-3**, 12 (1985)

Berman, B.L., Dahling, B.A., Datz, S., Kephart, J.O., Klein, R.K., Pantell, R.H., Park, H.: Channeling-radiation measurements at Lawrence Livermore National Laboratory. Nucl. Instrum. Methods B**10/11**, 611 (1985) (Proc. Eighth Conf. Application of Accelerators in Research and Industry)

Berman, B.L., Datz, S., Kephart, J.O., Klein, R.K., Pantell, R.H., Park, H., Swent, R.L., Alguard, M.J., Hynes, M.V.: Channeling radiation from LiH and LiD. Bull. Am. Phys. Soc. **30**, 373 (1985)

Datz, S., Berman, B.L., Dahling, B.A., Kephart, J.O., Klein, R.K., Pantell, R.H., Park, H.: Channeling radiation from damaged LiF crystals. Bull. Am. Phys. Soc. **30**, 373 (1985)

Kephart, J.O., Klein, R.K., Pantell, R.H., Park, H., Datz, S., Alguard, M.J., Swent, R.L., Berman, B.L.: Occupation lengths for electrons and positrons channeled in silicon crystals. Bull. Am. Phys. Soc. **30**, 374 (1985)

Park, H., Kephart, J.O., Klein, R.K., Pantell, R.H., Berman, B.L., Datz, S.: Channeling radiation from GaAs. Submitted to Appl. Phys. Lett.

Park, H., Kephart, J.O., Klein, R.K., Pantell, R.H., Berman, B.L., Datz, S., Swent, R.L.: Electron channeling radiation from diamonds with and without platelets. J. Appl. Phys. **57**, 1661 (1985)

Park, H., Kephart, J.O., Klein, R.K., Pantell, R.H., Hynes, M.V., Berman, B.L., Dahling, B.A., Datz, S., Swent, R.L., Alguard, M.J.: Temperature dependence of channeling radiation. Bull. Am. Phys. Soc. **30**, 374 (1985), and submitted to Appl. Phys. Lett.

8. Transition Radiation

G. B. Yodh

With 25 Figures

Transition radiation is emitted when a source, for example, a charged particle, is moving with a constant velocity in an inhomogeneous or a nonstationary medium. Basic formulas for X-rays emitted by an ultra-relativistic charged particle traversing a periodic medium are discussed. Experimental test of theoretical predictions are surveyed and current applications of transition radiation for particle identification, energy measurement, beam monitoring, and for providing a tunable source of hard X-rays are described.

8.1 Scope of the Review

Transition radiation (TR) is the electromagnetic radiation that is emitted when a charged particle traverses the boundary between two media of different dielectric or magnetic properties [8.1]. The process depends on the velocity of the particle and is a collective response of the matter surrounding the trajectory. The radiation is associated with the sudden change in the electromagnetic field of the particle in the transition. For ultra-relativistic particles, i.e., $E/Mc^2 = \gamma \gg 1$, the major part of the radiated energy is in X-rays and is sharply peaked in the forward direction. The number of photons emitted is small, of the order of the fine-structure constant .

Transition radiation differs from Čerenkov radiation and bremsstrahlung in its sizeable dependence on $\gamma = (1 - \beta^2)^{-\frac{1}{2}}$, which provides a new technique for distinguishing particles with the same charge but different masses at a given momentum, and provides a technique of energy measurement of particles whose charge and mass is known when conventional detectors become inoperative (e.g., at $\gamma \gtrsim 1000$). The smallness of the energy loss associated with TR makes these detectors nondestructive of the particles being identified and therefore offers the possibility of using TR for monitoring particle beams. The dependence of TR on collective properties of the medium makes it possible to determine plasma frequencies of materials. The coherent nature of TR from foils makes it possible to devise relatively monochromatic sources of hard X-rays.

There exists a vast amount of theoretical discussion on TR in the literature, by *Garibyan* [8.2,3], *Ter-Mikaelyan* [8.4,5], *Bass* and *Yakovenko* [8.6], *Durand* [8.7], *Artru* et al. [8.8], and *Harutyunian* and *Frangyan* [8.9]. It is not the purpose of this review to give yet another discussion of the theory. The main objectives of this review are:

i) to provide a concise summary of theoretically predicted yields of transition radiation from multilayered media, specially emphasizing the importance of interference effects (Sect.8.2), saturation due to the formation zone and nonzero gas density between layers, and to give a set of "universal" curves for yield integrated over angles in terms of scaled variables (Sect.8.3);

ii) to give a survey of experimental results which provide tests for the theoretical predictions (Sect.8.4);

iii) to discuss several different applications of TR that are currently being studied (Sect.8.5).

8.2 Characteristics of Transition Radiation

8.2.1 General Theoretical Formula for the Differential Yield

a) Two Semi-Infinite Media

For an ultra-relativistic particle with $\gamma = E/mc^2 \gg 1$ of charge Ze, the material medium through which it travels can be considered as an electron gas, with a "dielectric constant" given by

$$\varepsilon(\omega) = 1 - \omega_p^2/\omega^2 = 1 - \xi^2 , \qquad (8.2.1)$$

where ω_p is the plasma frequency of the medium:

$$\omega_p^2 = 4\pi\alpha n_e/m_e \qquad (8.2.2a)$$

(n_e = electron density, $\alpha = 1/137$ and we have set $\hbar = c = 1$).
A partial density ρ of an element $_Z X^A$ gives the contribution

$$\omega_p^2 = 4\pi\alpha(Z/A)N_0\rho/m_e \simeq 2(Z/A)(\rho/\text{gm cm}^{-3})(21 \text{ eV})^2 , \qquad (8.2.2b)$$

where N_0 is Avogadro's number. The frequency of the emitted photon is ω.

When the particle crosses the boundary between two media (media 1 and 2 in Fig. 8.1), radiation is emitted because the Coulomb field of the particle has to readjust itself [8.10]. For the emitted X-ray photons, one has $\omega \cong \gamma\omega_p$, and hence $\xi_1^2 = \omega_{p_1}^2/\omega^2$ and $\xi_2^2 = \omega_{p_2}^2/\omega^2$ are both $\ll 1$ and the emission angle $\theta \ll 1$. The energy radiated per unit solid angle per unit frequency interval can be approximated by [8.8,10]

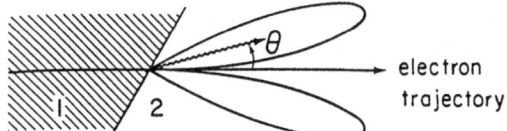

Fig.8.1. Schematic representation of the angular distribution of transition radiation emitted by a charged particle traversing a single interface

$$\frac{d^2W}{d\omega\,d\Omega} = \frac{\alpha}{\pi^2}\left|\frac{\theta}{\gamma^{-2} + \theta^2 + \xi_1^2} - \frac{\theta}{\gamma^{-2} + \theta^2 + \xi_2^2}\right|^2 \ . \tag{8.2.3}$$

This expression shows that the radiation is concentrated in a narrow cone, θ^2 being of order $\gamma^{-2} + \xi_1^2$ to $\gamma^{-2} + \xi_2^2$. The radiation falls off as θ^{-5} for large θ. As long as this cone is contained in the second medium there is no dependence on the angle of incidence.

b) Multilayered Medium

Consider the case of n parallel surfaces separating n + 1 different media, as shown in Fig.8.2. The total TR of frequency $\omega \gg \omega_p$ emitted at an angle θ is obtained by coherently adding the amplitudes from each surface taking into account absorption and phase retardation. We can neglect backward emission, reflections on the boundaries, and change in θ because of refraction. We can write for the radiation field [8.8]:

$$E(\omega,\theta) = \sum_{j=1}^{n} e^j(\omega,\theta)\,\exp\left[-\sum_{m \geqslant j}(\sigma_m + i\phi_m)\right] \ , \tag{8.2.4}$$

where $\mathbf{e}^j$ is the single-surface amplitude and $\boldsymbol{\theta}$ is a vector of length θ which is the difference between the unit vectors for the photon and particle directions. Up to a numerical factor,

$$e^j(\omega,\boldsymbol{\theta}) = \frac{\boldsymbol{\theta}}{\gamma^{-2} + \theta^2 + \xi_{j-1}^2} - \frac{\boldsymbol{\theta}}{\gamma^{-2} + \theta^2 + \xi_j^2} \ . \tag{8.2.5}$$

Absorption in the mth layer is given by $\exp(-\sigma_m)$ and the phase retardation due to the difference in the speed of the particle and that of the wave of an emitted X-ray is

$$\phi_m = \omega\ell_m/v - k_m \cdot \ell_m \ , \tag{8.2.6}$$

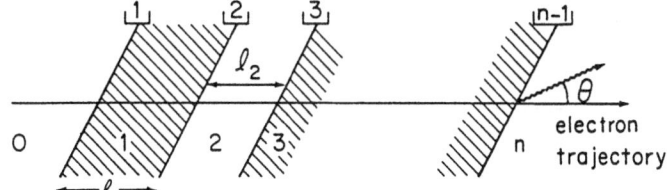

Fig.8.2. General schema of an electron traversing an n-interface radiator

197

where k_m is the wave vector in medium m and $\boldsymbol{\ell}_m$ the vector distance traversed by the particle in this layer. This phase difference can be written (for $\gamma \gg 1$ and $\xi^2 \ll 1$) as

$$\phi_m \simeq (\gamma^{-2} + \theta^2 + \xi_m^2)\omega\ell_m/2 \quad . \tag{8.2.7}$$

Here we have assumed that all $\boldsymbol{\ell}_m$ are collinear, i.e., we have neglected multiple scattering of the particle. It is useful to define a "formation length" $Z_m(\theta)$ given by

$$Z_m(\theta) = (\gamma^{-2} + \theta^2 + \xi_m^2)^{-1} 2/\omega \tag{8.2.8}$$

and to write

$$\phi_m = \ell_m/Z_m(\theta) \quad . \tag{8.2.9}$$

Note that if $\ell_m \ll Z_m$, then $\phi_m \cong 0$ and we can ignore the mth layer.

c) Single-Foil Yield (SF)

For a single sheet of material in a medium, the first and last media are the same, $\xi_0 = \xi_2$, and the TR yield is

$$\left(\frac{d^2W}{d\omega\,d\Omega}\right)_{SF} = 4\sin^2(\phi_1/2)\left(\frac{d^2W}{d\omega\,d\Omega}\right)_{Single\ Surface} \quad , \tag{8.2.10}$$

where $4\sin^2(\phi_1/2)$ represents single-foil "diffraction."

d) N-Foil Yield (NF)

For a stack of N foils of constant thickness and spacing, one can show that

$$\left(\frac{d^2W}{d\omega\,d\Omega}\right)_{NF} = 4\sin^2(\phi_1/2)\left(\frac{d^2W}{d\omega\,d\Omega}\right)_{SS} \times (I^N) \quad , \tag{8.2.11}$$

where I^N is the N-foil interference factor, given by

$$I^N = |(1 - C^N)/(1 - C)|^2 \quad , \quad \text{with} \tag{8.2.12}$$

$$C = \exp(i\phi_1 + i\phi_2 - \sigma_1/2 - \sigma_2/2) \quad . \tag{8.2.13}$$

The total phase shift for one foil and one gap is given by

$$\phi_{12} \equiv \phi_1 + \phi_2 = (\ell_1 + \ell_2)(\gamma^{-2} + \theta^2)\omega/2 + (\ell_1\omega_{P_1}^2 + \ell_2\omega_{P_2}^2)/2\omega \quad , \tag{8.2.14}$$

with $\sigma = \sigma_1 + \sigma_2$. We can rewrite (8.2.12) as

$$I^N = \exp[(1 - N)\sigma/2]\frac{\sin^2(N\phi_{12}/2) + \sinh^2(N\sigma/4)}{\sin^2(\phi_{12}/2) + \sinh^2(\sigma/4)} \quad . \tag{8.2.15}$$

Equation (8.2-3) for the single-surface yield, (8.2.10) for the single-foil yields, and (8.2.11,15) for the N-foil yield concisely describe the TR produced by ultra-relativistic particles for the experimental configurations to be discussed in later sections. They are sufficiently general to cover many other cases with appropriate modifications [8.11,12].

8.2.2 Interference Effects

For a fixed frequency ω and a fixed γ, the angular distribution of TR is the product of three terms representing single-surface yield, single-foil diffraction, and multifoil interference. The first term is peaked at $\theta \sim 1/\gamma$, the second oscillates with the distance $\Delta\theta_{SF} \sim (\pi/\omega)(1/\ell_{foil})$ between zeros, and the third has main maxima spaced by $(\Delta\theta)_{NF} = \pi/\omega(\ell_{foil} + \ell_{gap})$. The three terms are sketched in Fig.8.3 for a case where $\Delta\theta_{SF} \sim 2/\gamma$ and $\Delta\theta_{NF} \ll \Delta\theta_{SF}$ or $\ell_{gap} \gg \ell_{foil}$ [8.10]. For this case, the resultant yield is N times the single-foil yield without absorption. With absorption, when $\ell_{gap} \gg \ell_{foil}$ the yield is $[(1 - e^{-N\sigma})/(1 - e^{-\sigma})] \times$(single-foil yield).

An examination of (8.2.10) shows that if ϕ_1 is small, i.e., if $\ell_m \ll Z_m$, the single-foil yield is suppressed. It is also evident from the derivation of (8.2.11) that for the cases where absorption in one foil is small one can interchange the role of gap and foil without changing the total yield, so that as the gap thickness goes to zero the yield vanishes.

Finally, it is to be noted that, in order to see "interference" effects due to multiple foils, gap thickness and foil thickness should be of comparable size [8.8, 10,13].

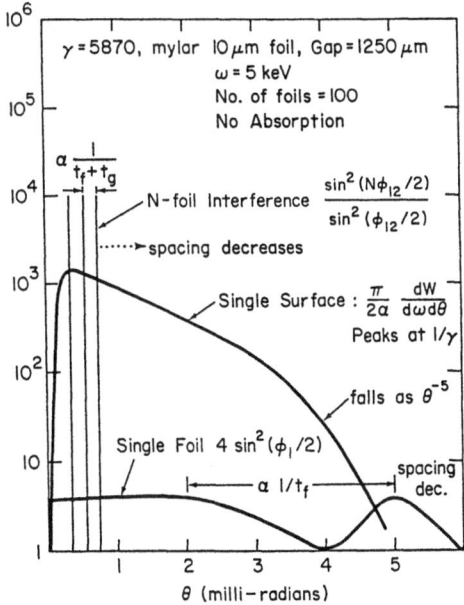

Fig.8.3. An illustration of contributions of the single-surface factor (8.2.3), the single-foil interference factor (8.2.10), and the N-foil interference factor (8.2.12) to the N-foil yield. Calculations are for the special case involving 100 10-μm mylar foils with a 1250-μm vacuum gap, ~3-TeV electrons, and X-ray frequency of 5 keV. Note that X-ray absorption in the foils has not been included

8.3 Yield Integrated Over Angle

8.3.1 General Remarks

In most practical devices one measures the TR yield integrated over angle θ. To study this, we introduce the variable $x = \gamma^{-2} + \theta^2$. We find [8.8] that

$$\frac{dW}{d\omega} = \frac{\alpha}{\pi} (\xi_1^2 - \xi_2^2) \int_{\gamma^{-2}}^{\infty} \frac{(x - \gamma^{-2})}{(x + \xi_1^2)^2 (x + \xi_2^2)^2} \, dx \quad \text{(interference factors)} \quad . \quad (8.3.1)$$

The interference factors are functions of x only (not of γ and θ separately) and are positive. Equation (8.3.1) defines a monotonic function of γ. If one of the media is vacuum, $dW/d\omega$ increases logarithmically with γ when $\gamma \to \infty$. If ω_{p_1} and $\omega_{p_2} \neq 0$, then $dW/d\omega$ reaches a finite limit when $\gamma \to \infty$. The yield saturates when $\gamma \gtrsim \omega/\omega_{p_2}$ (as the plasma frequency of the gap is less than ω_{p_1}). For normal air, $\omega_{p_2} = 0.7$ eV and $\omega/\omega_{p_2} = 10^4$ for keV X-rays. The functional dependence of the integrand in (8.3.1) on γ, ω_{p_1}, and ω_{p_2} is such that the yield for these parameters is the same as that for the vacuum case, where the parameters are

$$\omega_{p_1}' = \omega_{p_1} [1 - (\omega_{p_2}/\omega_{p_1})^2]^{-\frac{1}{2}} \quad , \quad \omega_{p_2}' = 0 \quad ,$$

$$\gamma' = \gamma(1 + \gamma^2 \omega_{p_2}^2 / \omega^2)^{-} \quad . \quad (8.3.2)$$

Note that $\gamma' < \gamma$ and $\omega_{p_1}' < \omega_{p_1}$; hence the yield in gas will always be smaller than that in vacuum.

Next, we discuss the main characteristics of TR produced in single-surface, single-foil, and multifoil (regularly spaced) arrangements.

8.3.2 Single-Surface Yield (SS)

The intensity of TR depends on particle velocity through the Lorentz factor $\gamma = (1 - \beta^2)^{-\frac{1}{2}}$. The differential radiated energy per unit frequency range, integrated over angle, depends logarithmically on γ and is given by

$$\left(\frac{dW}{d\omega} \right)_{SS} = \frac{\alpha}{\pi} \left[\left(\frac{\xi_1^2 + \xi_2^2 + 2\gamma^{-2}}{(\xi_1^2 - \xi_2^2)} \right) \ln \left(\frac{\gamma^{-2} + \xi_1^2}{\gamma^{-2} + \xi_2^2} \right) - 2 \right] \quad . \quad (8.3.3)$$

This formula can be restated in terms of a dimensionless scaled parameter η' given by [8.14]

$$\eta' = \omega/(\gamma' \omega_{p_1}') = (\eta^2 + r)(1 - r)^{-\frac{1}{2}} \quad , \quad (8.3.4)$$

where γ' and ω_{p_1}' are given by (8.3.2) and where

200

$$\eta = \omega/\gamma\omega_{p_1} \quad , \tag{8.3.5}$$

$$r = \omega_{p_2}^2/\omega_{p_1}^2 \cong \rho_2/\rho_1 \quad . \tag{8.3.6}$$

The yield is

$$\left(\frac{dW}{d\omega}\right)_{SS} = \frac{\alpha}{\pi} [(1 + 2\eta'^2)\ln(1 + \eta'^{-2}) - 2] \quad . \tag{8.3.7}$$

We make three remarks on single-surface yield:

i) The total energy radiated in medium-vacuum transitions obtained by integrating (8.3.1) over ω is [8.10]

$$W = 2\alpha\gamma\omega_{p_1}/3 \quad , \tag{8.3.8}$$

which is a yield increasing linearly with energy, a condition seldom encountered in experiments at high γ. Note also that the yield is small.

ii) Although the mean number of photons would diverge at low photon energies, this is no problem since we usually have some low-energy cutoff. In particular, the number of photons above $0.15\gamma\omega_p$ is quite small, being $\sim 0.5\alpha$. Thus one requires many surfaces to get sufficient yield (or many foils).

iii) Above a cutoff frequency given by $\omega_c \sim \gamma\omega_{p_1}$, the yield drops sharply.

8.3.3 Single-Foil Yield (SF)

The interference factor to be used in (8.3.1) is $4\sin^2[(\gamma^{-2} + \theta^2 + \xi_1^2)\omega\ell_1/4]$. This factor dominates in deciding many of the features of TR and hence it is necessary to discuss its ramification in some detail. We do this next.

a) Formation-Zone Effect

TR is a macroscopic process, so that the radiation yield tends to zero as the foil gets thinner. For each angle θ, we have defined the "formation length" $Z_m(\theta)$ in (8.2.8). Its value for $\theta = 0$ is referred to as the "formation zone" Z_1, given by

$$Z_1 = (2/\omega)(\gamma^{-2} + \xi_1^2)^{-1} \quad . \tag{8.3.9}$$

As pointed out in the previous section, if $\ell_1 \ll Z_1$ then the SF yield becomes small; this condition translates into the result that for $\gamma > \gamma_1$, the mean radiated energy W in vacuum increases only logarithmically with γ. The characteristic energy and frequency, γ_1 and ω_1, are respectively given by

$$\gamma_1 = \ell_1\omega_{p_1}/2 = 2.5(\omega_{p_1}/eV)(\ell_1/\mu m) \tag{8.3.10}$$

and

$$\omega_1 = \gamma_1 \omega_{p_1} \quad . \tag{8.3.11}$$

Notice that γ_1 and ω_1 depend on the foil material and thickness. It can be shown [8.8] that the TR spectrum from a single foil in vacuum is given by a universal function $G(\nu,\Gamma)$ of the dimensionless scaled variables

$$\Gamma = \gamma/\gamma_1 \quad , \tag{8.3.12}$$

$$\nu = \omega/\omega_1 \quad , \tag{8.3.13}$$

where

$$\left(\frac{dW}{d\omega}\right)_{SF} = \frac{2\alpha}{\pi} G(\nu,\Gamma) \quad . \tag{8.3.14}$$

For a foil in gas, we can transform γ, ω_{p_1} and ω_{p_2} according to (8.3.2,3), and thus obtain

$$G_{gas} = G(\nu',\Gamma') \quad , \tag{8.3.15}$$

where

$$\nu' = \nu(1 - r)^{-1} \quad , \tag{8.3.16}$$

$$\Gamma' = (\Gamma^{-2} + \nu^{-2})^{-\frac{1}{2}}(1 - r)^{-\frac{1}{2}} \quad . \tag{8.3.17}$$

The universal function $G(\nu,\Gamma)$ is shown in Fig.8.4 for the range of interest. In Fig. 8.5, the single-foil yield is compared with the incoherent sum (G_{inco}) of the yields from two surfaces. In Figs.8.4 and 5, absorption of photons in the foil has been neglected. The oscillations in $G(\nu,\Gamma)$ as a function of ν are due to single-foil interference effects. We describe the important features of $G(\nu,\Gamma)$:

i)   When $\nu \geqslant \Gamma^2$, interference is washed out and the yield begings to look like G_{inco} (see, e.g., the case for $\Gamma = 0.4$ in Fig.8.5).

ii) Constructive interference is shown by the maxima of $G(\nu,\Gamma)$. The function $G(\nu,\Gamma)$ has ridges at $\nu^{-1} = \pi, 3\pi,\ldots$ and valleys at $\nu^{-1} = 2\pi, 4\pi,\ldots$ for $\Gamma \geqslant \sqrt{\nu}$.

iii) The height of a ridge increases logarithmically with Γ:

$$G(ridge) \cong \ln(\Gamma^4/\nu^3) - 3 - 0.577\ldots \quad . \tag{8.3.18}$$

iv) For large values of Γ, formation-zone effects are seen for values of $\nu > 1$ where $G(\nu,\Gamma)$ falls below G_{inco}.

In the design of TR detectors, advantage may be taken of the ridge at the "magic value" $\nu = 1/\pi$ to optimize the total yield. One chooses the foil thickness and material (ω_{p_1}) such that $\bar{\omega} = \omega_1/\pi$ matches the X-ray detector. The other ridges are much less interesting because of their smaller bandwidth and greater absorption at lower frequency.

In the case when the foil is in a gas at high enough γ, so that ν is not longer greater than $r^{\frac{1}{2}}\Gamma$, the total yield saturates. As the main contribution to W comes from the last oscillation, i.e., $\nu \sim 1/\pi$, this occurs when

202

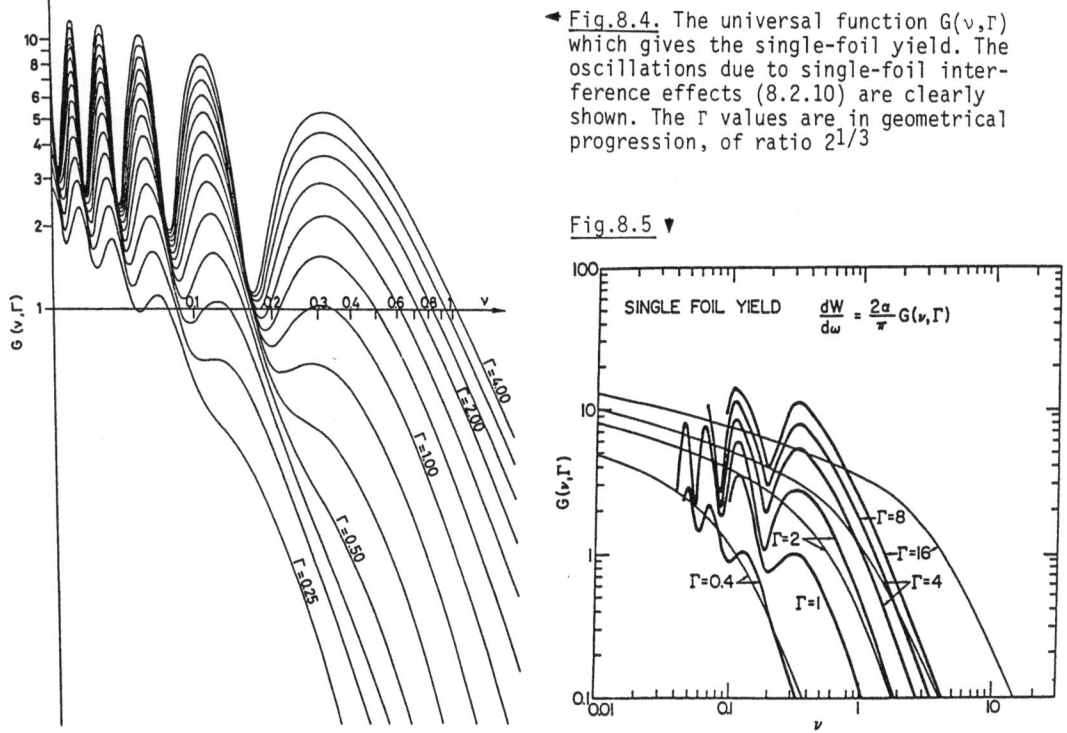

Fig.8.4. The universal function $G(\nu,\Gamma)$ which gives the single-foil yield. The oscillations due to single-foil interference effects (8.2.10) are clearly shown. The Γ values are in geometrical progression, of ratio $2^{1/3}$

Fig.8.5. ▼

Fig.8.5. The universal function $G(\nu,\Gamma)$ (——) compared with twice the single-surface yield divided by $2\alpha/\pi$ (———). Note that $G(\nu,\Gamma)$ falls below the incoherent addition of yield from two surfaces at $\nu > 1$ (the so-called formation-zone effect)

$$\Gamma = \Gamma_{sat,d} \sim \pi r^{-\frac{1}{2}} \quad . \tag{8.3.19}$$

b) Single-Gap Yield

For an isolated gap of length ℓ_2 in a medium, the yield saturates at a value γ_{sat} given by [8.8]

$$\gamma_{sat} = \min(\gamma_{sat,d} ; \gamma_{sat,FZ}) \quad , \tag{8.3.20}$$

where

$$\gamma_{sat,d} \cong \omega/\omega_{p_2} \quad , \tag{8.3.21}$$

$$\gamma_{sat,FZ} \cong 2.2(\ell_2\omega/2) \quad . \tag{8.3.22}$$

The yield is also negligible if $\ell_2 \ll Z_1$, a condition different from that of a single-foil formation-zone saturation, $\ell_1 \ll Z_1$. If $\ell_2 \gg Z_2$, then $G_{gap} \simeq G_{inco}$ and the interference factor averages out. Finally, contrary to the single-foil case, there is no oscillation in the ω spectrum.

8.3.4 N-Foil Yield

For this case, in addition to ω_{p_1}, ω_{p_2}, and ℓ_1, the yield depends on the ratio of gap thickness to foil thickness, $\tau = \ell_2/\ell_1$. Gap and foil play a symmetric role. Equations (8.3.20-22) can be written as

$$\Gamma_{sat} = \min\left(\Gamma_{sat,d}, \Gamma_{sat,FZ}\right) , \tag{8.3.23}$$

where

$$\Gamma_{sat,d} = \nu r^{-\frac{1}{2}} , \tag{8.3.24}$$

$$\Gamma_{sat,FZ} \cong 2.2(\tau\nu)^{\frac{1}{2}} . \tag{8.3.25}$$

As the major part of the radiation comes from the frequency region $\nu \sim 1/\pi$, we have typically

$$\bar{\gamma}_{sat} = \omega/(\pi\omega_{p_2}) \tag{8.3.26}$$

if

$$\tau r \simeq \rho_2\ell_2/\rho_1\ell_1 > 1/15 ,$$

and

$$\bar{\gamma}_{sat} \simeq 0.6\omega_{p_1}(\ell_1\ell_2)^{\frac{1}{2}} \tag{8.3.27}$$

if

$$\tau r < 1/15 .$$

The yield from N foils can be factorized into terms which account for absorption and an equivalent N-foil yield per foil:

$$\left(\frac{dN}{d\omega}\right)_{NF} = \frac{2\alpha}{\pi} N_{eff} G_{many} . \tag{8.3.28}$$

The interference factor G_{many} depends essentially on the three variables, ν, Γ, and τ for the vacuum case [8.8] (or ν', Γ', and τ in a gas). For each value of ν, Γ, and τ we can calculate "universal curves" $G_{many}(\nu,\Gamma,\tau)$. For large τ, $G_{many} \sim G(\nu,\Gamma)$ for the single-foil case. Figures 8.6 and 7 show the function $G(\nu,\Gamma,\tau)$ for a range of values of ν and Γ and for $\tau = 8$ and 63, respectively.

The new feature which appears for a stack of foils with regular spacing is the presence of sharp interference peaks at angles (for a fixed frequency) which satisfy the "resonance condition" [8.10]

$$\phi_{12} = 2p\pi , \tag{8.3.29}$$

where p is an integer. The spacing between the peaks is proportional to $1/(\ell_f + \ell_g)$. At each of these "resonances", the angular distribution of TR is sharply peaked, the position of the ring depending on γ and ω through the relation

$$\nu(1 + \tau)\Gamma_{thr}^{-2} + (1 + r\tau)\nu^{-1} = 2p\pi . \tag{8.3.30}$$

For fixed ω, these "thresholds", $\Gamma = \Gamma_{thr}$, are crossed as γ increases, the highest

204

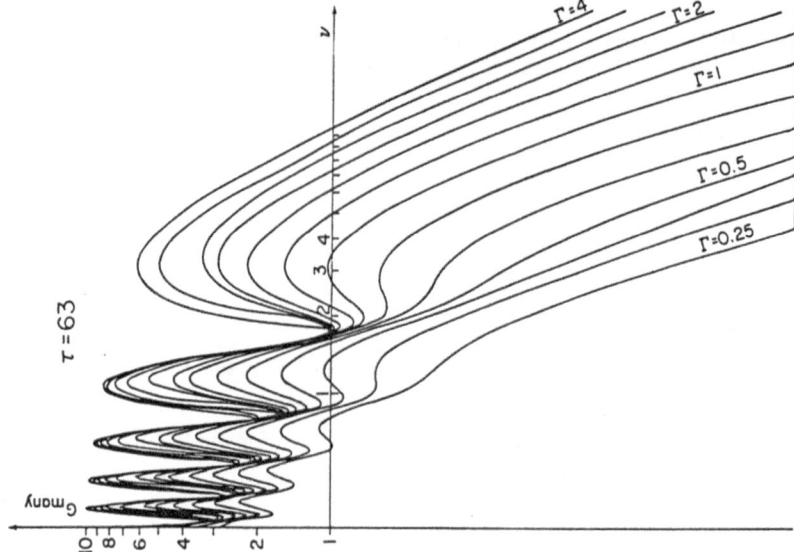

Fig.8.7. Same as Fig.8.6 but for $\tau = 63$, a large ratio of ℓ_g/ℓ_f. The dotted curves are omitted

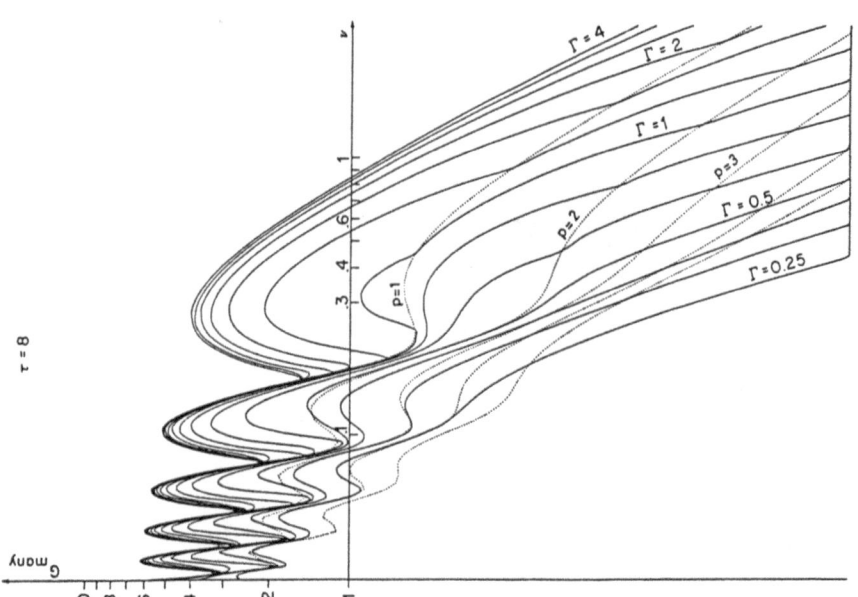

Fig.8.6. The function $G_{many}(\nu,\Gamma,\tau = 8)$ for different values of (same scales as in Fig.8.4). The quantities $G_{many}(\nu,\Gamma_{thr},\tau)$ are plotted with *dotted lines*, where Γ_{thr} is given by (8.3.30). They represent breaks in the $G(\nu,\Gamma)$ surface

205

threshold occurring for $p = 1$. At the thresholds, a break in $dW/d\omega$ curve occurs. In practical situations, one integrates over a bandwidth in ω and these threshold effects tend to be washed out. When the two terms in (8.3.30) are comparable, the threshold effect is not averaged out. A necessary condition for a "stationary" threshold is

$$\nu(1 + \tau)\Gamma^{-2} = (1 + r\tau)\nu^{-1} = p\pi \quad . \tag{8.3.31}$$

The threshold frequency and Lorentz factors are given by

$$\nu_{st} = (1 + r\tau)/p\pi \quad , \tag{8.3.32}$$

$$\Gamma_{sthr} = (1 + \tau)^{\frac{1}{2}}(1 + r\tau)^{\frac{1}{2}}/p\pi \quad . \tag{8.3.33}$$

In most experimental arrangements, $r\tau \ll 1$ and $p = 1$, so that $\nu_{st} = 1/\pi$ and $\Gamma_{sthr} = (1 + \tau)^{\frac{1}{2}}/\pi$. This $\nu \sim 1/\pi$ corresponds to a "magic value" for the single-foil yield and for reasonable τ (<100), Γ_{sthr} is within the experimental range.

In Fig.8.8, where the effective N-foil yield per foil for $\nu = 0.3$ is shown, all the features of N-foil yield discussed so far are visible. The figure shows: (a) sa-

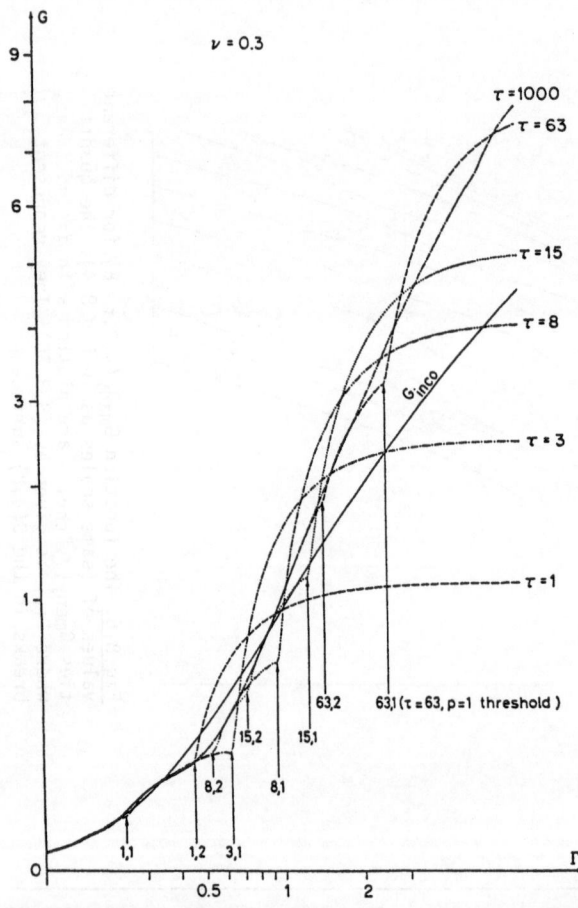

Fig.8.8. Energy dependence of the function $G_{many}(\nu = 0.3, \Gamma, \tau)$ for different values of τ. Only $p = 1$ and $p = 2$ thresholds (8.3.29) are clearly visible on these curves

turation due to the formation zone, $\Gamma_{sat,FZ} \simeq 2.2(\tau\nu)^{\frac{1}{2}}$, (b) occurrence of thresholds at $p = 1, 2, \ldots$, (c) approximation to G_{inco} for low Γ, and (d) enhancement over G_{inco} due to interference effects.

We summarize the conditions for neglecting some of the interference effects, leading to the particular cases:

$\ell_2 > 2\pi Z_2$, single foil ;

$\ell_1 > 2\pi Z_1$, integration in $\omega \rightarrow \sim$ single gap ;

$\left.\begin{array}{l} \ell_1 > 2\pi Z_1 \text{ , } \ell_2 > 2\pi Z_2 + \text{ integration in } \omega \\ \text{or} \\ \ell_1 \text{ and } \ell_2 > 2\pi Z_2 \end{array}\right\} \rightarrow$ single surface ; $\qquad$ (8.3.34)

$\Gamma_{sat} = \min(1.2\tau^{\frac{1}{2}}, r^{-\frac{1}{2}}\pi)$,

$\Gamma_{sthr} = (1 + \tau)^{\frac{1}{2}}/\pi$ for $r = 0$.

8.4 Experimental Tests of Transition-Radiation Theory

8.4.1 General Remarks

In order to test the predictions of the theory of transition radiation from ultra-relativistic particles it is necessary to study radiation from multilayered stacks of foils because of the intrinsic smallness of radiation from a single surface. For a practical radiator, N-foil interference effects cannot be neglected. The X-ray frequency range accessible to experiment is restricted by the absorption properties of the foils, gas, and detectors. The particle type and energy range that can be studied depends on the magnitude of the yield and the availability of controlled beams. In the last 20 years many experimental investigations have been done to investigate TR. Early experiments are described in a comprehensive review by *Harutyunian* and *Frangyan* [8.9] of the Institute for Physical Research of the Armenian SSR (1975), to which we refer the reader for details. The experiments which are discussed in this article cover a broad range of energies, materials, and particles. The relevant parameters of these experiments are summarized in Table 8.1.

As the major part of the radiated energy comes from the last oscillation of the X-ray frequency spectrum at $\nu = 1/\pi$, where $\nu = \omega/\omega_1$ with $\omega_1[keV] \simeq 10^4 \rho_f \ell_f [gm \ cm^{-2}]$, and as the gas is usually air ($\omega_g = 0.7$ eV and $r \approx 6 \times 10^{-4}$), we illustrate the coverage of representative experiments in the (Γ, τ) plane (Fig.8.9). Here, Γ is the Lorentz factor γ scaled by $\gamma_1 = \ell_f \omega_{p_1}/2$ and $\tau = \ell_g/\ell_f$. The yield is expected to saturate if $\Gamma > \min(\Gamma_{sat,d}, \Gamma_{sat,FZ})$ and the yield becomes low if $\Gamma < 2/\pi$. Figure 8.9 shows that experiments have covered the area of reasonable yield and energy dependence of TR from $2 < \tau < 800$ and $0.1 < \Gamma < 30$.

Table 8.1. A selected set of experiments on transition radiation

Particle	$\gamma/10^3$	Material	ω_p[eV]	ℓ_f[μm]	γ_1	ω_1[keV]	$\tau = \ell_g/\ell_f$	N_f	Method[a]	Det.	Remarks	Ref.
e^-	1.5 - 6	Al	33	25	2028	68.1	30	231	MS	Ge	Angular dist., energy spectra, FZ, W α γ	[8.15]
e^-	2.6 - 6	Mylar	24	4.4	264	6.34	360	100	SW	Ar,Kr	W(γ), FZ, SAT	[8.16]
				12.5	750	18.0	60	100				
e^-	2 - 30	Mylar	24	12.5	750	18	120	100-290	SW	Kr,Xe	W(γ), FZ, GAP, SAT	[8.17]
		Poly	18.6	25	1500	36	10-160	100-290				
				50	3000	72	30	100-290				
				12.5	580	10.8	120	188				
				12.5	1163	21.6	22	188				
e^-	1 - 4	Li	13.4	51	1700	22	11	1000	SW	Xe	W α γ	[8.18]
e^-	6 - 30	Poly	20	12.5	635	12.7	250	100	MS,SW	Kr,Xe	W(γ), SAT, dW/dω, Magic value	
		Mylar	24	25.4	1270	25.4	125	100	SW	Kr	W(γ), SAT	[8.19]
				50.8	2540	50.8	63	100				
				12.5	750	18	200	100				
e^-	10 - 18	Poly	20	25.4	1300	36	100	100	MS	X-ray spec.	dW/dω, oscillations, magic value	[8.20]
				16	812	16.3	87.5	1000				
e^-	2.65	Mylar	24	50	2538	51.8	28.0	250	MS	NaI(Tl)	dW/dω, oscillations, magic value	
				82	4162	83.2	17.0	200				
				11.5	701	16.8	264	28				
e^-	2.65	Li	13.4	25	1500	36.0	121	27	MS	Si	dW/dω, oscillations	
				50	3000	72.0	75	27				
				65	2177	29	5.1-46	28				
e^-	1.6 - 2.8	Mylar	24	25	1500	36	20	28	MS	Si	dW/dω, oscillations W(γ), N-foil threshold	

Table 8.1 (cont.)

Particle	$\gamma/10^3$	Material	ω_p[eV]	ℓ_f[μm]	Y_1	ω_1[keV]	$\tau = \ell_g/\ell_f$	N_f	Method[a]	Det.	Remarks	Ref.
π,e	0.7 – 1.8	Li	13.4	50	1701	22.8	4–10	500	SW	Ar,Xe	W α γ, independence of TR on particle type	[8.21]
π,p	0.04 – 1.4	Mylar	24	5	300	7.2	800	18	SW	Xe	Low γ detector, π/p identification	[8.22]
e	1.4 – 12	Li	13.4	53	1767	22.8	6	650	MS,SW	Si,Xe	W(γ), SAT, Spectra	[8.23]
π	1.0	Li	13.4	51	1700	22	6.7	1500	DC	Xe	W, π/e separation	[8.24]
e	>10	Mylar	24	5	300	7.2	28	1000	DC	Xe	W, π/e separation	[8.24]
π,e	1; >10	Li	13.4	30	1000	13	6.7	1500	DC	Xe	W(γ), π/e, π/K separation	[8.25]
π,K,p	1,0.5, 0.14	Mylar	24	5	300	7.2	28	960–420	PWC	Xe	W, π/K/p separation	[8.26]
e	0.2 – 2	Mylar	24	3	180	4.3	33	4000	SW	Xe	W, π/K/p separation	[8.27]
π,p	~5,0.7	Foam	~24	6	360	8.9	16	4000	SW	Ar	W, π/p separation, cosmic ray beam	[8.28]
				10	600	14.4	16	5000				
				17.5	1050	25	large	35				
π,K,p	1.4 – 2.9	Li	13.4	38	1267	17	20	1600	MS	Plastic scin.	Spectrum, angular separation, π/K/p separation, high int. beam	[8.29]
e	2 – 30	Various		Various				Various	SW	Kr,Xe	W(γ), SAT	[8.30]
e	2.7	Foam							MS	Si	W, spectrum	[8.12]

[a] MS magnetic separation; SW sandwich; DC drift chamber; PWC proportional wire chamber

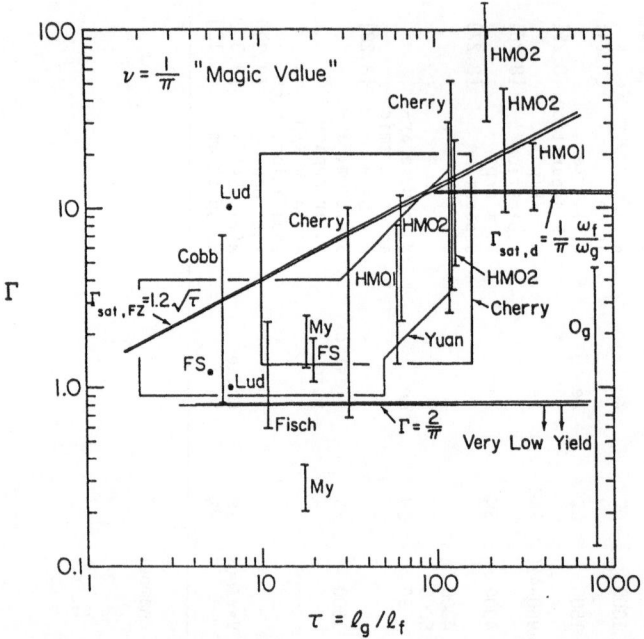

$$\tau = \ell_g / \ell_f$$

Fig.8.9. This figure shows the ranges of Γ and τ covered by various experiments. For the magic frequency $\nu = 1/\pi$, lines above which TR yield is saturated [$\Gamma_{sat,FZ}$ and $\Gamma_{sat,d}$ corresponding to (8.3.25 and 24)] are shown, as well as the line $\Gamma = 2/\pi$ below which the yield is very low. The names correspond to the following references: Cobb [8.23], Lud [8.25], FS [8.13], Fisch [8.18], My [8.29], Cherry [8.17], HMO1 [8.16], HMO2 [8.19], Yuan [8.15], and Og [8.22]

Experiments have verified the following predictions of TR theory:

1) The TR yield is a monotonically increasing function of the Lorentz factor γ, as discovered by *Yuan* et al. [8.15], *Cherry* et al. [8.17], *Fischer* et al. [8.18], *Ellsworth* et al. [8.19], *Bosshard* et al. [8.21], *Prince* et al. [8.30], *Alikhanyan* et al. [8.31].

2) The TR yield varies linearly with γ in regions of ν, Γ, and τ where incoherent addition is valid [8.15,17,18,21].

3) There are foil formation-zone effects, corresponding to decrease of yield with diminishing foil thickness [8.15,17,19,30,32].

4) Single-foil interference effects on the shape of the energy spectrum of X-ray photons, existence of enhancement at "magic" photon energies corresponding to $\nu = 1/\pi$, and its dependence of γ. Experiments were done by *Fabian* and *Struczinski* [8.13], *Ellsworth* et al. [8.19], *Cherry* et al. [8.20,32].

5) Variation of the TR yield with gap thickness [8.15,17,19].

6) Existence of thresholds in γ due to multifoil interference effects [8.20].

7) Saturation of the TR yield due to multifoil interference effects and to non-vacuum gaps [8.17,20,32].

210

It is to be noted that measured absolute yields generally agree with calculations and simulations of experimental conditions to better than 15%.

8.4.2 Discussion of Experimental Tests of Multifoil Theory

The experiments may be divided into three classes: (1) those which attempt to verify quantitatively the theory of TR, (2) those which apply TR to building particle identifiers for ultra-relativistic charged particles, and (3) those which use TR to measure the energy of charged particles in a nondestructive manner. The first class of experiments must use detectors with high efficiency over a wide frequency range and choose foil materials, thickness, and spacing to "isolate" the particular feature of TR to be studied. The second and third classes of experiments must deal with problems of maximizing signal-to-background ratio (e.g., TR X-rays to ionization loss), working with a high event-rate background and limited photon statistics. Generally speaking, the detected yield of TR depends on matching the "magic" frequency value, self-absorption in the foils, and the efficiency of the detector [8.23]. The challenge is to maximize the signal and differences between particles of different γ that have to be separated.

The nature of the problem is illustrated in Fig.8.10, where the interplay of the produced TR spectrum, the self-absorption in the foils, and the detection efficiency is shown for 25-μm mylar foils in vacuum with a large separation and using a 10-cm proportional counter filled with argon as X-ray detector. The maximum number of foils that can contribute at a given frequency is estimated by $1/\{1 - \exp[-\sigma(\omega)t]\}$, where $\sigma(\omega)$ is the X-ray absorption cross section and t is the foil thickness.

The linear dependence of the single-surface TR X-ray yield was first verified by *Yuan* et al. [8.15] using 231 aluminum foils of 25-μm thickness ($\omega_1 = 68.1$ keV and $\gamma_1 = 2062$) separated by 762-μm ($\tau = 30$). The TR photons were magnetically isolated from the charged particles and detected by a solid-state detector. This experiment observed three important features of TR X-rays: (1) that the angular distribution is forward peaked, $\theta \sim 1/\gamma$, (2) that the X-ray energy spectra show a sharp peak at the magic value of ~34 keV, and (3) that in a limited γ range, the linear dependence of the yield on γ is verified. Indeed the region in the ν, Γ, τ variables covered by this experiment happened to be that where G_{many} varies approximately linearly with γ.

The monotonic variation of TR yield with energy has been studied by many investigators. In Fig.8.11, the energy dependence of TR measured using mylar ($C_5H_4O_2$), lithium, and aluminum foils of varying thicknesses and arrangements is shown. The property that TR should only depend on charge and Lorentz factor of the particle and not on its mass is shown in Fig.8.11 where the yield from electrons at $\Gamma = 1.65$ ($\gamma = 3000$) is consistent with extrapolation of that measured using pions of lower γ.

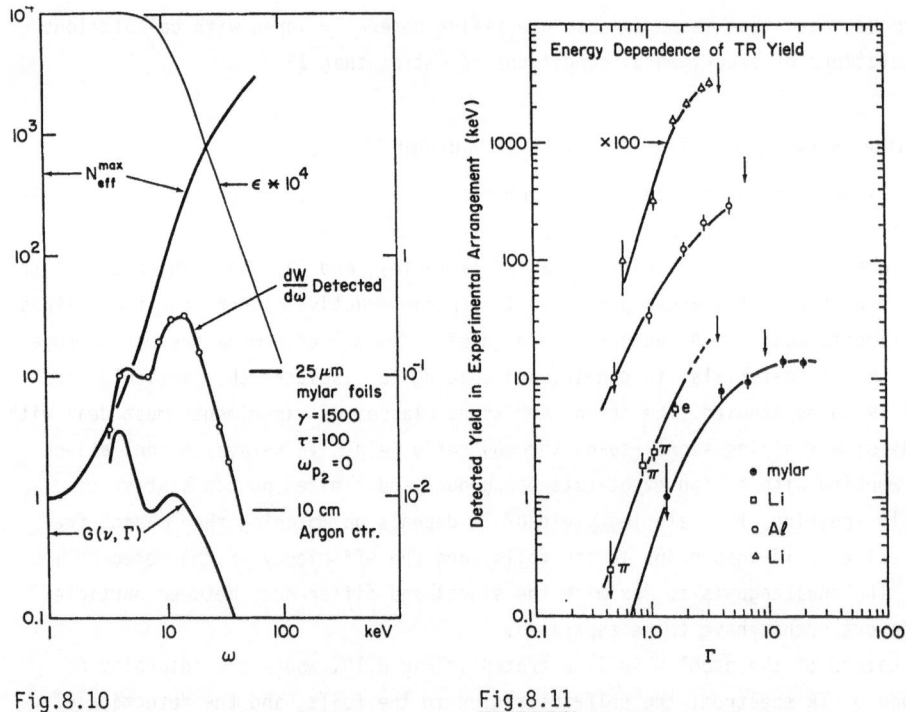

Fig.8.10   Fig.8.11

Fig.8.10. This figure shows the interplay of single-foil interference $G(\nu,\Gamma)$ absorption N_{eff}^{max}, and detection efficiency ϵ in determining the yield $dW/d\omega$ from N-foils when τ is large and the single-foil approximation is good for $G_{many}(\nu,\Gamma,\tau)$

Fig.8.11. Experimental data on γ dependence of TR yield for different foil materials. The results have been scaled by $\Gamma = \gamma/\gamma_1$, where $\gamma_1 = \omega_f \ell_f/2$ (8.3.10). The curves are linear below $\Gamma \sim 2$ after which saturation effects are expected to set in. The vertical arrows indicate the Γ above which saturation must take place. (● [8.17], □ [8.21], ○ [8.15], △ [8.18])

The approximate value of Γ where saturation of yield should set in is indicated by vertical arrows. The experimental results of *Cherry* et al. [8.17] clearly show this saturation.

In Figs.8.12 and 13, as set of representative curves are given which show formation-zone effects leading to a decrease of the TR yield at small foil thicknesses for fixed foil separation, and a decrease of the yield with decreasing gap thickness for fixed foil thickness, respectively [8.15,17]. In Fig.8.12, the horizontal axis is given in terms of foil thickness in μm as well as in terms of the location of the last oscillation of the TR spectrum due to single-foil interference at $\omega = \omega_1/\pi$, where $\omega_1 = \ell_f \omega_{p1}/2$. As most of the detected radiation comes from the region around ω_1/π, the detected yield should increase with foil thickness until a thickness is reached beyond which the efficiency of the detector starts to decrease. The data agree well with this expectation. The decrease in yield with gap spacing shown in

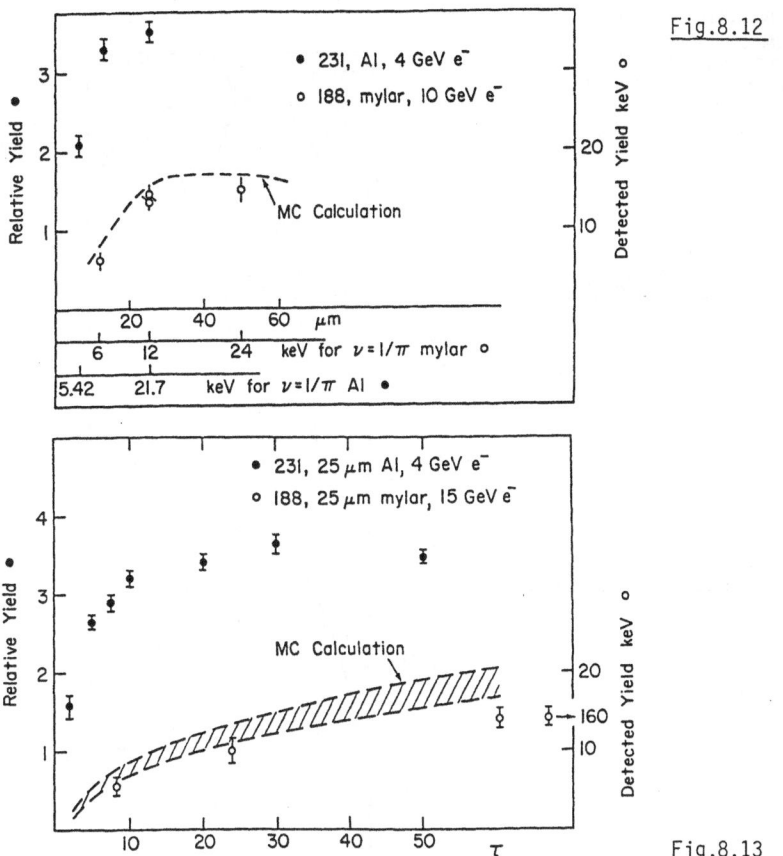

Fig.8.12

Fig.8.13

Fig.8.12. Experimental results on reduction in TR yield for small foil thickness (formation-zone effect). The frequency of the last oscillation in $G(\nu,\Gamma)$ corresponding to each foil thickness is shown in the lower two scales below the x axis. Comparison with a Monte Carlo calculation [8.17] for the case of mylar foils is quite satisfactory. The number of foils were kept fixed: 231 for Al [8.15] and 188 for mylar [8.17] respectively

Fig.8.13. Results of experimental investigations to study variation of TR yield with foil spacing for fixed foil thickness. The expected reduction in yield when $\tau < 5$ is clearly seen. The measurements were done using 231 25-μm Al foils [8.15] and 188 25-μm mylar foils [8.17]

Fig.8.13 [8.15,17] is a result of a formation zone in the gap, $\gamma_{sat,FZ} \approx 1.2\sqrt{\tau}$ (8.3.25). The horizontal axis is given in terms of the ratio $\Gamma_{sat,FZ}/\Gamma$, which decreases from 0.86 to 0.22, equivalent to a reduction in spacing from 4 mm down to 0.25 mm. The air formation zone becomes much larger than the foil spacing for 0.25 mm.

The most direct verfication of interference phenomena due to single-foil "diffraction" has been the study of energy spectra of TR X-ray photons, in which the emission from a small number of foils has been observed in an arrangement where the

213

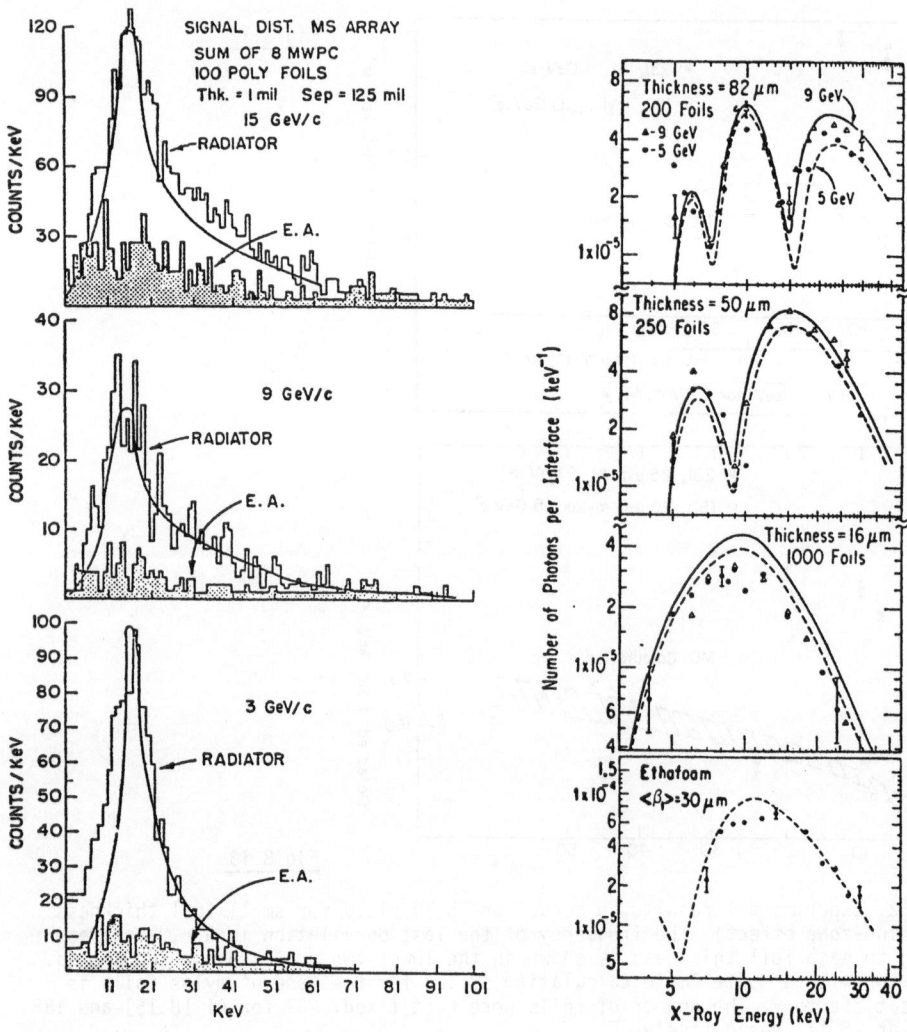

Fig.8.14   Fig.8.15

Fig.8.14. Experimental measurement of TR X-ray spectrum from 100 CH_2 foils [8.19].
Electrons of three different energies are seen to generate spectra which peak at
the same value of ω independent of γ, at the last oscillation of $G(\nu,\Gamma)$. The shaded curve shows the X-ray yield obtained by replacing the foils with a target of
solid polyethylene of the same total thickness as 100 25-μm Poly foils. The small
number of foils reduce the occurrence of multiple photon emission. The X-rays were
detected by 10 1.5-cm Xe-filled PWCs which also showed appropriate absorption
length for the TR X-rays

Fig.8.15. Frequency spectra of TR emerging from four different TR radiators as
measured by Cherry and Muller with a crystal spectrometer [8.20]. These beautiful
experimental results clearly show the oscillations due to interference effects
discussed in the text. The curves are Monte Carlo calculations for the experimental
conditions

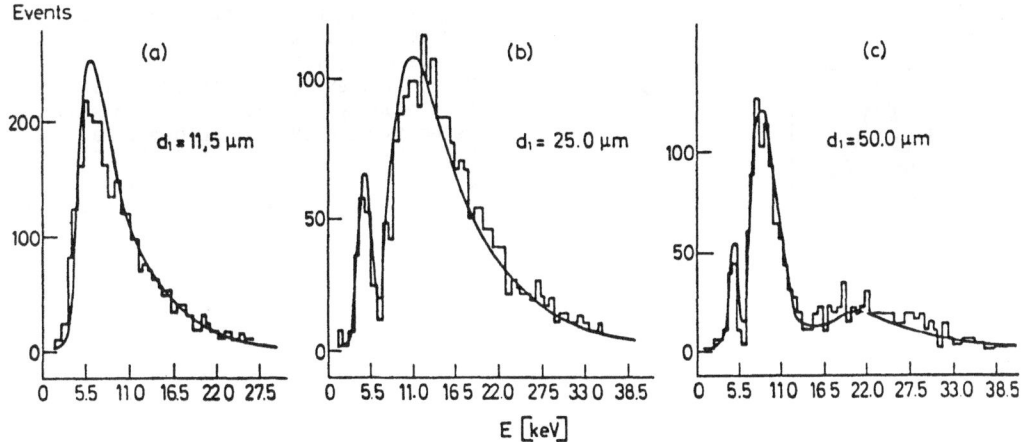

Fig.8.16a-c. Frequency spectra from mylar foils of different thickness compared with predictions [8.13]. Clear verficiation of oscillations is seen. Note the linear dependence of the frequency of $\nu = 1/\pi$ oscillation on foil thickness. The relative heights of maxima are determined by self-absorption in the foils and counter efficiency. The values of τ were 264, 121, and 75 for these runs

charged particles have been separated from the emitted X-rays by magnetic bending [8.13,19,20]. The X-ray spectrum has been detected using xenon multiwire proportional counters (MWPC), X-ray spectrometers, and solid-state detectors with good detection efficiency over the frequency range of the expected spectrum. The results of these measurements are shown in Fig.8.14 [8.19], Fig.8.15, [8.20], and Fig. 8.16 [8.13]. In each of these figures, data have been compared with Monte Carlo simulations based on TR "resonance" theory [8.3,8]. The data clearly show the following:

a) The location of the last maximum depends only on foil material and thickness, not on the Lorentz factor of the charged particle (Fig.8.14).
b) Clear evidence for oscillations and successive maxima for $\nu^{-1} = \pi$, 2π, 3π,... has been observed (Figs.8.15 and 16).
c) The absolute yield agrees with calculations to better than 10%. Generally, the observed yield is somewhat smaller than expected, but this may be due to the inherent difficulty of taking into account the multitude of effects which can cause a loss of X-ray photons through scattering, fluorescence escape, etc. [8.33].

Finally, we show in Fig.8.17 beautiful results of *Fabjan* and *Struczinkski* [8.13] on the observation of thresholds (Sect.8.3.4) due to multifoil interference effects. To observe the step in TR yield at $\gamma = 2300$, they studied the variation of X-ray yield in a narrow energy band, 9 keV $< \hbar\omega <$ 14.5 keV, using 28 25-μm mylar foils

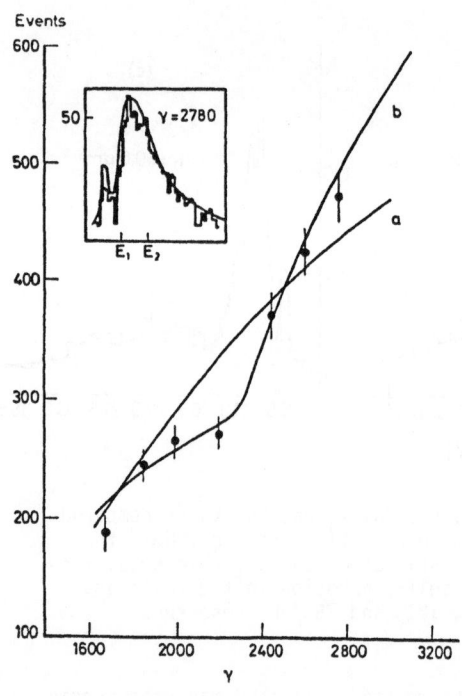

Fig.8.17. The figure shows beautiful confirmation of multifoil coherence effects using lithium foils and $\tau = 20$ [8.13]. The excitation function is for the frequency band $9.0 < \hbar\omega < 14.5$ keV. A step is seen at $\gamma \sim 2300$ due to a threshold according to (8.3.31). The curve labelled a is the prediction without multifoil effects. The curve labelled b includes them

with a gap spacing of 500-μm corresponding to a value of 20 for the parameter τ. The N-foil theory [8.8] predicts a step at $\Gamma \approx 1.5$ which corresponds to $\gamma \approx 2250$; such a step is clearly seen in the data.

   This brief survey provides convincing proof of the theory of transition radiation from multilayered media described in Sect.8.2 and 3 in a fairly detailed manner. Experiments which simultaneously measure the energy and angular distributions of transition radiation are needed to further test the theory. Attempts to do this using spark chambers and luminescence drift proportional chambers (*Alikhanyan* et al. [8.31], *Commichau* et al. [8.26]) offer exciting possibilities.

8.5 Current Applications of TR of High-Energy Particles, and New Developments

Transition radiation emitted by charged particles has been used for many different purposes in the last ten years: for particle identification, for energy determination, for beam optics, and as a monochromatic high-energy X-ray source. This progress has required better understanding of the design of radiators for TR, development of new techniques for X-ray and particle detection, and finding an optimal match between radiators and detector for each specific application. These developments are discussed briefly in this section.

8.5.1 Particle Identification

The TR technique offers a powerful method of identifying the mass of ultra-high-energy charged particles because of its dependence on the Lorentz factor, $\gamma = E/m$. The angular distribution of TR is peaked at an angle $\theta \sim 1/\gamma$ with respect to the particle direction, and the total yield of X-ray energy increases approximately linearly with γ until saturation sets in at high γ. At γ larger than ~ 1000, Čerenkov radiation becomes independent of γ, and this is where the TR technique becomes important. It offers a nondestructive technique for particle identification, when combined with measurement of momentum or energy of the charged particle. Just as is the case for Čerenkov counters, TR detectors must be specifically designed for each application. Note that the larger the mass difference between particles to be distinguished, the easier it is to design an identifier. The TR detector is designed so that the heavier particle produces a very small yield of TR X-rays, and the lighter particle gives a good TR signal above background. Such a detector is of the threshold type.

a) Optimization of TR Particle Identifiers

As the yield per foil is small, it is necessary to choose carefully the foil material, foil thickness, number of foils, and detector so as to achieve the best identification. The qualitative considerations involved can be succinctly described by examining (8.3.28) for the yield from N foils, which depends on the dimensionless parameters Γ, ν, and τ, where Γ is the reduced Lorentz factor, ν the reduced frequency, and τ the ratio of gap length to foil thickness [see (8.3.10-12,14) for their definitions]. From Sect.8.3, the following important qualitative features of the expected yield are summarized:

1) The last maximum in the radiation occurs at $\nu = \pi^{-1}$, above which the intensity decreases rapidly, becoming very small for $\nu > 1$ for any Γ (see Figs.8.6 and 7).
2) Most of the radiation comes from this "magic" region, the lower frequency photons being absorbed in most practical arrangements.
3) The range of Γ where the radiation increases rapidly is between $1/2 < \Gamma < 2$ to 3. At lower Γ values, the radiation becomes vanishingly small, and above ~ 3 the yield varies only logarithmically and then saturates.
4) There exist certain stationary thresholds in the amount of radiation due to multifoil effects. The corresponding value of Γ is given by $(1 + \tau)^{\frac{1}{2}}/\pi$.

A TR radiator designed to have optimum performance as a threshold detector should have $\Gamma = 1$, $\nu = 1/\pi$, and $\tau \simeq 5$ near its threshold. For a foil of a given material (ω_p), the threshold value of γ is obtained from $(1/2) \, \omega_p \ell_f = 2.5 \, (\omega_p/eV)(\ell_f/\mu m)$. Many materials, thicknesses of foil, and detectors have been used (see Table 8.1 for a summary from Li to Hg).

The detection of emitted X-ray photons separate from the particle ionization loss is difficult unless enough path length for magnetic bending is available, or high spatial resolution is possible in the detector. In many of the applications to date, one measures the sum of TR and ionization signal using proportional wire chambers (PWC). The most common gases used in these PWCs are argon, krypton, and xenon (with some quenching gas).

b) *Experimental Results*

The usual arrangement is the sandwich (SW) arrangement of foils and PWCs. The first experiment to show the separation of pions from electrons of 15 GeV/c using an SW array was one by the Chicago group [8.17]. They used several radiators consisting of 188 25-μm mylar foils separated by 1.5 mm each, and PWCs filled with xenon and krypton. Their beautiful results for the difference in detected signal for pions and electrons are shown in Figs.8.18 and 19 for 15-GeV/c and 3-GeV/c particles, respectively. A clear separation between pions and electrons is obvious.

Electron/hadron separation in an ISR experiment was achieved using only two layers of xenon PWCs and 50-μm Li foils [8.23]. The authors succeeded in rejecting protons at the 8% level with 90% efficiency for recording electrons.

The separation of pions from protons with a sandwich array is considerably more difficult, because the ratio of masses is only a factor of 7, rather than 280 for that of π to e. Identification of pions of a few hundred GeV of energy in a mixed cosmic-ray beam of pions and protons has been achieved using SW arrays [8.28,34]. In the experiment of *Ellsworth* et al. [8.28], 24 modules of 4.7-cm styrofoam radiators and 5.1-cm-thick PWCs filled with argon were used to identify pions from protons above 400 GeV. The fraction of pions in the beam was measured on a statistical basis, using likelihood analysis of all the PWC signals. The results of this experiment are illustrated in Fig.8.20, where the likelihood ratio for being a pion is shown for Monte Carlo pions and protons, and for the experimental data.

TR has been successfully used to tag pions from kaons and protons at energies above 200 GeV in the Fermilab M6-W beam [8.29]. A radiator consisting of 1600 38-μm Li foils with 750-μm gaps was placed inside the vacuum of the beam pipe. After traversal of the foils, the beam was deflected by 4 mrad with a 3-m-long dipole magnet, and the X-ray photons were detected by a 12-cm-long plastic scintillator placed in the path of the undeflected photons. In Fig.8.21, a photograph taken with an X-ray film located at the counter is shown with and without a copper adsorber to remove TR X-rays. A clear spot of TR X-rays next to the beam spot is seen. Using a negative beam of 300 GeV, it was possible to determine the number of K^- as compared to π^- by measuring signals with no TR. The K^-/π^- ratio determined in this way was in agreement with that measured in single-arm spectrometer studies of inclusive spectra. The use of a plastic scintillator to detect X-rays makes it feasible to operate such TR detectors in intense beams.

218

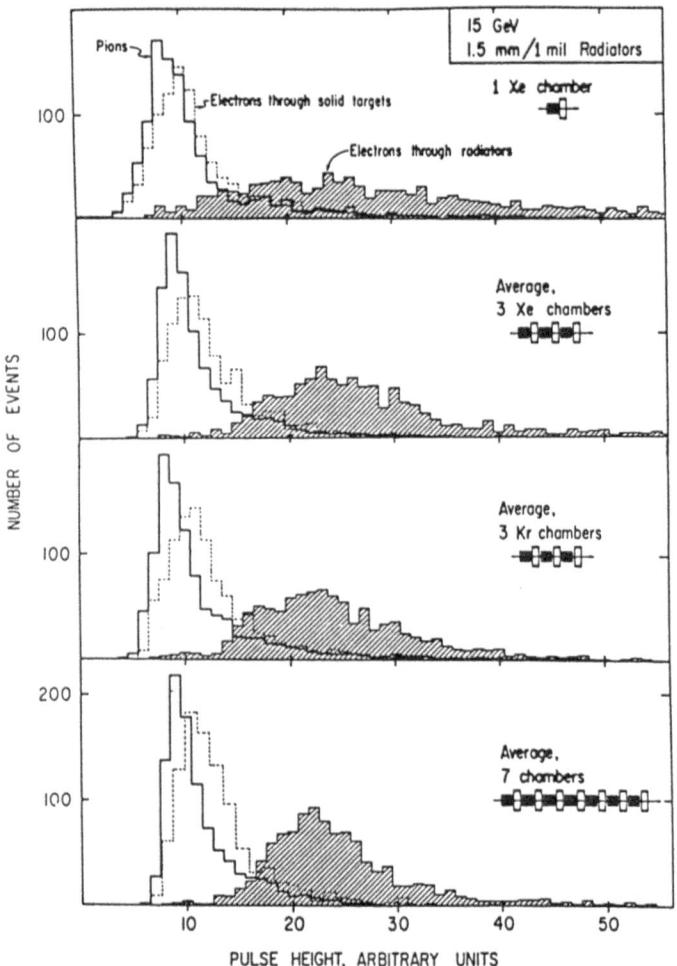

<u>Fig.8.18.</u> Results on pion-electron separation using a sandwich array of radiators and proportional chambers [8.30]. The proportional chambers are 4.3 cm thick and filled with 80% Xe and 20% CO_2, or 80% Kr and 20% CO_2 at 1 atmosphere pressure. Each radiator has 188 mylar foils. The dotted histogram is for electrons with equivalent absorber in place of TR radiators. It is clear that this arrangement can separate 15-GeV electrons from 15-GeV pions by using average pulse heights with 3 Xe-chamber or 7 Kr-chamber arrangements with good confidence

Particle identification has been achieved using the angular distribution of TR X-rays [8.24]. The method uses the time structure of signals in a drift chamber, as illustrated in Fig.8.22. One simultaneously measures the total charge Q and the total width W of the signal in a drift chamber placed downstream of a TR radiator. The percentage pion contamination for separating 140-GeV/c pions from kaons and 15-GeV/c pions from electrons is shown in Fig.8.23, using (Q + W) as a merit function. The method is seen to give good identification, but the aperture of the detec-

219

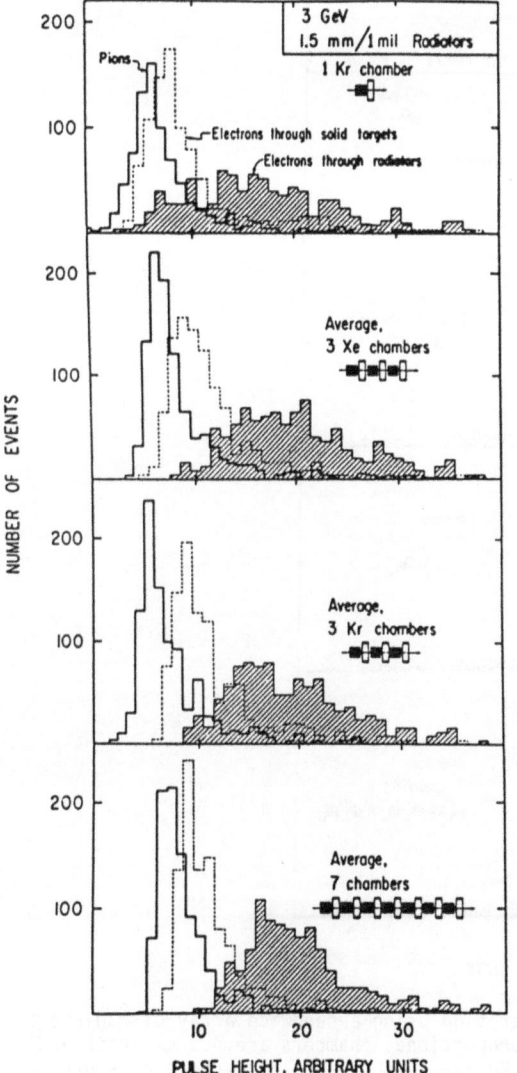

Fig.8.19. Same as in Fig.8.18 but for 3-GeV particles [8.30]

tor is limited by the requirement on the spacing between the radiator and drift chamber.

Another technique recently explored is to detect charge clusters due to TR photons along the track of a particle in a drift chamber [8.25]. A set of 12-radiator, MWPC modules was traversed by the beam. Charge clusters separated by 1 mm in space were measured by analyzing current pulses from each MWPC using ADCs operated by consecutive 40-ns gates. Only clusters above a fixed discriminator threshold were counted. These clusters indicate detection of TR X-rays and some fraction of δ rays. This technique enhances the signal-to-noise ratio by rejecting ordinary ionization loss. The cluster-counting technique improves the K-to-π separation by an order of

220

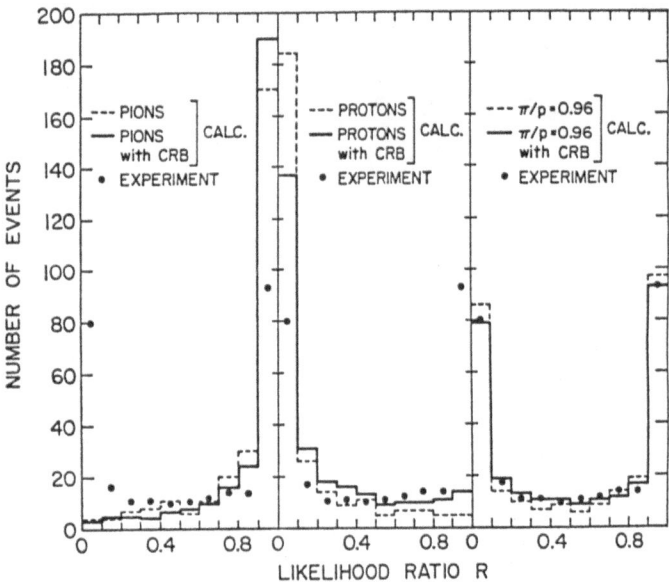

Fig.8.20. Application of TR SW-array technique to separate pions from protons in a cosmic-ray beam (CRB) for charged hadrons of 400-800 GeV energy [8.28]. The method calculates the likelihood ratio that the signals observed in a sequence of argon-filled proportional counters are due to a pion or due to a proton. The experimental data are compared with expectations from pure pion and pure proton beams. The method can identify particles on a statistical basis with good confidence

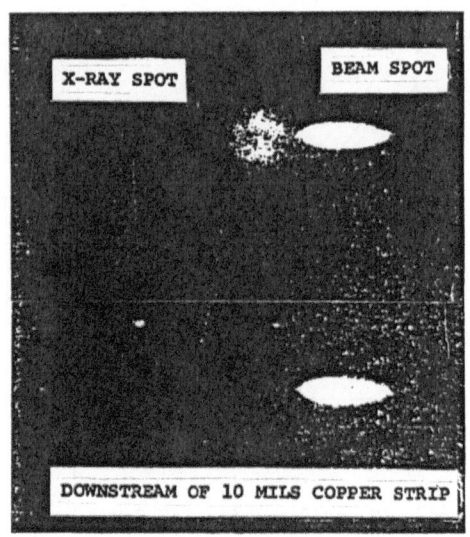

Fig.8.21. A photograph of a TR X-ray spot, as distinct from the beam spot from 300-GeV π^- in the FNAL M6 beamline [8.29]. The *upper figure* shows X-ray spot alongside beam spot, which is about 2.54 cm across, and in the *lower figure* X-rays are shown to be absorbed away by a 10-mil copper foil. The experiment used magnetic bending to deflect the beam and detected the X-rays using a plastic scintillator (see text)

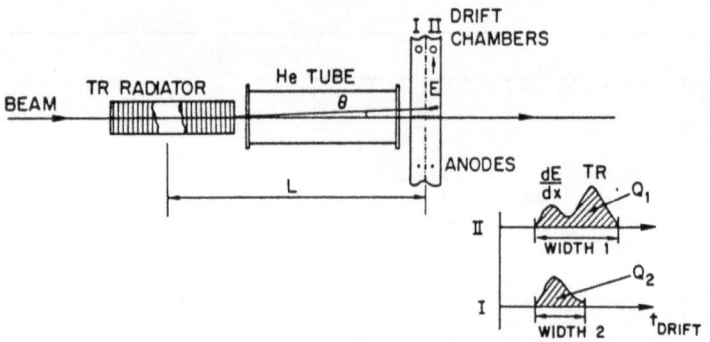

Fig.8.22. Experimental arrangement for the transverse drift method of detecting TR quanta using drift-chamber technique [8.27]. Two methods are indicated: one measures the total charge Q, and the other the time width W of the pulse

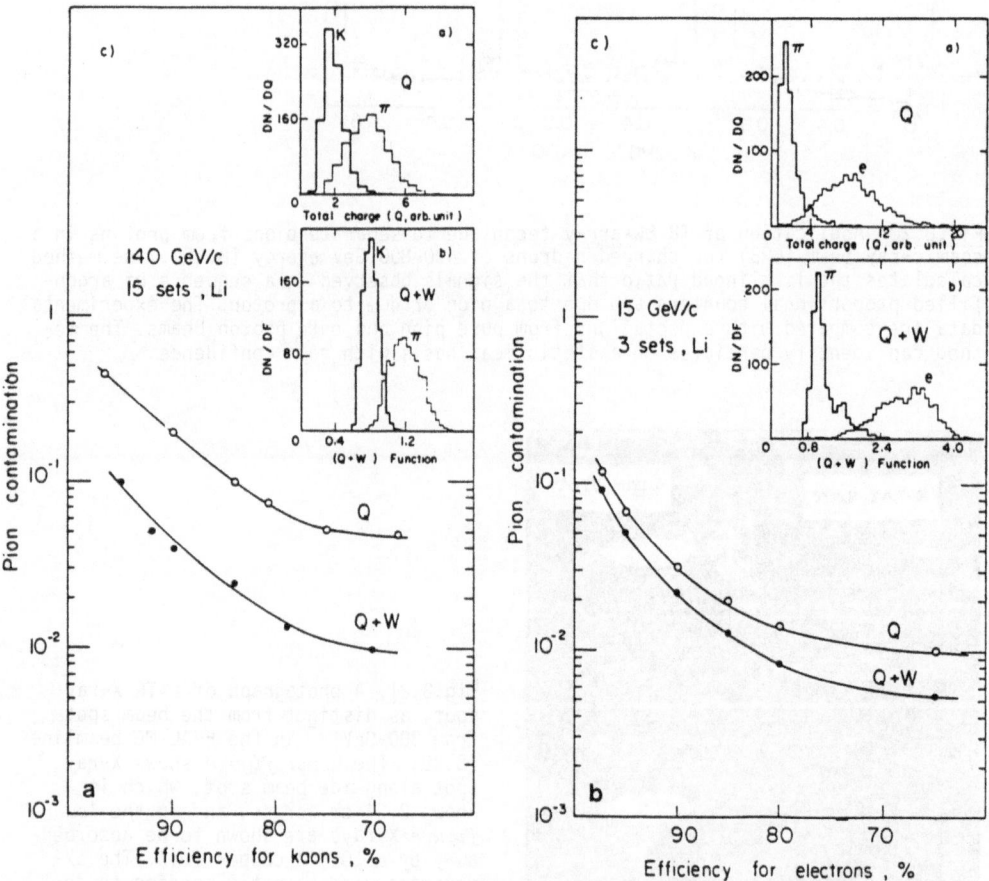

Fig.8.23. (a) Pion-kaon separation by two methods (Q and Q +W) for 140 GeV/c; (b) electron-pion separation at 15 GeV/c by Q and Q +W methods

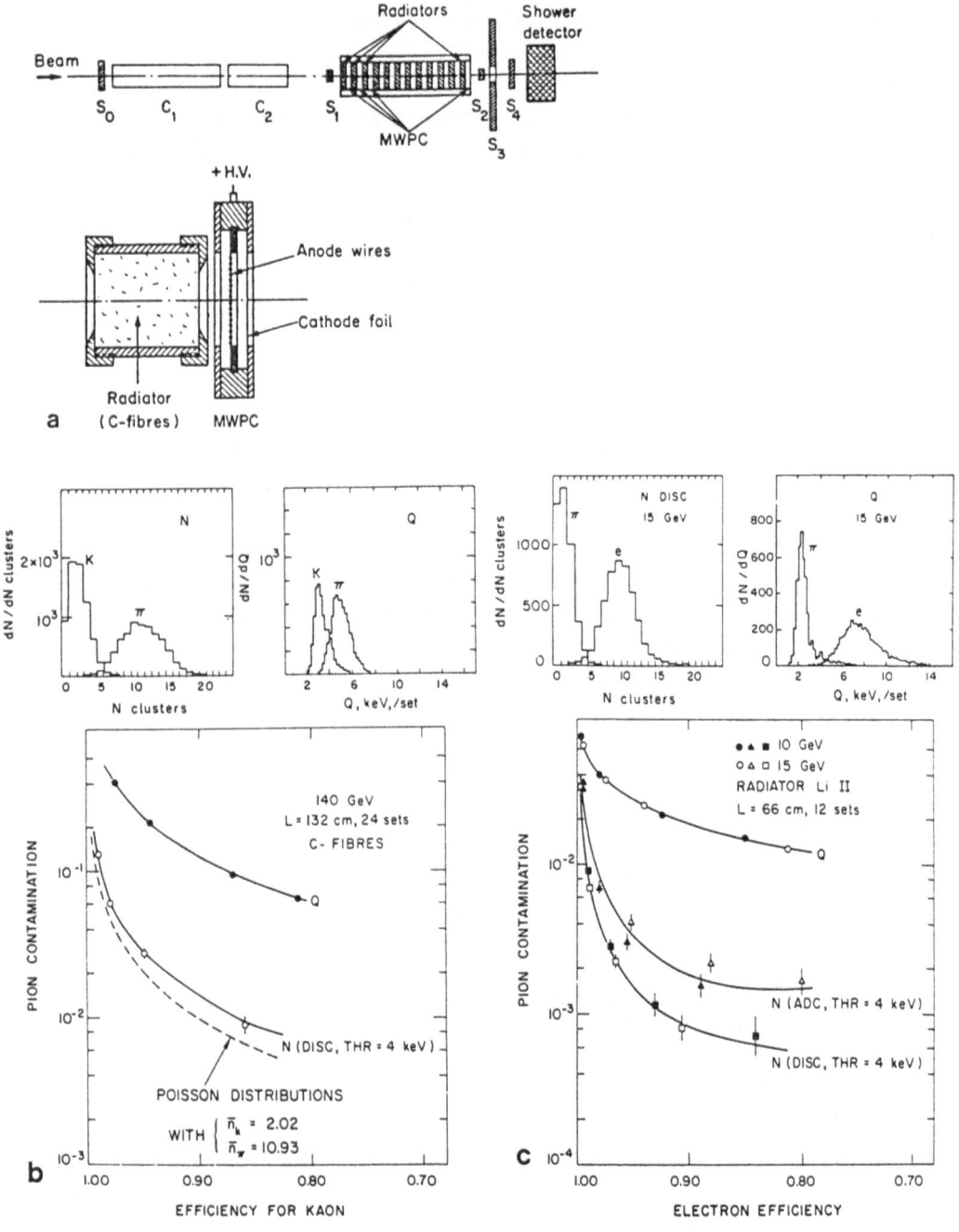

Fig.8.24a-c. Cluster-counting detector for particle separation [8.25]: (a) experimental arrangement —use of drift-chamber technique allows three measures for TR:
total charge Q, total width W, and total number of clusters N; (b) pion-kaon separation for 140 GeV/c by the three methods: (c) electron-pion separation at 15 GeV/c
with the three methods

magnitude. At 140 GeV/c, the pion contamination is less than 1% for a 90% kaon detection efficiency. This is shown in Fig.8.24.

An extension of the SW-array design to identify pions in the 100-GeV/c region has recently been successfully operated at CERN [8.27] and FNAL [8.22]. Correct identification of pions from kaons at the 85% level has been achieved at momenta down to 70 GeV/c.

8.5.2 Energy Measurement Using TR

Due to the smallness of the number of TR photons detected in most experimental arrangements, the energy resolution of TR detectors is very poor. For high-Z particles of high energy per amu (i.e., high Lorentz factor γ), the yield of TR photons increases as Z^2. It is, therefore, possible to reduce the fluctuations in detected energy by $1/Z$, and to construct an energy-measuring device in the region where the amount of radiation increases linearly with γ. Such a detector has been constructed by the University of Chicago group to measure the energy spectra of cosmic-ray nuclei with $Z \geqslant 4$ [8.35]. It is to be flown in the Space Shuttle in the near future. It uses carbon fibers as radiators and Xe PWCs as detectors of TR and ionization loss. A schematic drawing of the detector is shown in Fig.8.25. The group expects to measure the flux of iron nuclei up to energies of TeV/amu. The unavailability of accelerator beams of heavy ions at high energy makes it difficult to calibrate such instruments directly.

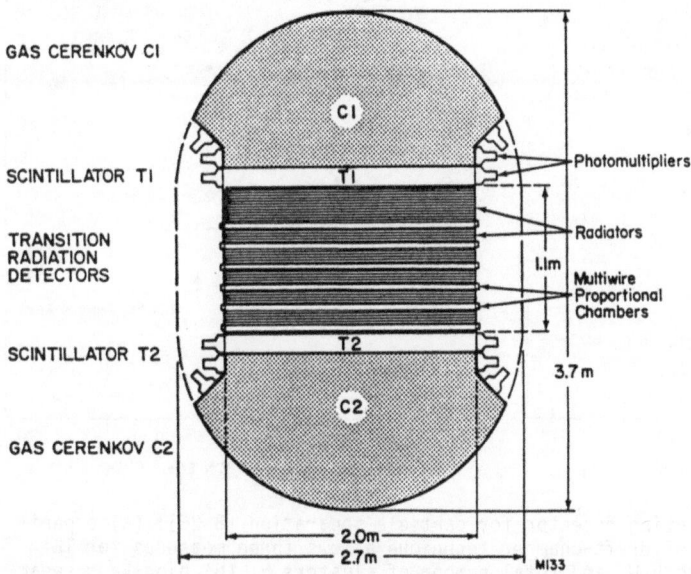

Fig.8.25. Application of the TR technique to measure the energy spectra of cosmic-ray nuclei with Z > 3. Schematic cross section of the University of Chicago cosmic-ray detector for Spacelab [8.35]

8.5.3 Transition Radiation Detectors as Accelerator Monitors

A novel application of TR has been made [8.36-38] for beam diagnostics and beam
monitoring at electron accelerators [AL 60 at Saclay ($\gamma \sim 100$) and LAL at Orsay
($\gamma \sim 2000$)] using TR produced by the beam in the optical region (2800 Å to 6000 Å).
Radiator samples, in the form of coatings of 3.5-μm of Al, Ag, Au, and Cu on mylar foils, were inserted into the beam. The angular distribution of TR reflected
from the radiators was observed. The angular distribution, polarization, and spectral distribution of the TR light was measured and found to be in good agreement
with theory. The angular distribution pattern was used (a) for accurate beam posi-
tioning on a positron target at LAL so as to optimize the positron yield, and (b)
for absolute energy determination (i.e., γ measurement) of the machine energy to
one percent. Future applications of this technique for diagnosis of accelerator
beams looks promising.

8.5.4 TR as a Tunable X-Ray Source

Single-foil interference gives rise to oscillations in the spectrum of transition
radiation from a stack of foils at $\nu^{-1} = \pi$, 2π, 3π, ... A group at SLAC [8.39,40]
has used this property of TR to set up a tunable X-ray source. They confirm the
predictions of theory with respect to expected frequencies and widths of oscillations. Stacks of mylar foils of varying thicknesses from 25-μm to 76-μm were placed
in a 4-GeV electron beam from SLAC. Their angle of inclination to the beam could be
varied to obtain different foil thicknesses in a continuous manner. A Li-drifted Si
detector was used to measure the X-ray spectrum. The authors showed that the resolution of the peaks of the oscillations improves with the order of the resonance, in
accordance with theory. The fractional width for the second-order oscillation is of
the order of 35-40%, while for the third order it improves by another factor of two.
This technique can provide relatively "monochromatic" hard X-ray beams for experi-
mentation.

8.5.5 Other Developments

Small grains (1 - 20-μm) of certain Type I superconductors such as In, Sn, Ta, Tl,
and Hg can remain in a metastable superheated superconducting state quasi-infini-
tely, unless some external energy causes them to go normal [8.41]. The energy required to 'flip' a grain of a few microns can be adjusted to be in the 1-keV to
1-MeV range. Individual grains can be flipped into normalcy by a traversal of the
grain by charged particles or by absorption of TR X-ray photons produced in the
traversal of the grain, or both. To make a TR detector, the "threshold" energy has
to be devised to give a good signal-to-noise ratio. A unique feature of grains as
detectors for TR photons is the high efficiency grains have for absorbing the TR
X-rays, because of their high Z and high density.

By making a colloidal suspension of these grains in a low-Z medium such as paraffin [called superheated superconducting colloids (SSC)], a TR radiator-detector device has been made [8.42]. The large plasma frequency of grain material and the small formation zone in paraffin make it possible to have a high filling factor of grains in the colloid without large loss of TR due to interference and absorption effects. The number of grains that go normal was measured by the frequency shift of a resonant circuit which contains a coil around the SSC detector. At large γ, the number of grains Δn chaning state must increase due to the monotonic increase of TR yield with γ; however, the number changing due to ionization loss should remain the same provided no showering occurs in the SSC. Experimental observations [8.42,43] show a linear dependence of Δn on γ below a value of γ of 4000, and preliminary results have indicated a quadratic dependence at higher γ (above 9000). Although the signal-to background ratio was quite good (~10), the time resolution was poor. Individual grain flips, however, have been successfully observed. The spatial resolution is limited by grain size and spacing and can be of the order of a few microns.

Although the method is promising, much work remains to be done to build practical detectors which have both high spatial and energy resolution as well as good particle identification capabilities.

Another technique to separate TR X-rays from particle ionization has been to use an electroluminescent drift chamber with 20-μm resolution [8.44]. Two adjacent high-pressure drift chambers (e.g., filled with 10 atmospheres of xenon, with a drift field of keV/cm and a 4-mm gap), each viewed by a photomultiplier were used to extract the TR signal. With this arrangement, it is possible to compare the pulse shapes from the two adjacent chambers to measure simultaneously the energy deposit and time shift of pulses due to detection of TR X-rays in one of the chambers. The time shift is a result of the angular spread of TR X-rays with respect to particle direction. The method offers a possibility of particle identification in a tight geometry.

We have seen that a large number of predictions of TR theory have been beautifully confirmed by experiments. There are, however, some theoretical aspects which still need to be investigated. One has to do with the effects of multiple scattering and bremsstrahlung on the emission of TR at still higher values of the Lorentz factor ($>10^5$). These phenomena appear to lead to enhancement of the production of hard X-rays [8.8,45-48] from ultra-relativistic electrons. The quadratic dependence on γ observed [8.42,49] may be due to this effect. Another has to do with emission of TR from monocrystals at certain well-defined "Bragg" angles [8.50], a phenomenon where experimental observation is rather difficult. Interference effects between transition radiation and Čerenkov radiation have been studied by several investigators: *DeRaad* et al. [8.51], *Zrelov* and *Ruzicka* [8.52], *Wu-Yang Tsai* et al. [8.53], and *Olsen* and *Kolbenstvedt* [8.54]. These coherence effects need to be studied experimentally. Emission of TR and occurrence of "transition scattering" in vacuum in the presence of strong magnetic fields have been studied theoretically [8.55].

Acknowledgments. This work has been supported in part by a grant from the National Science Foundation. It is a pleasure to acknowledge discussions with R.W. Ellsworth, L.C.L. Yuan, D. Muller, M. Cherry, C.W. Fabjan, W.J. Willis, M. Atach, A.G. Oganesian, and X. Artru during the course of my research activity on TR since 1969.

References

8.1 V.L. Ginzburg, I.M. Frank: Zh. Eksp. Teor. Fiz. **16**, 15 (1946);
V.L. Ginzburg, V.N. Tsytovich: Phys. Rep. **49**, 1 (1979)
8.2 G.M. Garibyan: Zh. Eksp. Teor. Fiz. **33**, 1403 (1957) [English transl.: Sov. Phys.-JETP **6**, 1079 (1958)]
8.3 G.M. Garibyan: Zh. Eksp. Teor. Fiz. **39**, 332 (1969) [English transl.: Sov. Phys.-JETP **12**, 237 (1961)];
G.M. Garibyan: Erevan Report No. EO-27 (1973)
8.4 M.L. Ter-Mikaelyan: Nucl. Phys. **24**, 43 (1961)
8.5 M.L. Ter-Mikaelyan: *High-Energy Electromagnetic Processes in Condensed Media* (Wiley-Interscience, New York 1972)
8.6 F.G. Bass, V.M. Yakovenko: Usp. Fiz. Nauk **86**, 189 (1965) [English transl.: Sov. Phys.-Usp. **8**, 420 (1965)]
8.7 Loyal Durand III: Phys. Rev. D**11**, 89 (1975)
8.8 X. Artru, G.B. Yodh, G. Mennessier: Phys. Rev. D**12**, 1289 (1975)
8.9 F.R. Harutyunian, A.S. Frangyan: Inst. Phys. Res., Acad. Sci. of the Armenian SSR, Report No. IPhR-75-22 (1975), Parts 1 and 2 (unpublished)
8.10 G.M. Garibyan: *Adventures in Experimental Physics* (World Science Education, Princeton 1972) p.120
8.11 G.M. Garibyan, L.A. Gevorgian, C. Yang: Zh. Eksp. Teor. Fiz. **66**, 552 (1974) [English transl.: Sov. Phys.-JETP **39**, 265 (1974)]
8.12 C.W. Fabjan: Nucl. Instrum. Methods **146**, 343 (1977)
8.13 C.W. Fabjan, W. Struczinski: Phys. Lett. **57**B, 483 (1975)
8.14 G.B. Yodh, X. Artru, R. Ramaty: Astrophys. J. **181**, 725 (1973)
8.15 L.C.L. Yuan, C.L. Wang, H. Uto, S. Prunster: Phys. Lett. **31**B, 603 (1970);
L.C.L. Yuan, C.L. Wang, H. Uto, S. Prunster: Phys. Rev. Lett. **25**, 1513 (1970)
8.16 F. Harris, T. Katsura, S. Parker, V.Z. Peterson, R.W. Ellsworth, G.B. Yodh, W.W.M. Allison, C.B. Brooks, J.H. Cobb, J.H. Mulvey: Nucl. Instrum. Methods **107**, 413 (1973)
8.17 M.L. Cherry, G. Hartmann, D. Muller, T.A. Prince: Phys. Rev. D**10**, 3594 (1973)
8.18 J. Fischer, S. Iwata, V. Radeka, C.L. Wang, W.J. Willis: Phys. Lett. **49**B, 393 (1974)
8.19 R.W. Ellsworth, J. MacFall, G.B. Yodh, F. Harris, T. Katsura, S. Parker, V. Peterson, L. Shiraishi, V. Stenger, J. Mulvey, B. Brooks, J. Cobb: In *Proc. XIII Int. Conf. on Cosmic Rays*, Denver, Vol.4 (Colorado Associated University Press, Boulder 1973) p.2819
8.20 M.L. Cherry, D. Muller: Phys. Rev. Lett. **38**, 5 (1976)
8.21 R. Bosshard, J. Fischer, S. Iwata, V. Radeka, C.L. Wang, M. Atac: Nucl. Instrum. Methods **127**, 141 (1975)
8.22 A.G. Oganesian, A.T. Sarkissian, M. Atac: In *Proc. Int. Symp. on Transition Radiation of High Energy Particles* (Erevan Phys. Inst., Erevan 1977) p.269; Fermilab Report No. FN-305 (1977) (unpublished)
8.23 J. Cobb, C.W. Fabjan, S. Iwata, C. Kourkoumelis, A.J. Lankford, G.C. Moneti, A. Nappi, R. Palmer, P. Rehak, W. Struczinski, W. Willis: Nucl. Instrum. Methods **140**, 413 (1977). Similar techniques have been studied by B. Merkel, J.P. Repellin, G. Sauvage, M. Gourquin, J.C. Chollet, J.M. Gaillard, A. Hrisoho, P.H. Jean: Nucl. Instrum. Methods **138**, 625 (1976)

8.24 M. Deutschmann, W. Struczinski, C.W. Fabjan, W. Willis, I. Gavrilenko, S. Mai-
     burov, A. Shmeleva, P. Vasiljev, V. Tchernyatin, B. Dolgoshein, V. Kantserov,
     P. Nevski, A. Sumarokov: Nucl. Instrum. Methods **180**, 409 (1981)
8.25 T. Ludlam, E. Platner, V. Polychronakos, M. Deutschmann, W. Struczinski, C.W.
     Fabjan, W. Willis, I. Gavrilenko, B. Dolgoshein, V. Kantserov, P. Nevski, A.
     Surnarokov: Nucl. Instrum. Methods **180**, 413 (1981);
     C.W. Fabjan, W. Willis, I. Giavrilenko, S. Maiburov, A. Shmeleva, P. Vasiljev,
     V. Chernyatin, B. Dolgoshein, V. Kantserov, P. Nevski, A. Sumarokov: Nucl.
     Instrum. Methods **185**, 119 (1981)
8.26 V. Commichau, M. Deutschmann, H. Goddeke, K. Hangartner, U. Putzhofen, R.
     Schulte, W. Struczinski: Nucl. Instrum. Methods **176**, 325 (1980)
8.27 C. Camps, V. Commichau, M. Deutschmann, H. Goddeke, K. Hangartner, W. Liesmann,
     U. Putzhofen, R. Schulte: Nucl. Instrum. Methods **131**, 411 (1975)
8.28 R.W. Ellsworth, A.S. Ito, J.R. MacFall, F. Siohan, R.E. Streitmatter, S.C.
     Tonwar, P.R. Vishwanath, G.B. Yodh: Phys. Rev. **D27**, 2041 (1983)
8.29 L.C. Myrianthopoulos, R.W. Ellsworth, R.G. Glasser, R.S. Holmes, H. Strobele,
     G.B. Yodh: Univ. of Maryland Techn. Report No. 80-106 (1980) (unpublished);
     Bull. Am. Phys. Soc. **A6**, 25 (1980)
8.30 T.A. Prince, D. Muller, G. Hartmann, M.L. Cherry; Nucl. Instrum. Methods
     123, 231 (1975);
     S.P. Swordy, J.L. Heureux, D. Muller, P. Meyer: Nucl. Instrum. Methods **193**,
     591 (1982)
8.31 A.I. Alikhanyan, K.M. Avakina, G.M. Garibyan, M.P. Lorikian, K.K. Shikhliarou:
     Phys. Rev. Lett. **25**, 635 (1979);
     A.I. Alikhanyan, S.A. Kankanian, A.G. Oganessian, A.G. Tamanian: Phys. Rev.
     Lett. **30**, 109 (1973)
8.32 M.L. Cherry: Phys. Rev. **D17**, 2245 (1978);
     M.L. Cherry, D. Muller, T.A. Prince: Nucl. Instrum. Methods **115**, 141 (1975)
8.33 P. Gorenstein, H. Burskyk, G. Garmire: Astrophys. J. **153**, 885 (1968)
8.34 V.V. Avakyan, A.T. Arundjian, L.S. Boyadjian, S.P. Kazarian, S.S. Kazariyan,
     E.A. Mamidjanian, R.M. Martirosov, M.M. Muradian, J.S. Oganezova, G.J. Oganian,
     G.G. Ovespian, M.P. Pleshko, S.O. Sohayan, O.M. Vinnitzski: *17th Int. Cosmic
     Ray Conf., Paris, 1981, Conference Papers*, Vol.5 (Centre d'Etudes Nucléaires,
     Saclay 1981) p.90
8.35 D. Muller: In *Proc. Workshop on Very High Energy Cosmic Ray Interactions*, ed.
     by M.L. Cherry, K. Lande, R.I. Steinberg (University of Pennsylvania, Phila-
     delphia 1982) p.448. Transition-radiation detectors have been used to identi-
     fy primary cosmic-ray electrons in balloon and satellite experiments, and to
     estimate their energy spectra. Reference to these may be found in
     T.A. Prince: Astrophys. J. **227**, 676 (1979);
     A.A. Gusev, G.I. Pugacheva, V.I. Zatsepin, A.F. Titenkov: Izv. Akad. Nauk.
     SSSR Ser. Fiz. **43**, 2491 (1979) [English transl.: Bull. Acad. Sci. USSR, Phys.
     Ser. **43**, No. 12, 24 (1979)]
8.36 L. Wartski, S. Roland, P. Brunet: In *Proc. Int. Symp. on Transition Radiation
     of High Energy Particles* (Erevan Phys. Inst., Erevan 1977) p.561
8.37 L. Wartski, S. Roland, J. LaSalle, M. Bolore, G. Filippi: J. Appl. Phys. **46**,
     3644 (1975)
8.38 L. Wartski, J. Marcou, S. Roland: IEEE Trans. Nucl. Sci. **NS-20**, 544 (1975)
8.39 A.N. Chu, M.A. Piestrup, P.F. Finman, R.H. Pantell, R.A. Gearhart: IEEE Trans.
     Nucl. Sci. **NS-29**, 336 (1982);
     A.N. Chu, M.A. Piestrup, T.W. Batbee, R.H. Pantell, F.R. Buskirk: Rev. Sci.
     Instrum. **51**, 597 (1980);
     A.N. Chu, M.A. Piestrup, T.W. Babee, R.H. Pantell: J. Appl. Phys. **51**, 1290
     (1980)
8.40 P.F. Finman, M.A. Piestrup, R.H. Pantell, R.A. Gearhart: IEEE Trans. Nucl.
     Sci. **NS-29**, 340 (1982)
8.41 A.K. Drukier: In *Proc. Int. Symp. on Transition Radiation of High Energy Par-
     ticles* (Erevan Phys. Inst., Erevan 1977) p.354; see also C.E.A. Saclay pre-
     print Phys. Sol. No. 74/21, 1974 (unpublished)
8.42 A.K. Drukier, C. Valette, G. Wavsand, L.C.L. Yuan, F. Peters: Nuovo Cimento
     Lett. **14**, 300 (1975);

L.C.L. Yuan: In *Proc. Int. Symp. on Transition Radiation of High Energy Particles* (Erevan Phys. Inst., Erevan 1977) p.344; see also
A.K. Drukier, L.C.L. Yuan: Nucl. Instrum. Methods **138**, 213 (1976)

8.43 A.K. Drukier: Nucl. Instrum. Methods **173**, 259 (1980)

8.44 A.I. Alikhanyan, V.I. Basakov, V.K. Chernyatin, B.A. Dolgoshein, V.M. Fedorov, I.L. Gavrilenko, S.P. Konoralov, O.M. Kozodaeva, V.M. Lebedenko, S.N. Majburov, S.V. Muravjec, V.P. Pustovetov, A.S. Romanjuk, A.P. Shmoleva, P.S. Vasiljev: Nucl. Instrum. Methods **158**, 137 (1979). For another application of drift-chamber techniques, see
R.A. Astabatyan, M.P. Lorikyan, G.A. Manukyan, K.Zh. Markaryan: Nucl. Instrum. Methods **187**, 477 (1981)

8.45 L.A. Avakian, C. Yang: *Proc. Int. Symp. on Transition Radiation of High Energy Particles* (Erevan Phys. Inst., Erevan 1977) p.592

8.46 V.E. Pofamov: Zh. Eksp. Teor. Fiz. **47**, 530 (1964) [English transl.: Sov. Phys.-JETP **20**, 353 (1965)]; Dokl. Akad. Nauk. SSSR **133**, 1315 (1960) [English transl.: Sov. Phys.-Dokl. **5**, 850 (1960)]

8.47 G.M. Garibyan, I.Ya. Pomeranchuk: Zh. Eksp. Teor. Fiz. **37**, 1828 (1959) English transl.: Sov. Phys.-JETP **10**, 1290 (1960)

8.48 I.I. Goldman: Zh. Eksp. Teor. Fiz. **38**, 1866 (1960) [English transl.: Sov. Phys.-JETP **11**, 1341 (1969)]

8.49 G.M. Garibyan, L.A. Varadanian, C. Yang: *Proc. Int. Symp. on Transition Radiation of High Energy Particles* (Erevan Phys. Inst., Erevan 1977) p.374

8.50 M.A. Aginian, C. Yang: *Proc. Int. Symp. on Transition Radiation of High Energy Particles* (Erevan Phys. Inst., Erevan 1977) p.193

8.51 L.L. DeRaad, Jr., Wu-Yang Tsai, T. Erber: Phys. Rev. D**18**, 2152 (1978)

8.52 V.P. Zrelov, J. Ruzicka: Nucl. Instrum. Methods **165**, 307 (1979); Nucl. Instrum. Methods **165**, 91 (1979); Nucl. Instrum Methods **160**, 327 (1979; Nucl. Instrum. Methods **151**, 395 (1978)

8.53 Wu-Yang Tsai, J. Nealing, L.L. DeRaad, Jr.: Phys. Rev. D**21**, 3428 (1980)

8.54 H.A. Olsen, H. Kolbenstvedt: Phys. Rev. A**21**, 1987 (1980)

8.55 V.L. Ginzburg, V.N. Tsytovitch: Zh. Eksp. Teor. Fiz. **74**, 1621 (1978) [English transl.: Sov. Phys.-JETP **47**, 845 (1978)]

Subject Index